Farmers' Crop Varieties and Farmers' Rights

Over the last 50 years there has been a growing appreciation of the important role that farmers play in the development and conservation of crop genetic diversity, and the contribution of that diversity to agro-ecosystem resilience and food security. This book examines policies that aim to increase the share of benefits that farmers receive when others use the crop varieties that they have developed and managed, i.e., 'farmers' varieties'. In so doing, the book addresses two fundamental questions. The first question is 'how do farmer management practices – along with other factors such as environment and the breeding systems of plants – affect the evolution and maintenance of discrete farmers' varieties?' The second question is 'how can policies that depend on being able to identify discrete plant varieties accommodate the agricultural realities associated with the generation, use and maintenance of farmers' varieties?' This focus on discreteness is topical because there are no fixed, internationally recognized taxonomic or legal definitions of farmers' varieties. And that presents a challenge when developing policies that involve making specific, discrete farmers' varieties the subject of legal rights or privileges.

The book includes contributions from a wide range of experts including agronomists, anthropologists, geneticists, biologists, plant breeders, lawyers, development practitioners, activists and farmers. It includes case studies from Asia, Africa, Latin America and Europe where, in response to a diversity of contributing factors, there have been efforts to develop policies that provide incentives or rewards to farmers as stewards of farmers' varieties in ways that are sensitive to the cultural, taxonomic and legal complexities involved. The book situates these initiatives in the context of the evolving discourse and definition of 'farmers' rights', presenting insights for future policy initiatives.

Michael Halewood is a senior scientist and leader of the Genetic Resources Policies, Institutions and Monitoring Group at Bioversity International, Rome, Italy.

Issues in Agricultural Biodiversity
Series editors: Michael Halewood and Danny Hunter

This series of books is published by Earthscan in association with Bioversity International. The aim of the series is to review the current state of knowledge in topical issues associated with agricultural biodiversity, to identify gaps in our knowledge base, to synthesize lessons learned and to propose future research and development actions. The overall objective is to increase the sustainable use of biodiversity in improving people's well-being and food and nutrition security. The series' scope is all aspects of agricultural biodiversity, ranging from conservation biology of genetic resources through social sciences to policy and legal aspects. It also covers the fields of research, education, communication and coordination, information management and knowledge sharing.

Published titles:

Crop Wild Relatives
A manual of *in situ* conservation
Edited by Danny Hunter and Vernon Heywood

The Economics of Managing Crop Diversity On-Farm
Case studies from the Genetic Resources Policy Initiative
Edited by Edilegnaw Wale, Adam Drucker and Kerstin Zander

Plant Genetic Resources and Food Security
Stakeholder perspectives on the International Treaty on Plant Genetic Resources for Food and Agriculture
Edited by Christine Frison, Francisco López and José T. Esquinas

Crop Genetic Resources as a Global Commons
Challenges in international law and governance
Edited by Michael Halewood, Isabel López Noriega and Selim Louafi

Community Biodiversity Management
Promoting resilience and the conservation of plant genetic resources
Edited by Walter S. de Boef, Abishkar Subedi, Nivaldo Peroni and Marja Thijssen

Diversifying Food and Diets
Using agricultural biodiversity to improve nutrition and health
Edited by Jessica Fanzo, Danny Hunter, Teresa Borelli and Federico Mattei

Community Seed Banks
Origins, evolution and prospects
Edited by Ronnie Vernooy, Pitambar Shrestha and Bhuwon Sthapit

Farmers' Crop Varieties and Farmers' Rights
Challenges in taxonomy and law
Edited by Michael Halewood

Farmers' Crop Varieties and Farmers' Rights

Challenges in taxonomy and law

Edited by Michael Halewood

First published 2016
by Routledge
2 Park Square, Milton Park, Abingdon, Oxon OX14 4RN

and by Routledge
711 Third Avenue, New York, NY 10017

Routledge is an imprint of the Taylor & Francis Group, an informa business

British Library Cataloguing-in-Publication Data
A catalogue record for this book is available from the British Library

Library of Congress Cataloging in Publication Data
Farmers' crop varieties and farmers' rights : challenges in taxonomy and law / edited by Michael Halewood.
pages cm. — (Issues in agricultural biodiversity)
Includes bibliographical references and index.
1. Plant varieties—Protection. 2. Agrobiodiversity conservation—Law and legislation. 3. Farmers—Legal status, laws, etc. 4. Agriculture and state. I. Halewood, Michael, 1963– editor.
K3876.F37 2016
343.07'6—dc23
2015036553

ISBN: 978-1-84407-890-5 (hbk)
ISBN: 978-1-84407-891-2 (pbk)
ISBN: 978-1-84977-566-3 (ebk)

Typeset in Bembo
by Apex CoVantage, LLC

Contents

Notes on contributors

Regine Andersen is executive director of Oikos – Organic Norway and previous senior research fellow of the Fridtjof Nansen Institute as director of the Farmers' Rights Project (on leave). She holds a doctoral degree in political science.

Enrico Bertacchini is assistant professor in the Department of Economics and Statistics, University of Torino, Italy.

Richard J. Blaustein is a science and environmental journalist based in Washington, DC. His article, 'Predicting Tipping Points,' appears in the spring 2015 *World Policy Journal*. He contributes articles to *BioScience* magazine, primarily on genetic resources, conservation and climate change.

Riccardo Bocci is general director of Rete Semi Rurali (Italian Rural Seed Network), Florence, Italy.

François Burgaud is international relations and public affairs manager for French seed organisation (GNIS). He has worked for 40 years on seed sector in Europe, Africa and Asia. He has been an expert on genetic resources matters for over 10 years.

Jorge Cabrera Medaglia is an international consultant on environmental law, professor at the University of Costa Rica and lead counsel on international biodiversity law at the Center for International Sustainable Development Law in Montreal, Canada.

Salvatore Ceccarelli is currently a freelance consultant after serving as barley breeder at the International Centre for Agricultural Research in the Dry Areas, Aleppo, Syria.

Jeremy Cherfas is a biologist, seed saver and freelance science writer.

Carlos M. Correa is director of the Center for Interdisciplinary Studies on Industrial Property and Economics, University of Buenos Aires, Argentina, and special advisor on intellectual property and trade, South Centre, Geneva, Switzerland.

Nguyen Van Dinh is a professor of agriculture at Vietnam University of Agriculture, Hanoi, Vietnam.

Lim Eng Siang is a retired officer of the Ministry of Agriculture and Agro-Based Industry, Malaysia.

Carlo Fadda is a senior scientist and theme leader, Intensification & Ecosystems Service Programme research area at Bioversity International, Addis Ababa, Ethiopia.

Gabrielle Gagné works at Trudel Johnston & Lespérance, a Canadian law firm specializing in class action. She has obtained a master's degree in common law and transnational law (LL.M.) as well as a master's degree in molecular biotechnology and law (M.B.M.D.).

Michael Halewood is a senior scientist and leader of the Genetic Resources Policies, Institutions and Monitoring Group at Bioversity International, Rome, Italy.

Nguyen Thi Ngoc Hue is a scientist in the field of agrobiodiversity conservation in Vietnam. Currently, she is senior consultant on theses and projects conducted in Vietnam relating to the conservation and utilization of PGRFA.

Devra Jarvis is principal scientist, Genetic Diversity, Productivity and Resilience at Bioversity International, Rome, Italy, and adjunct faculty, Crop and Soil Sciences, Washington State University, Pullman, USA.

Nguyen Ngoc Kinh is a professor in agriculture. He was dean of faculty of agronomy, Vietnam University of Agriculture, Hanoi, Vietnam, and head of the Department of Science, Technology and Environment, Ministry of Agriculture and Rural Development. He is vice-chairman and general secretary of the Vietnam Seed Association.

Isabel Lapeña is a legal specialist working on research projects related to the impacts of policies on actors' capacity to access, conserve and exchange crop diversity. At present she works as an independent consultant for Bioversity International and the FAO.

Isabel López Noriega is a scientist and policy specialist in the Genetic Resources Policies, Institutions and Monitoring Group at Bioversity International, Rome, Italy.

Niels Louwaars was at the time of writing his chapters senior scientist biopolicies, connected to the Centre for Genetic Resources the Netherlands and the Department of Law and Governance of Wageningen University, the Netherlands. Currently, he is director of Plantum.

Eid M. A. Megeed is an intellectual property rights policy expert, holds four US patents and earned the FAO Food Security Prize in 2007. He was leader of the GRPI-IPGRI Taskforce (Egypt) for Genetic Resources, Indigenous Knowledge Inventory and Documentation. He is the author of *Plant Genetic Resources Legislations Guide in Arab World.*

Alejandro Mejías is a Bioversity alumnus, a graduate of the New York University School of Law, USA, and an intellectual property and unfair competition practitioner based in Madrid, Spain.

Godfrey Mwila is deputy director of technical services at the Zambia Agriculture Research Institute in the Ministry of Agriculture in Zambia. He is a plant genetic resources conservation and use specialist and is involved with ITPGRFA development and implementation.

Dwijen Rangnekar is an associate professor of law at the School of Law, University of Warwick, UK. Among his current research projects is a monograph on the global political economy of plant breeders' rights.

Chutima Ratanasatien is a senior plant variety protection specialist at the Department of Agriculture, Thailand. Her main research interests are in the areas of community conservation of plant genetic resources, access benefit sharing and plant variety protection.

Manuel Ruiz Muller is director of the Biodiversity Program of the Peruvian Society for Environmental Law (SPDA). He teaches international environmental law at the Catholic University, Lima, Peru, and has been a consultant for Bioversity International, FIELD, UNCTAD, and other international organizations.

Juliana Santilli (posthumous) was a lawyer and researcher in environmental and cultural heritage law. She had a Ph.D. in environmental law and was also a federal prosecutor in Brasilia, Brazil.

Pratap Kumar Shrestha is a program specialist for seed systems and plant genetic resources, and coordinator of the scientific advisory team at USC Canada, Ottawa.

C. S. Srinivasan is associate professor of agricultural and development economics at the University of Reading, UK. His main research interests are the economic impacts of intellectual property rights on agricultural innovation and policy issues relating to plant genetic resources.

Raymond Sognon Vodouhe is a geneticist, plant breeder and seed specialist at Bioversity International, Benin. He works on adaptation to climate variation in humid tropics and dry land systems of West Africa.

Acknowledgements and dedication

We acknowledge the financial support of the Government of the Netherlands (through the Genetic Resources Policy Initiative). We also acknowledge the support of the CGIAR Research Program on Policies, Institutions and Markets and the CGIAR Research Program on Climate Change, Agriculture and Food Security.

Additional thanks are due to Stacy Beldon for language editing, Evelyn Clancy for coordinating the last rounds of author revisions, and to Tim Hardwick and Ashley Wright at Routledge and Jennifer Bonnar at Apex CoVantage for their support and unending patience. Finally, thanks very much to all of the colleagues in the Bioversity Policy Unit with whom I have had the pleasure of working over the years that this book was under development.

For My Yung, Matteo and Linh

1 Farmers' varieties and farmers' rights[1]

Challenges at the crossroads of agriculture, taxonomy and law

Michael Halewood and Isabel Lapeña

This book is about crop plant varieties developed by local farmers – commonly referred to as farmers' varieties – and policies to increase the share of benefits farmers receive from the use of those varieties. These are not new subjects. Over the course of the last 50 years there has been a growing appreciation on the part of different stakeholders including biologists, activists and policy makers of the important role that farmers have played in the development and conservation of crop varieties and crop genetic diversity generally. Over successive generations of seed selection (or cutting or bud selection), exchange, and replanting across a range of environments, farmers exert selection pressures contributing to the evolution of plant populations. Farmers have domesticated wild species – indeed, they continue to do so (Scarcelli et al. 2006; Vodouhe et al. 2011) – and are largely responsible for the extraordinary genetic diversity within species (intraspecific diversity) that exists today (Brush 2004). By way of corollary, farmers' selection can also contribute to the maintenance of a variety or population under environmental circumstances that would otherwise contribute to its gradual disappearance or extinction (Louellan 1999).[2]

Farmer crop diversity management, also referred to as on-farm management, has been recognized as an important aspect of food security (Smale 2006; De Schutter 2009, 2011; Lenné and Wood 2011). The genetic heterogeneity of farmers' varieties contributes to production system resilience in response to biotic and abiotic stresses lowering the risk of overall crop failures (Ceccarelli 2009; Altieri and Toledo 2011). Climate change is increasing the potential importance of crop diversity (and agrobiodiversity generally) for farmers' capacity to adapt to increasingly unpredictable and dramatic changes in climate (Burke, Lobell and Guarino 2009; PAR 2010; Fujisaka, Williams and Halewood 2011; Bedmar Villanueva, Halewood and López Noriega 2015; Jarvis et al. 2015). Farmers can use crop diversity, including farmers' varieties, to decrease the incidence and impact of plant pests and diseases (Zhu et al. 2000; Jarvis et al. 2011; Hajjar, Jarvis and Gemmill-Herren 2008; Mulumba et al. 2012; Tooker and Frank 2012). Farmers' varieties sometimes outperform formal sector improved varieties, especially when deployed in difficult environments, and in systems where farmers cannot afford inputs that are recommended to boost the performance of formal sector improved materials (Burdon and Jarosz 1990; Ceccarelli 1996; Balcha and Tanto 2008; Kumemi et al. 2012). They can constitute important sources of nutrition in diversified food systems (Johns and Eyzaguirre 2006; Frison et al. 2011; Powel et al. 2015), and often play an important role in reinforcing cultural identity and continuity (Argumedo and Pimbert 2008; Nabham 2009; Gentilcore 2010).[3]

Farmers' varieties do not exist on their own; they are epiphenomena of the farmer-centred innovation systems, which create and continuously maintain them (or alternatively allow them to fall into disuse and disappear). For some crops, in some parts of the world, the entire development chain consisting of diversity conservation, variety enhancement, seed multiplication, exchange and use is entirely farmer-led, without any involvement of formal sector organizations, and entirely unregulated. On the other hand, for many crops, in many parts of the world, farmers' crop and diversity management cuts across both formal and informal seed systems[4] (Van der Ploeg 2010), for example, accessing seed of the formal sector improved varieties through local markets or neighbours and mixing those varieties with their own varieties, or contributing their own farmers' varieties to be used in formal sector research or breeding programs (Louwaars and De Boef 2012). It is estimated that smallholder farmers in developing countries access up to 90 percent of their seed through informal mechanisms, including from their own harvests, from neighbours and local markets (Badstue et al. 2006; Hodgkin et al. 2007; Thomas et al. 2012; Pautasso et al. 2013). In addition, based on data gathered from six countries covering forty crops, McGuire and Sperling (2016) estimate that half of this informally sourced seed is obtained by farmers through local markets. There is a danger that the spread in scope and application of the commercial seed sector supported by favourable policies, including subsidization (and concomitant reduction in public research and plant breeding and extension) will place informal seed systems under increasing pressure, with the result that farmers' ability to innovate and contribute to development and maintenance of diversity will be lost (De Schutter 2009).

The increased recognition of farmers' dynamic role in the evolution and conservation of crop diversity, and the utility of that diversity, has been attended by calls for institutional and policy support (Jarvis et al. 2011; Santilli 2011). Recognizing and promoting farmers' rights to save, use, exchange and sell reproductive material is the most directly linked, and most frequently called for policy support (Robinson 2007; Pelegrina and Salazar 2011; Da Via 2012; Braunschweig et al. 2014). If farmers choose (or are required) to buy seed each season, then on-farm, across farm, and across agroecosystem levels of farmer-managed crop evolution will be interrupted. Promoting farmers' right to benefit from others' use of their varieties, for example, by allowing the sale of seeds of their varieties on the open market (Thijssen et al. 2008; Lipper, Anderson and Dalton 2010), or receiving benefits when others use them for research, breeding and further commercialization subject to prior informed consent (Humphries et al. 2012), are some options. Increasing farmers' access to quality seed of a wider, more diverse range of varieties – including farmers' varieties from their own agroecosystems – to introduce, mix and manage on their farms (Jarvis et al. 2011) is also essential. Such access can be increased through a variety of measures to strengthen informal and mixed formal and informal seed systems (Beck 2011; Gill et al. 2013; Kloppenburg 2014). Land tenure security is an important prerequisite for farmers to be willing to invest in diversity management (Dennis et al. 2003; Lapeña, Turdieva and López Noriega 2014), also enhancing access to markets for products of niche crops (Kuntashula et al. 2011; Giuliani et al. 2012). Participatory plant breeding (Halewood et al. 2007; Ceccarelli, Guimarães and Weltizien 2009; IPES FOOD 2015), variety selection and other forms of research (De Schutter 2009) with mixed teams of farmers and formal sector researchers can result in new and useful crop diversity being developed and deployed in farmers' fields (Howeler, Lutaladio and Thomas 2013). Recognizing farmers' autonomy to organize collective actions related to seed exchange, diversity conservation, participatory crop evaluation and improvement – all aspects of what is called community biodiversity management – is also critically important (Argumedo et al. 2011; De la Perrière and Kastler 2011; De Boef et al. 2013; Poudel, Shrestha and Sthapit 2015). Community-led seed enterprises (Afari-Sefa et al. 2013) and community seed banks are gaining increased attention as interventions that can create network linkages between farmers and experts from national gene banks, research organizations and so forth who are interested in conserving, using and evolving crop diversity in farmer-managed innovation systems (Vernooy et al. 2014; Vernooy, Shrestha and Sthapit 2015).

Ultimately, there is a continuum between policies and institutions directly related to development and conservation of farmers' varieties per se and those that are related to the maintenance and strengthening of the broader systems of farmer-centred innovation to produce and sustain those varieties in the first place. Farmer-centred innovation, community empowerment, biological diversity conservation, sustainable development and food security are interlinked. In the end, the range of matters that need to be addressed cut across property, social, economic, environmental and human rights issues.

The linked challenges addressed in this book

It is beyond the scope of this book to examine the full range of benefits associated with the use of farmers' varieties and supportive policies mentioned earlier. Instead, the book focuses in on two interlinked questions. The first question is, 'How do farmer management practices (along with other factors such as environment and the breeding systems of plants) affect the evolution and maintenance of discrete farmers' varieties?' The second question is, 'How can policies that depend on being able to identify discrete plant varieties accommodate the agro-cultural realities associated with the generation, use and maintenance of farmers' varieties?'

This focus on discreteness is topical because there are no fixed, internationally recognized taxonomic or legal definitions of farmers' varieties. And that presents a common challenge to policies that involve making specific farmers' varieties the object of legal rights or privileges. One such policy is to create *sui generis* intellectual property (IP) laws for farmers' varieties, to give farmers the means to control others' use of those varieties. Another policy is to create space for including farmers' varieties in national variety release lists so they can be sold on the open market. A third closely related policy strategy for promoting (or at least defending) farmers' interests in their varieties is to defensively publish information about those varieties to prohibit others from making them the subject of intellectual property rights. The purpose of the exercise is to defeat intellectual property applicants' claims that the variety for which they are claiming protection is not discrete (or new or novel or distinct or nonobvious).

In the past, when confronted with demands to recognize property protection for sexually producing plants, policy makers drew on taxonomists' criteria for distinguishing and naming cultivars (i.e. distinctness, uniformity and stability) as set out in the International Code of Nomenclature for Cultivated Plants (Sherman 2008; Parry 2012). Those criteria have been adopted by the UPOV Convention (International Convention for the Protection of New Varieties of Plants) establishing minimum standards for plant breeders' rights, and a number of regional agreements to harmonize seed trade regulations. Farmers' varieties, however, beg the question that is answered through reference to the International Code, because they frequently do not satisfy the criteria of distinctness, uniformity and stability that it embraces (Halewood et al. 2006). In this way, farmers' varieties present the next generation of linked taxonomic and policy challenges, at least in the context of developing and implementing policies that make discrete varieties the object of legal rights or privileges. This book attempts to respond, at least in part, to these challenges.

Before continuing, it is important to highlight that most of the policy options described in the preceding subsection are not directly affected by the lack of a taxonomic or legal definition of farmers' varieties. For example, it is not necessary to be able to identify discrete farmers' varieties (and distinguish them from other farmers' varieties), as part of asserting farmers' rights to save, use, exchange and sell farm-saved seed, or to land tenure security, or to participate in decision making relevant for the conservation and use of these resources. The fact that the book focuses on policy options that make particular varieties the objects of rights or privileges does not mean that we feel that such policies have the most potential to ensure and increase the distribution of benefits to farmers. As already stated, farmers' right to save, use, exchange and sell farm-saved seed is a sine qua non for their contributions to the ongoing evolution of crop diversity and to enjoy benefits associated with its use. However, because the right to save, use, exchange and sell seed does not turn on the ability to define discrete farmers' varieties, it does not fit within the main scope of enquiry of this book.

All of these disclaimers aside, ultimately, farmers' rights in relation to plant genetic resources for food and agriculture will need to be promoted from local to global levels through combinations of policy initiatives. Events in recent years have demonstrated that policy actors around the world are interested in exploring a range of options, including those which require the ability to identify discrete farmers' varieties.

Policy development context

One would be justified in wondering how it has come to pass that countries are involved in developing such laws that turn on the ability to identify discrete farmers varieties when there is still uncertainty, in taxonomy and in law, about the definition of farmers' varieties. Perhaps the most accurate answer is that these policy initiatives are instances of political good will – often stimulated by civil society and farmers' organizations – getting ahead of the policy makers' appreciation of the scientific basis upon which the policies must be made to operate. These are not grounds for indictment. In fact, a lot of policies are developed in this way, with politicians and high-level policy makers – often responding to outside pressures – opening up small spaces for interested actors (including stakeholders and technical specialists) to go off the beaten path to explore, with the possibility that, once more fully investigated, the facts will justify further investment in further policy development and implementation strategies (Lapeña 2012). Indeed, as discussed later and particularly in Chapter 4 of this volume, this process describes the origin and evolution of the concept of farmers' rights (Mooney 2011). At this stage, what is already clear is that appropriate policies linked to seed and farmers' varieties are necessary to create an enabling environment for smallholder agriculture, that promote the use of plant genetic resources for food and agriculture and integrate it better in seed sector development (FAO 2015).[5]

This book is meant to be a contribution to this process. The authors are part of a virtual troupe of stakeholders, practitioners and technical experts squeezing through relatively narrow policy space that have been opened up for farmers' rights, examining practical options for expanding and implementing those rights in meaningful ways that reflect the reality of how farmers conserve, use, exchange and improve their varieties, and how those uses influence the diversity of those varieties. It is our hope that by focusing on the technical issue of 'what is a farmers' variety?' in taxonomy and in law, the book will help policy makers assess their options, taking into account the opinions of stakeholders and experts and lessons learned from initiatives in different countries.

A brief summary of the relevant policies, laws and related strategies

Variety registration

As part of their national seed regulation, many countries have established a range of conditions that plant varieties must satisfy before they can be included in national variety lists, and the seed of those varieties can be sold on the open market. The objective of these laws is to ensure that seeds sold to farmers are of high quality, and perform in predictable ways. In order to be included on the national lists of varieties that can be commercialized, varieties must meet the criterion of distinctness, uniformity and stability (DUS), and have demonstrated additional value above other varieties of the same species. European countries were the first to develop such laws, with developing countries following suit. Such regulations were developed with the market for formally bred seed in mind, not seeds of varieties developed and managed by local farmers over generations. The premise behind this book, and behind the widespread use of the term farmers' varieties, is that farmers' varieties are often different from varieties developed through the formal plant-breeding sector. And one of the differences is that they frequently are not as distinct, uniform and stable as the varieties arising from the formal breeding sector. As a result, farmers' varieties – particularly those of open-pollinated species – do not satisfy these registration requirements, and therefore cannot be legally sold in the country, even if they have demonstrated commercial potential (Munyi and De Jonge 2015).

There are a number of possible ways to attempt to address this problem. One relatively indirect way is to exempt exchanges and sales of farm-saved seed between smallholder farmers from the scope of national variety registration and release regulations. Ethiopia, for example, includes such an exemption in its national seed regulations.[6] This opens up some limited space for commercialization of farmers' varieties, for sales between farmers, but not on a larger scale, with companies taking an active role. Another more direct way would be to exempt farmers' varieties entirely from mandatory registration requirements as a precondition for commercialization. This possibility is built into national seed regulations that are completely voluntary, such as in the United States. However, as far as the authors are aware, no countries that have adopted seed laws that include mandatory variety registration have created a wholesale exemption for farmers' varieties.

Another option that an increasing number of countries (and at least one region) are experimenting with is to adapt variety registration regulations and practices, to make them more flexible, with different criteria for registration, reflecting the special nature of farmers' varieties (IFOAM 2004; FAO 2006; Paavilainen 2009; Lapeña 2012; Mahop, De Jonge and Munyi 2013b). Some of the countries that have created alternative registration lists with alternative registration criteria for farmers' varieties (or for categories of varieties that would at least partially include what we are calling farmers' varieties in this publication) are Peru, France, Italy, the Netherlands, Costa Rica, Benin, Nepal,[7] Finland,[8] Switzerland,[9] Republic of Korea[10] and Ecuador.[11] A number of these national regulations are examined in this book.

It makes sense that introduction of revised registration options for farmers' varieties would be accompanied by rules allowing farmers, and farmers' organizations, to register varieties in their own names. In some countries, registrants must be companies, or organizations with at least one formally trained breeder; other countries have revised their regulations to address this issue (e.g. Nepal and Ecuador). In Ecuador, this reform – brought about by decree in 2012[12] – is already considered quite successful in terms of its outreach to farmers and farmers' organizations.

Most countries' seed regulations establish criteria for regulating and registering seed multipliers, to ensure that seed is grown under conditions that promote its genetic integrity and health. Often, the farmers and farming communities that have played key roles in the development and maintenance of farmers' varieties are unable to meet those criteria. The result is that, in the event that a farmers' variety is registered, other organizations – perhaps without substantial connection to the farmers concerned – would need to take responsibility for the maintenance of the variety and multiplication and marketing of the seed. Ultimately, this creates another layer of disincentive for farmers, farming communities and research and civil society organizations that may be assisting them to pursue farmer variety registration.

In 1993, the Food and Agriculture Organization of the UN (FAO) introduced the concept of 'quality declared seed' (QDS) to respond to the combined facts that (1) many countries did not have the capacity to implement and enforce full-scale, centralized seed quality control regulations, (2) there are already a range of potential seed multipliers/distributers who can provide and are providing seed through unregulated, informal systems, and (3) less ambitious, more incremental improvements to those seed systems were possible (FAO 1993, 2006). In principle, the QDS system is not meant to compete or enter into conflict with more stringent seed regulations. It was designed for use in countries, and for crops, where the formal seed sector (including highly demanding seed quality standards and regulations) has not expanded in scope (Bishaw and Louwaars 2014). Zambia's Plant Varieties and Seeds Act,[13] for example, has a two-tiered system of standards to promote seed quality: one with stricter standards conforming to the standard model described earlier, including DUS as preconditions for variety registration, and a second tier that establishes a QDS system. QDS involves a centralized system of registering varieties for which QDS can be produced and marketed, a centralized system or producer registration, and checks by national authorities of seed multiplication cropping and seed prepared for sale. FAO's QDS guidelines established three kinds of varieties that can be registered: bred varieties, local varieties and varieties developed through participatory plant breeding. For a local variety, an applicant for registration must submit its name, its origin, a simple morphological description and its value for cultivation and use, with an indication of the agroecological zone for which the variety is suited, and information about the procedures to be followed for maintaining the variety (FAO 2006). This system potentially has flexibility for registration of farmers' varieties, and commercial sale of quality declared seed of those varieties. It also has flexibility to empower smaller organizations – like farmers' organizations (or consortia of organizations that include farmer organizations) – to produce and trade in farmers' variety seeds. A number of countries have included QDS as part of their seed regulation, including Tanzania, Peru, Ecuador, Rwanda, Ethiopia and Zambia.

In an effort to increase the size of seed markets through cross-border trade, some regions have developed (or are developing) agreements to harmonize their seed laws and variety registration and seed certification requirements (Rohrbach, Minde and Howard 2003). As part of these agreements, member countries agree to recognize varieties that are registered in other member countries, thereby reducing transaction costs associated with regulatory compliance for companies that want to sell their seeds in more than one country. Harmonization may ultimately contribute to farmers' getting faster access to quality seed of new varieties (Gisselquist 2013, Setimela et al. 2009) produced through formal sector mechanisms, even in countries with small seed markets.

Most of these regional harmonization agreements, however, reproduce the standard DUS and VCU (value for cultivation and use) conditions for registration, thereby further entrenching and expanding the scope of rules that do nothing to address the special situation of farmers' varieties. For example, the Economic Community of West African States (ECOWAS) adopted harmonization regulations in 2008, which included the creation of the West African Catalogue of Plant Species and Varieties.[14] This regional catalogue is comprised of the national catalogues of each member state. The rules specify that to be included in a national list of varieties that can be sold within the region, varieties must be DUS with demonstrated VCU (FAO 2008). Likewise, the Common Market for Eastern and Southern Africa (COMESA) Seed Trade Harmonization Regulations, 2014,[15] establishes a catalogue of all varieties that can be commercialized within the East and Southern Africa subregion, requiring that they be distinct, uniform and stable, with demonstrable VCU. The Memorandum of Understanding for the harmonization of seed regulations in the Southern African Development Community (SADC) came into force in 2013; it also includes DUS and VCU as criteria for inclusion in a regional catalogue. All three regional seed law harmonization initiatives have been subject to considerable criticism by civil society organizations on these grounds (GRAIN 2005; ACB 2012; AFSA 2013). Some of these harmonization rules leave room for the maintenance of looser standards and alternative registration lists for varieties whose seeds will only be marketed within the country concerned.

The European Common Catalogue, established in 1972, also includes the DUS criteria.[16] However, in 2008 and 2009 EU directives[17] created exceptions for the registration and limited sales of 'conservation varieties', defined as those landraces and varieties that have been traditionally grown in particular localities and regions and are threatened by genetic erosion, which were not held to the same DUS standards. While an important step forward in principle, this approach has nonetheless been the subject of considerable criticism for the limitations it places on farmers and farmers' varieties. The European derogation and more recent attempts to introduce new legislation are addressed in later chapters in this book. It is not clear why other regions' seed law harmonization efforts are not actively exploring registration criteria exceptions for farmers' varieties that build (and expand) upon the EU treatment of conservation varieties.

Clusters of questions that this book explores are: What was the rational and historical evolution of mandatory variety registration as a precondition for commercialization? Why were the DUS criteria in particular adopted? What variations of registration criteria are some countries or regions experimenting with? In countries that have developed alternative systems, who is allowed to actually register farmers' varieties? Who is driving the process of considering policy alternatives in national settings, and what challenges have they encountered?

Ultimately, policy makers will need to feel comfortable that suggested modifications to the 'default' systems of DUS variety registration are balanced with competing policy goals, that is that consumer safety is adequately promoted while simultaneously promoting farmer innovation, income generation, and genetic diversity conservation. One possibility is that the balance should be struck differently in different situations, taking into consideration the state of the markets and competition between suppliers of the crops in question, the institutional capacity of the country concerned to implement and enforce such a policy and so on. Here, we are optimistic about the possibility of national laws being able to accommodate the kinds of changes that are necessary to allow the inclusion of farmers' varieties in national lists, as long as they are complemented by other measures to ensure farmers-as-seed-buyers are not exposed to too much risk when purchasing seed on the open market. In some countries and for some crops (depending upon their breeding system), the market operates relatively well in this regard. Where this is not so, other measures have to be built in to minimize consumers' risk.

Sui generis *plant variety protection*

An increasing number of countries have developed plant variety protection laws that allow owners of plant varieties to exclude others' use of those varieties for a number of proscribed purposes, including production, multiplication, conditioning for the purpose of propagation, selling, importing, exporting or stocking for any of the purposes mentioned. The objective of such laws is to create incentives for breeders to engage in innovative plant breeding activities, by providing them a time-limited exclusive right to commercially exploit their varieties. Most countries' plant variety protection laws stipulate that varieties must be novel, distinct, uniform and stable to qualify for intellectual property protection (Kanniah and Antons 2012; Koonan 2014; Lertdhamtewe 2014). The reappearance of DUS in this context is not a coincidence. Variety registration in national seed laws and plant variety protection laws both adopted criteria and practices from taxonomists for identifying and naming cultivars (Sherman 2008; Parry 2012). Problems related to the reliance on these criteria in plant variety protection laws chart very closely those problems associated with their use in variety commercialization regulations. If farmers' varieties don't satisfy these criteria, then they can't be the subject of plant variety protection laws, and farmers are thereby precluded from the possibility of exploiting would-be exclusive rights over the use of the varieties they develop and maintain (Bishaw and Van Gastel 2009, p. 582 in relation to the protection of PPB varieties; Mahop, De Jonge and Munyi 2013a; De Jonge 2014).

One possible solution, to potentially provide farmers with increased benefits derived from others' use of their varieties would, again, be to alter the DUS criteria for protection so the more heterogeneous farmers' varieties could be recognized and protected (Leskien and Flitner 1997; IPGRI 1999; Crucible II Group 2000; Robinson 2007; Koonan 2014; ISSD 2015). Some countries have experimented with introducing alternative conditions for protecting farmers' varieties. Malaysia was one of the earliest countries to attempt to do this;[18] its plant variety protection (PVP) law establishes that farmers' varieties can be protected if they are novel, distinct and identifiable. India and Thailand[19] have dropped the novelty requirement for certain classes of varieties to qualify for forms of intellectual property protection. India has also passed a regulation in furtherance of the implementation of the 2001 Protection of Plant Varieties and Farmers' Rights (PPV&FR) Act, which specifies that, for farmers' varieties, uniformity standards could be relaxed to allow double the number of off-types as otherwise permitted for registration of other categories of varieties under that Act.[20] There are still relatively few examples of laws including alternative (or formally relaxed) criteria for farmers' varieties, and the few that do exist have not yet established proven track records, with the numbers of protected farmers' varieties remaining very low.

Directly related issues concern the impact of such policies on the very farmers they are meant to reward and incentivize. Farmers have been innovating through selection, exchange, and experimentation for millennia in largely informal, unregulated systems. Presumably they have been engaged for millennia in the innovative behaviours that have given rise to the crop variety diversity that already exists. There is the possibility that layering on a new set of incentives would encourage farmers to act in ways that would counteract their diversity-creating behaviours. There is also the very real possibility that individual private property rights are culturally at odds, and therefore inappropriate, with more open systems of farmer innovation and shared interests in farmers' varieties (Correa 2000; Robinson 2007).

Some recent national policy initiatives involve an amalgam of intellectual property rights and the kind of rights that are created by ABS (access and benefit sharing) laws. For example, under the Indian PPV&FR Act, parties seeking intellectual property protection for new varieties that incorporate registered farmers' varieties need to prove that they have obtained consent from the registrees as one of the conditions precedent for the grant of the right. Indonesia's Plant Variety Protection law, 2000,[21] also provides for the registration of local varieties (by public officials on behalf of local people); anyone who wants to use registered varieties must then negotiate an agreement with the local authorities (Kanniah and Antons 2012).

Efforts to promote harmonized plant variety protection at regional and global scales have reinforced countries' adoption of novelty, distinctness, uniformity and stability (NDUS) as the standards for national plant variety protection laws. The number of countries joining the International Union for the Protection of Plant Varieties (UPOV)[22] and implementing plant variety protection laws incorporating NDUS has increased considerably in recent years. UPOV had 72 member states as of 10 June 2014. While the WTO TRIPS[23] agreement does not actually require countries to implement UPOV (Leskien and Flitner 1997), it appears to have provided countries with motivation to do so. So too have bilateral trade agreements, which require the implementation of UPOV compliance standards (Robinson 2007; De Schutter 2009). Through regional agreements like the Organisation Africaine de la Propriété Intellectuelle (OAPI) revised Bangui Agreement (Annex X),[24] the African Regional Intellectual Property Organization (ARIPO) 2015 Arusha Protocol for the Protection of New Varieties of Plants,[25] and the 2012 version of the Draft Protocol for the Protection of New Varieties of Plants (Plant Breeders' Rights) in the Southern African Development Community Region,[26] entire subregions have committed themselves to reproducing the same standards. These initiatives have also been subject to considerable criticism from civil society (Berne Declaration 2014; AFSA and GRAIN 2015) and academics (Mahop et al. 2013a), who indicate that such standards are inappropriate for developing countries, and suggest ways of implementing them that take advantage of whatever flexibilities exist (De Jonge, Louwaars and Kinderlerer 2015; De Jonge and Munyi 2015).

This book examines a cluster of questions regarding the inclusion of NDUS criteria in national plant variety protection laws that are very similar to those which it examines concerning seed laws. What is the origin of these criteria in plant variety protection laws? What variations are countries (and commentators) considering? What actors are involved in promoting the exploration of alternatives, and what factors – institutional and otherwise – are challenging their implementation?

Prohibiting third parties' claims of control over farmers' varieties

The corollary of seeking intellectual property rights over farmers' varieties is seeking to prohibit others from doing so, with the result that they remain in the public domain. One way to do this is to attempt to defeat others' claims that the varieties over which they seek intellectual property protection are either distinct (in the case of plant variety protection) or novel and nonobvious (in the case of patents). In recent years, there have been a number of cases wherein patents and plant variety protection (PVP) claims over traditional crop varieties (or traditional uses of plants) have been attacked or defeated on these grounds (Comisión Nacional Contra la Biopiratería 2014).

Building on these cases, this book also investigates what kinds of information, in what form, can potentially be used to defeat intellectual property claims, and how that information can be systematically assembled to provide a basis for fending off such claims in the future. In this context, it will consider a range of initiatives to document or register biological diversity (including farmers' varieties) and traditional knowledge and the extent to which they do or can provide a basis for defensive publication.

Farmers' rights

Everything discussed earlier, and in this book generally, is situated within the larger political context of promoting farmers' rights in relation to plant genetic resources for food and agriculture. The concept of farmers' rights was first introduced in the context of a formal intergovernmental meeting[27] in 1985, by a Mexican delegate during a hotly contested session of the FAO's Commission on Plant Genetic Resources (Bjørnstad 2004; Mooney 2011).

The concept of farmers' rights – fuzzy though it was – gained a foothold over time. It was included in a resolution adopted by the FAO Council related to the nonbinding International Undertaking on Plant Genetic Resources.[28] The idea, simply stated, was that farmers' contributions to the creation and conservation of biological diversity should be rewarded through newly legal and or political rights, called farmers' rights, with financial support for continued stewardship of crop diversity. International political negotiations regarding the content of farmers' rights reached their zenith (or nadir, depending on your perspective) (Egziabher, Matos and Mwila 2011) at 3 a.m. on 3 November 2001, when exhausted negotiators from 116 countries agreed to the text of Article 9 of the International Treaty on Plant Genetic Resources for Food and Agriculture (the 'Treaty').

Article 9 states:

9.1 The Contracting Parties recognize the enormous contribution that the local and indigenous communities and farmers of all regions of the world, particularly those in the centres of origin and crop diversity, have made and will continue to make for the conservation and development of plant genetic resources which constitute the basis of food and agriculture production throughout the world.

9.2 The Contracting Parties agree that the responsibility for realizing Farmers' Rights, as they relate to plant genetic resources for food and agriculture, rests with national governments. In accordance with their needs and priorities, each Contracting Party should, as appropriate, and subject to its national legislation, take measures to protect and promote Farmers' Rights, including:

a. protection of traditional knowledge relevant to plant genetic resources for food and agriculture;
b. the right to equitably participate in sharing benefits arising from the utilization of plant genetic resources for food and agriculture; and
c. the right to participate in making decisions, at the national level, on matters related to the conservation and sustainable use of plant genetic resources for food and agriculture.

9.3 Nothing in this Article shall be interpreted to limit any rights that farmers have to save, use, exchange and sell farm-saved seed/propagating material, subject to national law and as appropriate.[29]

While the inclusion of farmers' rights in the Treaty is certainly important, the Treaty leaves a lot undone in terms of spelling out the content of farmers' rights and how they should be implemented. Very significantly (and this is one of the sorest points for people who had hoped for more), the Treaty leaves these issues to be sorted out by national governments (Egziabher et al. 2011; South Centre 2015). With nowhere to turn for precedents, national policy makers and farmers are struggling to find creative, practical measures to promote the interests of farmers in ways that are connected to their role as conservers and promoters of plant genetic diversity.

The whole range of policies mentioned in the introductory paragraphs of this chapter – including the narrower subset of those policies focused on in this book – can be fairly characterized as potential ways to promote farmers' rights. The inclusion of farmers' rights in the Treaty provides incentives and momentum within countries for exploring such options. So far, the Governing Body of the Treaty (comprised of 136 countries as of November 2015) has not had many opportunities to focus much attention on the implementation of Article 9. Over the coming years, the Governing Body will function as a de facto clearing house for collection and dissemination concerning domestic policies, laws and administrative procedures to promote farmers' rights. Consequently, this book should also have an international audience, in addition to be immediately of interest (we hope) to national policy makers and advocates.

The logic and structure of this book

Part 1 of the book provides a technical basis for the analysis that follows. Chapters 2 and 3 address the combined genetic, environmental and human cultural complexities involved in answering the questions, 'What is a farmers' variety?' and 'How can you tell them apart?' Jeremy Cherfas (Chapter 2) takes readers through a range of interdependent variables that affect how farmers' varieties evolve, and when they may (or may not) develop distinctive traits. Carlo Fadda (Chapter 3) focuses in particular on how farmers' choices affect the evolution and maintenance of certain traits. Since most people reading this book probably are not scientists, we have provided extra space for the explanation of key scientific concepts.

The case studies in Part 2 are designed to provide readers with an appreciation of the kinds of situations in which farmers and their research and development partners come face-to-face with the kinds of policy challenges that this book addresses. The three studies (presented in Chapters 4, 5 and 6), focus on farmer variety enhancement efforts in Nepal, Vietnam and Syria, and the ways in which those efforts were helped or hindered by national policies and laws. The authors are Pratap Shrestha, Nguyen Thi Ngoc Hue, Devra Jarvis and Michael Halewood, and Salvatore Ceccarelli, respectively.

Part 3 of the book situates the concept of farmers' rights in relation to plant genetic resources for food and agriculture in a broader international policy landscape. In Chapter 7, Regine Andersen analyzes the birth and development of international policies and organizations which are concerned with the conservation and use of plant genetic resources for food and agriculture in general. In Chapter 8, she narrows her focus on the concept of farmers' rights, as it has appeared and evolved at the international level, and where it has 'found legs' in national policies and practices.

It is in Part 4 that the book gets into the history, rationale, and criticisms of seed and plant variety protection laws, defensive publication strategies, and consideration of institutional capacities. Chapter 9 provides a historical account of the evolution of plant variety protection laws. Carlos Correa provides an analysis of the kinds of discussions that have permeated, for years, public debate about developing *sui generis* versions of PVP laws for farmers' varieties, for and against, and he considers the last 30 years of countries implementing, or planning to implement, such laws. In keeping with the theme of the book, the author dedicates considerable attention to the DUS standards and the possibilities of different standards being relied upon in *sui generis* approaches.

Chapter 10, by Niels Louwaars and François Burgaud, provides an historical account of the development of variety registration policies and standards, at both the national and more recently the regional levels. They examine the manner in which those laws, first created in developed countries, have been duplicated without much adaptation in developing countries. They examine the policy rationale for potential differences that can be structured into the design in variety registration standards and associated rights and privileges. It devotes considerable attention to the 'historicity' of the inclusion and implementation of DUS standards. And it looks to those exceptional cases where some deviation of those standards has been used, such as, for example, in the recently adopted European derogation.

In Chapter 11, Isabel López Noriega considers options for defensive publication in light of the evolution of plant variety protection and patent law, and actual cases wherein interested parties have succeeded or failed to defeat intellectual property claims over farmers' varieties, or traditional uses of plants, on the basis of existing prior art. She considers a range of biodiversity traditional knowledge registration projects with an eye to both understanding their primary objectives (which often have little or nothing to do with intellectual property concerns), and the extent to which the kinds of information documented could be useful in defeating intellectual property rights claims. Ultimately, she identifies kinds of information, and ways in which it can be shared, that can potentially be used as a basis for systematically collecting and publishing information to defeat intellectual property claims.

Much of the writing to date on developing *sui generis* intellectual property rights regimes, adapting variety registration requirements and promoting defensive publication schemes has paid little attention to challenges associated with the institutional capacity of countries to effectively implement them. In Chapter 12, C.S. Srinivasan analyzes the different kinds of institutional challenges that developing countries in particular face in implementing these three approaches to promoting farmers' rights.

Part 5 includes a number of country case studies focusing on district or national level policy initiatives related to variety registration and intellectual property laws, with a view to how those initiatives address (or don't address) the special situation of farmers' varieties. Part 5 also includes case studies concerning how governments have proactively developed means to defensively publish information about farmers' varieties. In each case, the author(s) describe the content of the law or policy in question and provide some context relating to the actors who promoted its development, whether or not it was successfully adopted, and its current state of implementation. Many of these case studies are cross referenced in Part 4, so some readers will likely flip ahead to read the case study while they are reading the chapter addressing the same subject matter in Part 4.

These case studies are organized by theme. The first group of case studies concerns intellectual property rights. Dwijen Rangnekar writes about the Indian Protection of Plant Varieties and Farmers' Rights Act, 2001, analyzing the extent to which it introduces novel standards of protection for farmers' varieties and the recent history of its implementation. Lim Eng Siang writes about the Malaysian Protection of New Plant Varieties Act, 2004. Eid M. A. Megeed provides details about the Egyptian law on the Protection of Intellectual Property Rights, Book 4, 2002. Gabrielle Gagné and Chutima Ratanasatien analyze the Thai Plant Varieties Protection Act, and Godfrey Mwila describes efforts (which have not yet been successful) to introduce adapted forms of intellectual property protection included in the Zambian Plant Variety and Seeds Act.

The second group of case studies concerns seed laws and variety registration. Pratap Shrestha writes about the Nepalese seed law and changes to implementation guidelines to accommodate farmers' varieties. The US regulatory approach, which does not include mandatory variety registration as a precondition for commercialization, is introduced by Richard Blaustein. Juliana Santilli provides a commentary on the Brazilian seed law with a particular focus on its treatment of creole varieties. Alejandro Mejias, Enrico Bertacchini and Riccardo Bocci describe the range of approaches that Italian provinces have taken to developing alternative lists of crop varieties, as part of regionalized efforts to promote conservation and sustainable use of local diversity. Godfrey Mwila describes the Zambian seed law and the manner in which it incorporates the concept of quality declared seed. Nguyen Van Dinh and Nguyen Ngoc Kinh highlight relevant aspects of Vietnam's Regulation on Production Management of Farm Households' Plant Varieties, 2008. Manuel Ruiz Muller analyzes Peru's initiative to create a registry for native crop varieties under Law No. 28477. Jorge Cabrera Medaglia comments on ongoing efforts to develop a draft seed law in Costa Rica that will – if successful – create space for the registration and commercialization of farmers' varieties. Finally, Raymond Vodouhe and Michael Halewood write about the conditions for registering farmers' varieties in the Benin seed law.

Notes

1 The authors thank Juanita Chaves and Bram De Jonge for their comments on earlier drafts of this chapter.

2 Other related factors that contribute to the extent to which plant populations evolve, or don't, in farmer-managed systems of crop innovation include the breeding systems of the plants themselves, the variability of the environmental conditions where farmers are planting, the reliability and geographic range of seed exchange systems, the kinds of new materials that are available to farmers to introduce, and demands for the harvested products from those varieties (Halewood et al. 2006; Cox 2009; Döring et al. 2011).

3 The importance of some crops to national cultural identity is reflected in the 2014 Costa Rica Decree No. 38.538/C/MAG, which declares its diversity of maize varieties and uses as cultural patrimony of Costa Rica, and in Guatemala Legislative Decree No. 13–2014 which states that maize forms part of the intangible cultural heritage of the nation and creates August 13 as the 'National Day of Maize', and in Peru Supreme Resolution No. 009-2005-AG that declares May 30 as the 'National Day of Potato' to recognise the country as the centre of origin and diversity of the potato, with high concentration of native potatoes and wild relatives diversity.

4 Formal seed supply systems are characterised by a vertically organised production and distribution of tested seed and approved varieties, while informal, local or traditional seed systems are referred to local reproduction of the seed by farmers themselves, using 'local' seed selection, production and conditioning practices (Almekinders, Louwaars and de Bruijn 1994).

5 Voluntary Guide for National Seed Policy Formulation, endorsed by the Commission at its 15th Regular Session in Rome, 19–23 January 2015. CGRFA-15/15/Report. CGRFA-15/15/Inf.25.

6 Seed Proclamation No. 782/2013 exempts 'the exchange or sale of farm-saved seed among smallholder farmers or agro-pastoralists' from the coverage of the Proclamation, thereby allowing farmers to sell and exchange materials, among themselves, without first satisfying variety registration, and seed producer and distributor standards. Of course, this exemption also creates space for farmers to exchange and sell seed of registered varieties as well (amongst themselves).

7 The relevant laws of Peru, France, Italy, the Netherlands, Costa Rica, Benin and Nepal are examined in Chapters 10, 18, 21, 24, 25 and 26 of this volume.

8 Seed Trade Act of 2000 (728/2000).

9 Ordonnance du Département fédéral de l'économie, de la formation et de la recherche (DEFR) sur les semences et plants du 7 décembre 1998 (Etat le 1er janvier 2013).

10 Seed Industry Law (Act No. 5024 of 6 December 1995, as last revised by Act No. 6374 of 26 January 2001).

11 Acuerdo No. 494 Normativa para la Aplicación de la Ley de Semillas, 30 November 2012.

12 Ibid.

13 Plant Varieties and Seeds Act, 1998, Chapter 236 of the Laws of Zambia.

14 Regulation C/REG.4/05/2008 on Harmonization of the Rules Governing Quality Control, Certification and Marketing of Plant Seeds and Seedlings in ECOWAS Region, Sixtieth Ordinary Session of Minsters, Abuja, 17–18 May 2008, available at www.coraf.org/wasp2013/wp-content/uploads/2013/07/Regulation-seed-ECOWAS-signed-ENG.pdf (last accessed 29 May 2015).

15 COMESA Seed Trade Harmonization Regulations, 2014, adopted by the COMESA Council, at its 32nd meeting held at Kinshasa, Democratic Republic of Congo, available at http://foodtradeesa.com/wp-content/uploads/2013/06/COMESA-Seed-Harmonisation-Regulations.pdf (last accessed 29 May 2015). A distinction between the ECOWAS and COMESA regulation may be possible on the basis that the latter (at paragraphs 26, 27 and 28) appears to set standards only for varieties to be marketed on the regional basis, and not for registration and marketing strictly within the borders of member states.

16 EU Council Directive 2002/53/EC of 13 June 2002 on the common catalogue of varieties of agricultural plant species.

17 Commission Directive 2008/62/EC of 20 June 2008 providing for certain derogations for acceptance of agricultural landraces and varieties which are naturally adapted

to the local and regional conditions and threatened by genetic erosion and for marketing of seed and seed potatoes of those landraces and varieties; Commission Directive 2009/145/EC of 26 November 2009 providing for certain derogations, for acceptance of vegetable landraces and varieties which have been traditionally grown in particular localities and regions and are threatened by genetic erosion and of vegetable varieties with no intrinsic value for commercial crop production but developed for growing under particular conditions and for marketing of seed of those landraces and varieties.

18 Protection of New Plant Varieties Act No. 634 (2004). Available at: www.wipo.int/wipolex/en/text.jsp?file_id=128880 (last accessed 4 March 2015).

19 Thailand Plant Variety Protection Act B.E. 2542 (1999). Available at: www.wipo.int/wipolex/en/text.jsp?file_id=129781 (last accessed 4 March 2015).

20 Protection of Plant Varieties and Farmers Rights (Criteria for Distinctiveness, Uniformity, and Stability for Registration) Regulation, 2009, GSR 452(E), 29 June 2009, available at www.plantauthority.gov.in/pdf/gnotifi376.pdf (last accessed 4 March 2015).

21 Laws of Republic of Indonesia No. 29 of 2000 on Plant Variety Protection, Art. 7. www.wipo.int/wipolex/en/text.jsp?file_id=181116 (last accessed 15 May 2015).

22 UPOV Secretariat (1978; 1991) International Convention for the Protection of New Varieties of Plants. Geneva: UPOV Secretariat. www.upov.int (last accessed 15 May 2015).

23 World Trade Organization (1994) Agreement on Trade-Related Aspects of Intellectual Property Rights – Annex 1C of the Final Act of the Uruguay Round Agreement. Geneva: WTO. https://www.wto.org (last accessed 15 May 2015).

24 OAPI joined UPOV on 10 June 2014. The 1977 Bangui Agreement was revised in 1999 to align with UPOV 1991 and entered into force in 2006.

25 The Arusha Protocol for the Protection of New Varieties of Plants was adopted by the Diplomatic Conference held in Arusha, Tanzania, on 6 July 2015. The Draft Protocol (ARIPO/CM/XIV/8) was found to be in conformity with the UPOV 1991 Act by the UPOV Council in April 2014. UPOV Council (2014). C(Extr.)/31/2; Examination of the Conformity of the Draft ARIPO Protocol for the Protection of New Varieties of Plants with the 1991 Act of the UPOV Convention. Available at: www.upov.int/edocs/mdocs/upov/en/c_extr_31/c_extr_31_2.pdf (last accessed 15 May 2015).

26 Southern African Development Cooperation (SADC). (2012). Draft Protocol for the Protection of New Varieties of Plants (Plant Breeders' Rights) in the Southern African Development Community Region, November 2012. www.ip-watch.org/weblog/wp-content/uploads/2013/04/SADC-Draft-PVP-Protocol-April-2013.pdf (last accessed 15 May 2015).

27 Mooney (2011), Santilli (2011) and Anderson (2005) note that the term "farmers' rights" was first coined in the 1980s by a nongovernmental organization – Rural Advancement Foundation International (RAFI), more recently renamed Erosion, Technology and Control (ETC) – that was active monitoring genetic resources issues being addressed under the aegis of FAO.

28 Resolution 5/89 (Adopted 29 November 1989),

> Endorses the concept of Farmers' Rights (Farmers' Rights mean rights arising from the past, present and future contributions of farmers in conserving, improving, and making available plant genetic resources, particularly those in the International Community, as trustee for present and future generations of farmers, for the purpose of ensuring full benefits to farmers, and supporting the continuation of their contributions, as well as the attainment of the overall purposes of the International Undertaking) in order to:
>
> a ensure that the need for conservation is globally recognized and that sufficient funds for these purposes will be available;
> b assist farmers and farming communities, in all regions of the world, but especially in the areas of origin/diversity of plant genetic resources, in the protection and conservation of their plant genetic resources, and of the natural biosphere;

c allow farmers, their communities, and countries in all regions, to participate fully in the benefits derived, at present and in the future, from the improved use of plant genetic resources, through plant breeding and other scientific methods.

29 International Treaty on Plant Genetic Resources for Food and Agriculture, Report on the Conference of FAO, Thirty-first session, Rome, 2–13 November 2001, c. 2001/REP, Appendix D at article 9. As of 15 May 2004, 52 states had ratified, accepted, approved or acceded to the Treaty. UN FAO, International Treaty on Plant Genetic Resources for Food and Agriculture, available at www.fao.org/Legal/TREATIES/033s-e.htm (last accessed 15 May 2004).

References

Afari-Sefa, V., T. Chagomoka, D. K. Karanja, E. Njeru, S. Samali, A. Katunzi, H. Mtwaenzi, L. Kimenye, K. Hannweg and M. Penter (2013). 'Private Contracting Versus Community Seed Production Systems: Experiences from Farmer-led Seed Enterprise Development of Indigenous Vegetables in Tanzania,' International Society for Horticultural Science (ISHS), Leuven, Belgium, Acta Horticulturae 1007: 671–80.

African Centre for Biosafety (ACB) (2012). *Harmonization of Africa's Seed Laws: A Recipe for Disaster. Players, Motives and Dynamics*, ACB, Melville.

AFSA and GRAIN (January, 2015). *Land and Seed Laws under Attack. Who Is Pushing Changes in Africa?* AFSA and GRAIN.

Alliance for Food Sovereignty in Africa (AFSA) (2013). *COMESA Approval of Seed Trade Regulations Spells Disaster for Small Farmers and Food Sovereignty in Africa.* GRAIN. Available at: www.grain.org/bulletin_board/entries/4795-comesa-approval-of-seed-trade-regulations-spells-disaster-for-small-farmers-and-food-sovereignty-in-africa (last accessed May 29 2015).

Almekinders, C., Louwaars, N., and de Bruijn, G.H. (1994). Local seed systems and their importance for an improved seed supply in developing countries. *Euphytica 78*(3), 207–216.

Argumedo, A., and M. Pimbert (2008). *Protecting Farmers' Rights with Indigenous Biocultural Heritage Territories: The Experience of the Potato Park*, International Institute for Environment and Development, IIED, Asociación ANDES.

Argumedo, A., K. Swiderska, M. Pimbert, Y. Song and R. Pant (2011). *Implementing Farmers' Rights Under the FAO International Treaty on PGRFA: The Need for a Broad Approach Based on Biocultural Heritage.* Paper prepared for the Fourth Governing Body of the International Treaty on PGRFA, Bali, 14–18 March 2011.

Badstue, L. B., M.R. Bellon, J. Berthaud, X. Juárez, I.M. Rosas, A.M. Solano and A. Ramírez (2006). 'Examining the Role of Collective Action in an Informal Seed System: A Case Study from the Central Valleys of Oaxaca, Mexico,' *Human Ecology* 34(2): 249–73.

Balcha, G., and T. Tanto (2008). 'Conservation of Genetic Diversity and Supporting Informal Seed Supply in Ethiopia,' in M.H. Thijssen, Z. Bishaw, A. Beshir and W.S. de Boef (eds.), *Farmers, Seeds and Varieties: Supporting Informal Seed Supply in Ethiopia*, Wageningen International, Wageningen, pp. 141–49.

Beck, R. (2011). 'Farmers Rights and Open Source Licensing,' *Arizona Journal of Environmental Law & Policy* 1(2): Marquette Law School Legal Studies Paper No. 10–28.

Bedmar Villanueva, A., M. Halewood and I. López Noriega (2015). *Agricultural Biodiversity in Climate Change Adaptation Planning: An Analysis of the National Adaptation Programmes of Action. CCAFS Working Paper No. 95.* CGIAR Research Program on Climate Change, Agriculture and Food Security (CCAFS). Copenhagen, Denmark. Available at: http://ccafs.cgiar.org/ (last accessed 29 January 2016).

Berne Declaration (2014). *Owning Seeds, Accessing Food.* A Human Rights Impact Assessment of UPOV1991. Based on Case Studies in Kenya, Peru and the Philippines.

Bishaw, Z., and N. Louwaars (2014). 'Evolution of Seed Policy and Strategies and Implications for Ethiopian Seed Systems Development,' in A. Wold, A. Fikre, D. Alemu, L. Desalegn and A. Kirub (eds.), *The Defining Moments in Ethiopian Seed System, Ethiopian Institute of Agriculture Research*, pp. 31–60. Available at: www.future-agricultures.org/publications/research-and-analysis/books (last accessed 29 January 2016).

Bishaw, Z., and A.J.G. van Gastel (2009). 'Variety Release and Policy Options,' in S. Ceccarelli, E.P. Guimarães and E. Weltizien (eds.), *Plant Breeding and Farmer Participation*, Food and Agriculture Organisation of the United Nations, Rome, pp. 565–87.

Bjørnstad, S.B. (2004). *Breakthrough for 'The South'? An Analysis of the Recognition of Farmers' Rights in the International Treaty on Plant Genetic Resources for Food and Agriculture*, p. 35, 92. Available at: www.fni.no/pdf/FNI-R1304.pdf (last accessed 5 February 2016).

Braunschweig, T., F. Meiienberg, C. Pionetti and S. Shashikant (2014). *Owning Seeds, Accessing Food.* A Human Rights Impact Assessment of UPove 1991 based on Case Studies in Kenya, Peru, and the Philippines. Berne Declaration. Zurich.

Brush, S. (2004). *Farmers' Bounty: Locating Crop Diversity in the Contemporary World.* Yale University Press, New Haven and London.

Burdon, J.J., and A.M. Jarosz (1990). 'Disease in Mixed Cultivars, Composites, and Natural Plant Populations: Some Epidemiological and Evolutionary Consequences,' in H.D. Brown, M.T. Clegg, A.L. Kahler and B.S. Weir (eds.), *Plant Population Genetics, Breeding and Genetic Resources.* Sinauer Associates, Sunderland, MA, pp. 215–28.

Burke, M.B., D.B. Lobell and L. Guarino (2009). 'Shifts in African Crop Climates by 2050, and the Implications for Crop Improvements and Genetic Resources Conservation,' *Global Environmental Change* 19: 317–25.

Ceccarelli, S. (1996). *Positive Interpretation of Genotype by Environment Interactions in Relation to Sustainability and Biodiversity.* The International Center for Agricultural Research in the Dry Areas (ICARDA), Aleppo, Syria.

Ceccarelli, S., E.P. Guimarães and E. Weltizien (eds.) (2009). *Plant Breeding and Farmer Participation.* FAO, Rome. Available at: ftp://ftp.fao.org/docrep/fao/012/i1070e/i1070e.pdf.

Comisión Nacional Contra la Biopiratería (2014). *Comisión Nacional contra la Biopiratería, única en el mundo, cumple 10 años. Combatiendo casos que afectan la biodiversidad del Perú.* Available at: www.biopirateria.org/comision-nacional-contra-la-biopirateria-unica-en-el-mundo-cumple-10-anos/ (last accessed 29 January 2016).

Cox, S. (2009). 'Crop Domestication and the First Plant Breeders,' in S. Ceccarelli, E.P. Guimarães and E. Weltizien (eds.), *Plant Breeding and Farmer Participation.* FAO, Rome, pp. 1–26.

Crucible II Group (2000). *Seeding Solutions.* Volume 1, IDRC, IPGRI and DHF, Rome.

Da Via, E. (2012). 'Seed Diversity, Farmers' Rights, and the Politics of Re-peasantization,' *International Journal of Sociology of Agriculture and Food* 19(2): 229–42.

De Boef, W., A. Subedi, N. Peroni, M. Thijssen and E. O'Keeffe (eds.) (2013). *Community Biodiversity Management. Promoting Resilience and Conservation of Plant Genetic Resources.* Routledge, Abingdon, Oxon.

De Jonge, B. (2014). 'Plant Variety Protection in Sub-Saharan Africa: Balancing Commercial and Smallholder Farmers' Interests,' *Journal of Politics and Law* 7(3): 100–111. Available at: www.ccsenet.org/journal/index.php/jpl/article/view/39778 (last accessed 29 January 2016).

De Jonge, B., N.P. Louwaars and J. Kinderlerer (2015). 'A Solution to the Controversy on Plant Variety Protection in Africa,' *Nature Biotechnology* 33(5): 487–8.

De Jonge, B. and P. Munyi (2015). *A Differentiated Approach to Plant Variety Protection in Africa.* Wageningen Working Paper Law and Governance 2015/04. Available at: http://ssrn.com/abstract=2619763 (last accessed 29 January 2016).

De la Perrière, R.A.B., and G. Kastler (2011). *Seeds and Farmers' Rights. How International Regulations Affect Farmers Seeds.* Biodiversity Exchange and Diffusion of Experience (BEDE) & Reseau Semences Paysannes (RSP), Brens, Montpellier.

Dennis, E., J. Ilyasov, E. Van Dusen, E. Treshkin, M. Lee and P. Eyzaguirre (2003). *The Role of Local Institutions in the Conservation of Plant Genetic Diversity, IFPRI, University of California at Berkeley.* Paper Presented at the CAPRi-IPGRI Workshop 'Property Rights, Collective Action and Local Conservation of Genetic Resources,' Rome, 29 September 2003.

De Schutter, O. (2009). *The Right to Food. Seed Policies and the Right to Food: Enhancing Agro-biodiversity and Encouraging Innovation.* Interim Report of the Special Rapporteur on the Right to Food, Olivier De Schutter, Submitted to the General Assembly. (A/64/170), 23 July 2009.

Dhar, B. (2002). *Sui Generis Systems for Plant Variety Protection: Options under TRIPS*, Quaker United Nations Office, Geneva.

Döring, T.F., S. Knapp, G. Kovacs, K. Murphy and M.S. Wolfe (2011). 'Evolutionary Plant Breeding in Cereals-into a New Era,' *Sustainability* 3(10): 1944–71.

Dutfield, G. (2011). *Food, Biological Diversity and Intellectual Property: The Role of the International Union for the Protection of New Varieties of Plants (UPOV). Intellectual Property Issue Paper No. 9.* QUNO, Geneva. Available at: www.quno.org/resource/2011/2/food-biological-diversity-and-intellectual-property (last accessed 29 January 2016).

Egziabher, T.B.G., E. Matos and G. Mwila (2011). 'The African Regional Group. Creating Fair Play between North and South,' in C. Frison, F. López and J.T. Esquinas-Alcázar (eds.), *Plant Genetic Resources and Food Security: Stakeholder Perspectives on the International Treaty on Plant Genetic Resources for Food and Agriculture*, Earthscan, London, pp. 41–57.

FAO (1993). 'Quality Declared Seed System,' *Technical Guidelines for Standards and Procedures.* FAO, Rome.

FAO (2006). 'Quality Declared Seed System,' *FAO Plant Production and Protection Paper 185.* Available at: ftp://ftp.fao.org/docrep/fao/009/a0503e/a0503e00.pdf.

FAO (2008). 'West African Catalogue of Plant Species and Varieties,' FAO, Rome.

FAO (2015). 'Draft Guidelines to Support the Integration of Genetic Diversity into National Climate Change Adaptation Planning (Revised Version),' Fifteenth Regular Session of the Commission on Genetic Resources for Food and Agriculture (CGRFA), Rome, 19–23 January 2015. Information document CGRFA-15/15/Inf.15. Available at: www.fao.org/3/a-mm507e.pdf (last accessed 8 February 2016).

Fonseca, M.F. (2004). *Alternative Certification and a Network Conformity Assessment Approach*, IFOAM. Available at: www.ifoam.bio/sites/default/files/page/files/alternativecertification andanetworkconformityassessmentapproach.pdf (last accessed 29 January 2016).

Frison, E.A., J. Cherfas and T. Hodgkin (2011). 'Agricultural Biodiversity Is Essential for a Sustainable Improvement in Food and Nutrition Security,' *Sustainability* 3(1): 238–53. http://dx.doi.org/10.3390/su3010238 (last accessed 9 February 2016).

Fujisaka, S., D. Williams and M. Halewood (eds.) (2011). 'The Impact of Climate Change on Countries' Interdependence on Genetic Resources for Food and Agriculture,' *Background Study No 48.* FAO, Rome. Available at: ftp://ftp.fao.org/docrep/fao/meeting/017/ak532e.pdf.

Gentilcore, D. (2010). *Pomodoro! A History of the Tomato in Italy*, Columbia University Press, New York.

Gill, T.B., R. Bates, A. Bicksler, R. Burnette, V. Ricciardi and L. Yoder (2013). 'Strengthening Informal Seed Systems to Enhance Food Security in Southeast Asia,' *Journal of Agriculture, Food Systems, and Community Development* 3(3): 139–53. http://dx.doi.org/10.5304/jafscd.2013.033.005 (last accessed 29 January 2016).

Giuliani, A., F. Hintermann, W. Rojas and S. Padulosi (eds.) (2012). *Biodiversity of Andean Grains: Balancing Market Potential and Sustainable Livelihoods*, Bioversity International, Rome, Italy.

Gisselquist, D., C.E. Pray, L. Nagarajan and D.J. Spielman (May 9, 2013). *An Obstacle to Africa's Green Revolution: Too Few New Varieties*. Available at: http://papers.ssrn.com/sol3/papers.cfm?abstract_id=2263042 (last accessed 29 January 2016).

GRAIN. (July, 2005). *Africa's Seed Laws: A Red Carpet for the Corporations,* Seedling. Available at: www.grain.org (last accessed 29 January 2016).

Hajjar, R., D.I. Jarvis and B. Gemmill-Herren (2008). 'The Utility of Crop Genetic Diversity in Maintaining Ecosystem Services,' *Agriculture, Ecosystems and Environment* 123: 261–70.

Halewood, M., J.J. Cherfas, J.M.M. Engels, T.H. Hazekamp, T. Hodgkin and J. Robinson (2006). 'Farmers, Landraces, and Property Rights: Challenges to Allocating Sui Generis Intellectual Property Rights to Communities over their Varieties,' in S. Biber-Klemm, T. Cottier and D.S. Berglas (eds.), *Rights to Plant Genetic Resources and Traditional Knowledge: Basic Issues and Perspectives*, CAB International, World Trade Institute, Berne, Switzerland, pp. 173–202. ISBN:0851990339.

Halewood, M., P. Deupmann, B.R. Staphit, R. Vernooy and S. Ceccarelli (2007). *Participatory Plant Breeding to Promote Farmers' Rights*, IPGRI, Rome. Available at: www.bioversityinternational.org/e-library/publications/detail/participatory-plant-breeding-to-promote-farmers-rights (last accessed 29 January 2015).

Hodgkin, T., R.B. Rana, J. Tuxill, D. Blama, A. Subedi, I. Mar, D. Karamura, R. Valdivia, L. Collado, L. Latournerir, M. Sadiki, M. Sawadogo, A.H.D. Brown and D.I. Jarvis (2007). 'Seed Systems and Crop Genetic Diversity in Agroecosystems,' in D.I. Jarvis, C. Padoch and H.D. Cooper (eds.), *Managing Biodiversity in Agricultural Ecosystems*, Bioversity International and Columbia University Press, SDC, IDRC, UNU and CBD, New York, NY, pp. 77–116.

Howeler, R., N. Lutaladio and G. Thomas (2013). *Save and Grow: Cassava, a guide to sustainable production intensification*, Food and Agriculture Organization of the United Nations, Rome.

Humphries, S., J. Jimenez, O. Gallardo, M. Gomez, F. Sierra and Associations of Local Agricultural Research Committees of Yorito, Victoria and Sulaco (2012). 'Rights of Farmers and Breeders' Rights in the New Globalizing Context,' in M. Ruiz and R. Vernooy (eds.), *The Custodians of Biodiversity: Sharing Access to and Benefits of Genetic Resources*, Earthscan, London, pp. 79–93.

International Plant Genetic Resources Institute (IPGRI) (1999). *Key Questions for Decision-makers: Protection of Plant Varieties Under the WTO Agreement on Trade-related Aspects of Intellectual Property Rights*. IPGRI, Rome, Italy. Available at: www.bioversityinternational.org/uploads/tx_news/Key_questions_for_decision-makers_41.pdf (last accessed 29 January 2016).

IPES FOOD. (May, 2015). *The New Science of Sustainable Food Systems. Overcoming Barriers to Food System Reform*. International Panel of Experts on Sustainable Food Systems.

Jarvis, A., H. Upadhyaya, C.L.L. Gowda, P.K. Aggarwal, S. Fujisaka and B. Anderson (2015). 'Plant Genetic Resources for Food and Agriculture and Climate Change,' *Coping with Climate Change – The Roles of Genetic Resources for Food and Agriculture*, FAO, Rome. Available at: www.fao.org/3/a-i3866e.pdf (last accessed 29 January 2016).

Jarvis, D.I., T. Hodgkin, B.R. Sthapit, C. Fadda and I. Lopez-Noriega (2011). 'An Heuristic Framework for Identifying Multiple Ways of Supporting the Conservation and Use of Traditional Crop Varieties within the Agricultural Production System,' *Critical Reviews in*

Plant Sciences 30(1–2): 125–76. ISSN:0735–2689. Available at: http://dx.doi.org/10.1080/07352689.2011.554358 (last accessed 29 January 2016).

Johns, T., and P.B. Eyzaguirre (2006). 'Linking Biodiversity, Diet and Health in Policy and Practice,' *Proceedings of the Nutrition Society* 65(2): 182–9.

Kanniah, R., and C. Antons (2012). 'Plant Variety Protection and Traditional Agricultural Knowledge in Southeast Asia,' *Australian Journal of Asian Law* 13(1): 1–23.

Keneni, G., E. Bekele, M. Imtiaz and K. Dagne (2012). 'Genetic Vulnerability of Modern Crop Cultivars: Causes, Mechanism and Remedies,' *International Journal of Plant Research* 2(3): 69–79.

Kloppenburg, J. (2014). 'Re-purposing the master's tools: the open source seed initiative and the struggle for seed sovereignty,' *The Journal of Peasant Studies* 41(6): 911–31. http://dx.doi.org/10.1080/03066150.2013.875897 (last accessed 29 January 2016).

Koonan, S. (2014). 'Developing Country Sui Generis Options. India's Sui Generis System of Plant Variety Protection,' QUNO's Briefing Paper, *Food, Biological Diversity and Intellectual Property* 4: 1–6.

Kuntashula, E., E. Wale, J.C.N. Lungu and M.T. Daura (2011). 'Consumers' Attribute Preferences and Traders' Challenges Affecting the Use of Local Maize and Groundnut Varieties in Lusaka: Implications for Crop Diversity Policy,' in E. Wale, A.G. Drucker and K.K. Zander (eds.), *The Economics of Managing Crop Diversity On-farm. Case Studies from the Genetic Resources Policy Initiative*. Bioversity International and Earthscan, Washington, DC and London, pp. 93–110. .

Lapeña, I. (2012). 'La nueva legislación de semillas y sus implicancias para la agricultura familiar en el Perú,' *Serie de Política y Derecho Ambiental (SPDA) No. 26*, octubre de 2012, Lima, Perú, pp. 1–20. Available at: http://registroagrobio.com/download/biblioteca/publicaciones/pequena%20agricultura/p5_lanuevalegis.pdf (last accessed 29 January 2016).

Lapeña, I., M. Turdieva and I. López Noriega (2014). 'Conservation of Fruit Tree Diversity in Central Asia: An Analysis of Policy Options and Challenges,' in I. Lapeña, M. Turdieva, I. López Noriega and W.G. Ayad (eds.), *Conservation of Fruit Tree Diversity in Central Asia: Policy Options and Challenges*. Bioversity International, Rome, Italy, pp. 1–31.

Lenné, J.M., and D. Wood (2011). 'Utilization of Crop Diversity for Food Security,' in J.M. Lenné and D. Wood (eds.), *Agrobiodiversity Management for Food Security: A Critical Review*. CAB International, Wallingford, UK, pp. 189–211.

Lertdhamtewe, P. (2014). 'Developing Country Sui Generis Options. Thailand' s Sui Generis System of Plant Variety Protection. QUNO's Briefing paper. *Food, Biological Diversity and Intellectual Property* 3: 1–6.

Leskien, D., and Flitner, M. (1997). 'Intellectual Property Rights and Plant Genetic Resources: Options for a Sui Generis System,' *Issues in Genetic Resources No. 6*. IPGRI, Rome.

Lipper, L., C.L. Anderson and T. Dalton (2010). *Seed Trade in Local Markets: Implications for Crop Diversity and Agricultural Development*, Earthscan, London.

Louwaars, N.P., and W.S. De Boef. (2012). 'Integrated Seed Sector Development in Africa: A Conceptual Framework for Creating Coherence between Practices, Programs and Policies,' *Journal of Crop Improvement* 26: 39–59.

Mahop, M.T., D.B. De Jonge and P. Munyi (2013a). 'Plant Variety Protection Regimes and Seed Regulations in Sub-Saharan Africa: Current Trends and Implications,' *BioScience Law Review* 13(3): 101–11.

Mahop, M.T., D.B. De Jonge and P. Munyi (2013b). *Seed Systems and Intellectual Property Rights: An Inventory from Five Sub Saharan African Countries*. Available at: www.narcis.nl/publication/RecordID/oai%3Alibrary.wur.nl%3Awurpubs%2F439674 (last accessed 29 January 2016).

Malaysia (2004, art. 14). *Protection of New Plant Varieties Act No. 634 of 2004*. Available at: www.wipo.int/wipolex/en/details.jsp?id=3143 (last accessed 29 January 2016).

McGuire, S. and L. Sperling (2016). 'Seed Systems Smallholder Farmers Use', *Food Security*. Available at: http://link.springer.com/article/10.1007%2Fs12571-015-0528-8 (last accessed 29 January 2016).

Mooney, P. (2011). 'International Non-governmental Organizations the Hundred Year (or so) Seed War – Seeds, Sovereignty and Civil Society – A Historical Perspective on the Evolution of 'The Law of the Seed,' in C. Frison, F. López and J.T. Esquinas-Alcázar (eds.), *Plant Genetic Resources and Food Security: Stakeholder Perspectives on the International Treaty on Plant Genetic Resources for Food and Agriculture*, Earthscan, London, pp. 135–48.

Mulumba, J.W., R. Nankya, J. Adokorach, C. Kiwuka, C. Fadda, P. De Santis and D.I. Jarvis (2012). 'A Risk-minimizing Argument for Traditional Crop Varietal Diversity Use to Reduce Pest and Disease Damage in Agricultural Ecosystems of Uganda,' *Agriculture, Ecosystems and Environment* 157: 70–86. Available at: http://dx.doi.org/10.1016/j.agee.2012.02.012 (last accessed 29 January 2016).

Munyi, P. and B. De Jonge (2015). 'Seed Systems Support in Kenya: Consideration for an Integrated Seed Sector Development Approach,' *Journal of Sustainable Development* 8(2): 161–73.

Nabham, G.B. (2009). *Where Our Food Comes From. Retracing Nikolay Vavilov's Quest to End Famine*, Island Press, Washington.

Paavilainen, K. (2009). 'National Policies and Support Systems for Landrace Cultivation in Finland,' in M. Veteläinen, V. Negri and N. Maxted (eds.), *European Landraces on Farm Conservation, Management and Use. Bioversity Technical Bulletin No. 15*. Bioversity International, Rome, Italy, pp. 296–99.

PAR (2010). *The Use of Agrobiodiversity by Indigenous Peoples and Rural Communities in Adapting to Climate Change*. A Discussion Paper Prepared by the Platform for Agrobiodiversity Research. Available at: www.agrobiodiversityplatform.org/blog/wp-content/uploads/2010/05/PAR-Synthesis_low_FINAL.pdf (last accessed 29 January 2016).

Parry, B. (2012). 'Taxonomy, Type Specimens and the Making of Biological Property in Intellectual Property Rights Law,' *International Journal of Cultural Property* 19(3): 251–68.

Pautasso, M., G. Aistara, A. Barnaud, S. Caillon, P. Clouvel, O.T. Coomes, M. Delêtre, E. Demeulenaere, P. De Santis, T. Döring, L. Eloy, L. Emperaire, E. Garine, I. Goldringer, D. Jarvis, H.I. Joly, C. Leclerc, S. Louafi, P. Martin, F. Massol, S. McGuire, D. McKey, C. Padoch, C. Soler, M. Thomas and S. Tramontini (2013). 'Seed Exchange Networks for Agrobiodiversity Conservation. A Review,' *Agronomy for Sustainable Development*, 33(1): 151–75.

Pelegrina, W.R., and R. Salazar (2011). 'Farmers' Communities: A Reflection on the Treaty from Small Farmers' Perspectives,' in C. Frison, F. López and J.T. Esquinas-Alcázar (eds.), *Plant Genetic Resources and Food Security: Stakeholder Perspectives on the International Treaty on Plant Genetic Resources for Food and Agriculture*, FAO, Bioversity International and Earthscan, New York and London, pp. 175–81.

Perales, H., S.B. Brush and C.O. Qualset (2003). 'Landraces of Maize in Central Mexico: An altitudinal transect,' *Economic Botany* 57: 7–20.

Poudel, D., P. Shrestha and B. Sthapit (2015). 'An Analysis of Social Seed Network and Its Contribution to On-Farm Conservation of Crop Genetic Diversity in Nepal,' *International Journal of Biodiversity*, Article ID 312621. Available at: http://dx.doi.org/10.1155/2015/312621 (last accessed 29 January 2016).

Powel, B., S.H. Thilsted, A. Ickowitz, C. Termote, T. Sunderland and A. Herforth (2015). 'Improving Diets with Wild and Cultivated Biodiversity from Across the Landscape,' *Food Security* 7: 535–54. http:// link.springer.com/article/10.1007/s12571-015-0466-5/fulltext.html (last accessed 9 February 2016).

Rangnekar, D. (2002). *Access to Genetic Resources, Gene-Based Inventions and Agriculture*. Commission on Intellectual Property Rights. Integrating Intellectual Property Rights and

Development Policy. Study paper 3a. Available at: www.iprcommission.org/graphic/documents/study_papers.htm (last accessed 29 January 2016).

Robinson, B.D. (2007). *Exploring Components and Elements of Sui Generis Systems for Plant Variety Protection and Traditional Knowledge in Asia.* ICTSD Programme on IPRs and Sustainable Development.

Robinson, B.D. (2008). *Towards a Balanced 'Sui Generis' Plant Variety Regime.* Guidelines to Establish a National PVP Law and an Understanding of TRIPS-plus Aspects of Plant Rights. UNDP.

Rohrbach, D., I.J. Minde and J. Howard (2003). 'Looking Beyond National Boundaries: Regional Harmonization of Seed Policies, Laws and Regulations,' *Food Policy* 28: 317–33.

SADC Secretariat (2008). *Technical Agreements on Harmonization of Seed Regulations in the SADC Region.* Seed Variety Release Seed Certification and Quality Assurance Quarantine and Phytosanitary Measures for Seed, Gaborone, Botswana, 2008.

Scarcelli, N., S. Tostain, C. Mariac, C. Agbangla, O. Da, J. Berthaud and J.-L. Pham (2006). 'Genetic Nature of Yams (Dioscorea spp.) Domesticated by Farmers in W. Africa (Benin),' *Genetic Resources and Crop Evolution* 53: 121–30.

Setimela, P.S., B. Badu-Apraku and W. Mwangi (2009). *Variety Testing and Release Approaches in DTMA Project Countries in Sub-Saharan Africa.* Harare, Zimbabwe, CIMMYT. Available at: http://repository.cimmyt.org/xmlui/bitstream/handle/10883/807/93477.pdf?sequence=1 (last accessed 29 January 2016).

Sherman, B. (2008). 'Taxonomic Property,' *Cambridge Law Journal* 67(3): 560–84.

Smale, M. (2006). *Valuing Crop Biodiversity: On-farm Genetic Resources and Economic Change,* CABI Publishing, Wallingford, Oxfordshire, UK.

South Centre (March, 2015). Towards a More Coherent International Legal System on Farmers' Rights: The Relationship of the FAO ITPGRFA, UPOV and WIPO. *Policy Brief No. 17.* Geneva. Available at: www.southcentre.int/wp-content/uploads/2015/04/PB17_More-Coherent-International-Legal-System-on-Farmers'-Rights_EN.pdf (last accessed 29 January 2016).

Thijssen, M.H., Z. Bishaw, A. Beshir and W.S. de Boef (eds.) (2008). *Farmers, Seeds and Varieties: Supporting Informal Seed Supply in Ethiopia,* Wageningen, Wageningen International.

Thomas, M., E. Demeulenaere, J.C. Dawson, A.R. Khan, N. Galic, S. Jouanne-Pin, C. Remoue, C. Bonneuil and I. Goldringer (2012). 'On-Farm Dynamic Management of Genetic Diversity: The Impact of Seed Diffusions and Seed Saving Practices on a Population-variety of Bread Wheat,' *Evolutionary Applications* 5(8): 779–95.

Tooker, J.F., and S.D. Frank (2012). 'Genotypically Diverse Cultivar Mixtures for Insect Pest Management and Increased Crop Yields,' *Journal of Applied Ecology* 49(5): 974–85.

Van der Ploeg, J.D. (2010). 'The Peasantries of the Twenty-first Century: The Commoditisation Debate Revisited', *The Journal of Peasant Studies* 37(1): 1–20.

Vernooy, R., P. Shrestha and B. Sthapit (eds.) (2015). *Community Seed Banks: Origins, Evolution and Prospects,* Routledge, Abingdon, Oxon.

Vernooy, R., B. Sthapit, G. Galluzzi and P. Shrestha (2014). 'The Multiple Functions and Services of Community Seedbanks,' *Resources* 3(4): 636–56. Available at: www.mdpi.com/2079-9276/3/4/636 (last accessed 9 February 2016).

Vodouhe, R., A. Dansi, H.T. Avohou, B. Kpèki and F. Azihou (2011). 'Plant Domestication and its Contributions to In Situ Conservation of Genetic Resources in Benin,' *International Journal of Biodiversity and Conservation* 3(2): 40–50.

Zhu, Y.Y., H. Chen, J. Fan, Y. Wang, Y. Li, J. Chen, J. Fan, S. Yang, L. Hu, H. Leung, T.W. Mew, P.S. Teng, Z. Wang and C.C. Mundt (2000). 'Genetic Diversity and Disease Control in Rice,' *Nature* 406: 718–22.

Part I

Dynamism in the field

Factors affecting the evolution and maintenance of distinct traits in farmers' varieties

2 Technical challenges in identifying farmers' varieties[1]

Jeremy Cherfas

Introduction

Discussions around the policy questions associated with farmers' varieties and their possible protection have generally tended to treat farmers' varieties as a monolithic entity. This notion is somewhat surprising as it denies the very diversity that is at the heart of farmers' varieties. This chapter will examine through biological and cultural lenses the various attempts to define farmers' varieties in an effort to isolate, if not clarify, some of the technical difficulties that beset a seemingly simple question such as, 'Can we identify a landrace?'

Definitions

Nomenclature in this area is confused at best. One sees terms such as landrace, farmers' variety, traditional variety, farmer selection and others used more or less interchangeably. One also sees fine distinctions being drawn among the various names to denote slightly different concepts. This chapter will generally refer to landraces and farmers' varieties as if they were interchangeable, because usually they are. It is, however, worth looking at the historical development of the ideas that these various terms embrace, because it gives interesting insights into the role of the farmer in the process and, thus, may point to future changes.

The full name of a living thing is essentially a hierarchy of ever-smaller categories. Thus, durum wheat is *Triticum durum*; the genus *Triticum* indicates that it is closely related to other wheats, for example, *T. dicoccoides* (emmer wheat). The species *durum* indicates that durum wheat is not emmer wheat, which is a different species. One generally accepted definition (with exceptions, of course) is that even if they can interbreed, members of different species do not usually produce fertile offspring. Landraces and improved modern varieties alike fall below the rank of species. They can interbreed and produce fertile offspring. The rules that govern the characteristics that indicate the various different ranks are agreed by international bodies that publish the International Codes of Nomenclature.

There are four different codes. The International Code of Nomenclature for Bacteria was originally part of the International Code of Nomenclature for

Botany and was separated in 1975.[2] The International Code of Botanical Nomenclature, in addition to plants, also covers organisms traditionally studied by botanists, such as blue-green algae and fungi, which are no longer considered to be plants.[3] The International Code of Zoological Nomenclature is responsible for animals,[4] and the International Committee on Taxonomy of Viruses manages the naming of viruses.[5] Each of these bodies was established by agreement among the members of various international scholarly bodies, and each is responsible for setting the standards for naming and for adjudicating competing claims, which each does by virtue of its agreed constitution. Their chief purpose is to ensure that organisms have an agreed scientific name that is accepted worldwide.

Of particular concern to this chapter, the International Codes of Nomenclature also establish the rules governing the different ranks of a taxonomy – for example, the genus and species – and, thus, might be expected to shed light on the question of what constitutes a farmers' variety. In botany and zoology, recognizably different populations of the same species that can interbreed and produce fertile offspring are known as subspecies. For cultivated plants, the problem is more complex and is addressed in part by the International Code of Nomenclature for Cultivated Plants (ICNCP).[6] The ICNCP states that 'the cultivar is the primary category of cultivated plants whose nomenclature is governed by this Code' (Article 2.1) and it defines a cultivar as 'an assemblage of plants that has been selected for a particular attribute or combination of attributes and that is clearly distinct, uniform and stable in its characteristics and that when propagated by appropriate means, retains those characteristics' (Article 2.2). The ICNCP goes on to explain that while cultivars 'differ in their mode of origin and reproduction . . . only those plants which maintain the characteristics that define a particular cultivar may be included within that cultivar' (Article 2.5). The ICNCP recognizes clones of different kinds, and states that

> an assemblage of individual plants grown from seed derived from uncontrolled pollination may form a cultivar when it meets the criteria laid down in Art 2.2 and when it can be distinguished consistently by one or more characters even though the individual plants of the assemblage may not necessarily be genetically uniform.
>
> (Article 2.11)

Similarly, 'an assemblage of plants grown from seed that is repeatedly collected from a particular provenance and this is clearly distinguishable by one or more characters (a topovariant) may form a cultivar' (Article 2.15). Finally (for our purposes), the ICNCP notes that 'in considering whether two or more plants belong to the same or different cultivars, their origins are irrelevant. Cultivars that cannot be distinguished from others by any of the means currently adopted . . . are treated as one cultivar' (Article 2.17).

Crucial points to consider are that the ICNCP's definitions are important largely in terms of the intellectual property rights protection that can be afforded to a recognized cultivar, which by definition will be distinct from all other cultivars. 'Cultivar' thus is often treated as equivalent to 'plant variety' within the meaning of the International Union for the Protection of New Varieties of Plants (UPOV Convention), and the recognition of a cultivar or variety may be essential to provide its breeder (or other designated entity) with some legal protection.[7] And while the ICNCP does not define a landrace or farmers' variety on the basis of its definitions and exegesis, some farmers' varieties might well qualify as cultivars, while others definitely would not because they fail to meet one or more of the essential qualities of distinctness, uniformity and stability.[8]

The ICNCP is essentially an instrument drawn up by and for plant breeders, not farmers. The natural habitat of the ICNCP's accepted definition of cultivar is the breeders' fields, and the 'appropriate means' of propagation are those that ensure that the cultivar is indeed 'distinct, uniform, and stable' (DUS) and that it remains so. Landraces exist in farmers' fields, and unless one knows about the use, cultivation, and management of a given landrace, it is impossible to decide whether it is likely to be DUS. This notion is expanded on later in this chapter, but, for example, a farmer may be exercising balancing selection on a variety in one part of her farm with the result that the characteristics of the 'assemblage of plants' remains stable from year to year, while in another part of the farm she is exerting directional selection such that the assemblage changes over time. Use, cultivation and management directly influence DUS characteristics.

What is a landrace?

The use of landrace to denote a biological entity associated with agriculture emerged first in the 1890s. A.C. Zeven (1998) provides a review of definitions since that time. In its original incarnation, a landrace was viewed as a source of material for plant breeding and little more. Landraces were recognized as genetically variable populations – which was one reason for their interest to breeders – with generally lower yields than improved varieties. The main use of the term was to distinguish some cultivated populations from wild species and from the products of scientific breeding. Early definitions also often involved the idea of endemism: a landrace was associated with a particular place (although usually not with the people who lived there). How long the landrace had to have been there varied from 'time immemorial' to 'a generation or two.'

Stability of yield, which is often associated explicitly with adaptability and genetic diversity, is another characteristic often used to identify landraces and to distinguish them from scientifically bred and genetically uniform cultivars. Such stability can arise in two nonexclusive ways. First, genetic heterogeneity confers the wide adaptability that allows a population to yield under a wide range of environmental conditions during any one growing season. Second, the same adaptability based on genetic heterogeneity allows the population to respond to shifts in conditions from year to year, with concomitant shifts in the most favoured genotypes. Scientific breeding tends to minimize the genetic heterogeneity of a cultivar, but something else clearly maintains the variability of a landrace.

Early authors dismissed the link between the activities of farmers in a particular place and the maintenance of landraces. Indeed, von Rümker (1908, cited by Zeven 1998) said that no human selection was involved. A landrace, he thought, was the result of unselected adaptation to growing conditions and would maintain its distinguishing characteristics even when grown outside its region of origin. This view persisted until the 1970s, when J.R. Harlan (1975) was one of the first to suggest that landraces depend on cultivation and, hence, a measure of artificial selection is essential to their survival. The importance of farmers in the maintenance of landraces quickly gained ground over the following decades, with several contributions based on theoretical and practical considerations. Hodgkin, Ramanatha and Riley (1993) state that 'the most important feature' of landraces is that human intervention is needed to create and maintain them. Brush (1995) agrees that landraces owe their existence to farmer selection. Louette (2000) demonstrates that selection by farmers is crucial to the maintenance of local maize varieties in Cuzalapa, Mexico, while Teshome and his colleagues (1999) show that farmers not only maintain local landraces but are also consistent in recognizing and distinguishing sorghum landraces with an accuracy that 'approximates the accuracy of standard scientific taxonomic approaches.' Prain and Campilan (1997, p. 325) demonstrate that in upland Irian Jaya certain sweet potato landraces (which they call cultivars) are culturally important in land consecration and rituals associated with first planting and that without this 'cultural saliency' these landraces might have vanished years ago.

Despite this more recent work, Zeven (1998, p. 127) concludes his review by defining a landrace as follows: '[A]n autochthonous landrace is a variety with a high capacity to tolerate biotic and abiotic stresses resulting in a high yield stability with an intermediate yield under a low input agricultural system.' The insertion of 'autochthonous' (i.e. endemic) ignores the question of whether farmers outside the landrace's area of origin can maintain that landrace. It also requires that one know about a plant population's place of origin and agronomic performance in order to decide whether it is a landrace. And as Halewood and his colleagues (2006, p. 175) note, 'it is curious that after providing such a full account of the development of the appreciation of farmers' roles in landrace development, conservation and use, Zeven's own definition underplays the element of dynamic farmer selection/maintenance.'

In this chapter, we attempt to rectify what we see as Zeven's omission. Yet before we do so, we should acknowledge that there is a distinct temptation to avoid grasping the nettle of attempting a definition, given that it could well be picked over and found wanting in the future. Nevertheless, there are certain characteristics of landraces or farmers' varieties that must be taken into account. Landraces require the activities of farmers for their maintenance. In this necessity, they are like modern cultivars, which also require propagation by 'appropriate means' to retain their characteristics. Excepting individuals of a clonally propagated species, which are multiplied by vegetative methods rather than by sexual selection (see the following discussion) and which, as a result, are essentially genetically identical to one another, however, landraces represent a population of individuals that are genetically much more heterogeneous than the individuals of a modern cultivar. This characteristic reflects the necessarily less controlled practices of farmers compared to scientific plant breeders, who can take great pains to ensure that their selections are genetically uniform, both in order to deliver agronomic performance that is highly predictable and to ensure themselves of the protection afforded by a 'cultivar' that can be registered for plant breeders rights if so desired.[9] A landrace is thus made up of individuals with different genotypes. The frequency distribution of any particular trait (or allele) will depend on the particular landrace. Of more interest is the fact that the frequency of different traits is likely to vary from year to year and from place to place, even in the absence of farmer selection. Since different genotypes will perform differently in any given season, the proportions of the various genotypes are extremely unlikely to be constant from year to year. (See also Box 2.1 on page 39.)

Consider two extremes. At one extreme is a farmer who buys a specific named and registered variety (cultivar) from a merchant in the formal sector every year. The farmer never saves seeds from his harvest and relies on the merchant to supply the same entity – the DUS cultivar reproduced by 'appropriate means' – each season. If growing conditions change, the farmer may notice a declining trend in yield until such a time as he decides to ask the merchant or seed company for a different cultivar, one that will perform better under the changed conditions. This farmer is definitely not growing a landrace or a farmers' variety. If he decided to stop buying seed, saved all his own seed and was diligent about keeping it DUS it would still not be a farmers' variety, as it would be identical to the cultivar that was originally obtained. A single sample in any field in any year would be sufficient to define the population.

At the other extreme is the farmer who never buys seed from the formal sector. Each year, she selects and saves seed from her harvest to plant the next season. She may additionally make other selections from different parts of the farm, perhaps to be sure of having some seed lots that perform well under wetter conditions. Even if she does not make a conscious selection, her variety will to some extent track changing growing conditions, especially if the trend is unidirectional year after year, because those individuals with genotypes better suited to the conditions will produce more seed than those less suited, and so will dominate the harvest to some extent. In between these two extremes are farmers who save much of their own seed, who exchange seed with neighbours, who sometimes obtain seed from further afield through informal channels and who may also experiment with small quantities of DUS cultivars obtained through the formal sector and through the efforts of government or nongovernmental organizations.

For the second kind of farmer, no snapshot based on a single season or a single locality will be able to capture a genetic or phenotypic description of her landraces. The shifting genotypes may cycle back so that over a period of a few seasons there will be an average composition. Alternatively, the average may be moving in response to some trend. As with the problem of defining palaeontological species, where the ability to interbreed is a matter of conjecture, this raises the issue of when the changes to a landrace over time require outsiders or farmers to recognize it as a new landrace. The genetic variability that characterizes landraces also underpins their most important characteristics for farmers: the fact that they are adaptable and resilient.

In the earlier discussion, we explored some definitions of landraces or farmers' varieties and some of the difficulties of placing those definitions in a modern population biology framework. Our own view, that landraces are generally genetically more heterogeneous than cultivars and thus more adaptable and resilient, and that they require the activities of farmers (as opposed to breeders) for their maintenance, is not a definition but rather a description. Even so, there are undoubtedly exceptions that might be classified as farmers' varieties or landraces but that do not fit the description.

Berg (2009) offers a more nuanced view that distinguishes between landraces and 'folk varieties' largely on the basis of how farmers select, sow, harvest and handle the crop in question. He points out that the rise of the use of the term landrace roughly coincides with the period in European agricultural history when modern cereal varieties, most notably wheat and barley, were beginning to be adopted. Landraces, Berg says, were 'adapted to local growing conditions through natural selection, usually with no intentional selection. However, the term was quickly adopted as a generic for all farmers' varieties, including those bred and maintained by active seed selection on-farm' (2009, p. 423). Berg prefers to call the latter folk varieties. He points out that typical European cereals, such as wheat and barley, are sown broadcast, harvested in bulk, threshed soon after harvest and stored as bulk commodities. This process makes selection difficult. Crops such as rice, millet, sorghum and even maize tend to be handled individually at many stages, from planting through to threshing and even cooking, making it much easier for farmers to observe, select and retain individuals that possess desirable traits. Berg notes that these characteristics of the interaction between people and crops, what he calls 'affordance,' makes the emergence of a wide range of diversity more likely. He 'define[s] folk variety as a farmers' variety that is selected and maintained for one or more distinctive properties. It may be fairly uniform for the selected traits, but otherwise diverse and therefore responsive to new selection' (Berg 2009, p. 426). He further notes that 'farmers who have folk varieties exchange seeds as a routine' (p. 427).

Perhaps the difficulty of arriving at a suitable all-encompassing definition reflects the fact that different crop species exhibit diverse reproductive systems, while the reproductive system has a crucial influence on the genetic heterogeneity and adaptability of plant populations. Reproductive systems also interact with agricultural practices. Thus, it makes sense to consider different crop species at least to some extent separately.

Landraces and farmers' varieties as open and closed systems

The three major breeding systems in plants are cross-pollination, self-pollination and clonal propagation, although these categories are not always absolutely distinct. There is a continuum from completely clonal, with no evidence for any production of seed, through to obligated out-breeders that possess genetic and biochemical mechanisms to ensure self-sterility. From a simplistic point of view, species that are multiplied clonally (including potatoes, cassava, dates and olives) would maintain their distinctive features over time because no gene flow normally occurs among different selections and, hence, the only source of change is somatic mutation. Such mutations would first have to be noticed and then propagated to create a new variety, but establishing the new variety is extremely easy. This method of establishment is, in fact, a relatively common source of varieties, especially among long-lived perennials such as fruit trees, where the pink grapefruit, Shamouti orange and Red Delicious apple are well-known examples.

In-breeders (including beans, rice, wheat and barley) generally self-pollinate and change relatively slowly. However, any new characteristics that do arise in an in-breeding variety will be reasonably easy to select, and, thus, new varieties will be relatively easy to create. Out-breeders (including maize, pearl millet and the brassicas) cross-pollinate and change most rapidly.[10] Thus, out-breeders would be the easiest and fastest crops for which to develop new and distinctive traits, but these traits will be difficult to maintain. It is also difficult for farmers to maintain an out-breeding variety in the face of the easy exchange of genes from other varieties of the same and closely related species.

This view, however, ignores the influence of human societies and the environment, which also vary in ways that could be described as open and closed. Commerce and trade serve to spread varieties and genes. Hence, societies that are more open to commerce and trade will probably be in receipt of genetic diversity that is likely to influence the stability and integrity of 'their' landraces. Landraces associated with more closed societies are likely to remain more distinct. The environment will also exert a similar influence. Crops growing in isolated environments, such as high, deep valleys and dense tropical forests, are likely to change less rapidly than those in open environments such as windswept plains and areas well served by road networks. Crops in areas where the growing conditions are very different from year to year will change more rapidly than those in more stable environments. The nature of the society and the nature of the environment in which it is embedded will also interact in ways that are largely unpredictable. For example, one could argue a priori that a society confined to a closed environment, such as a mountain valley, might be extremely open to new species and varieties that do somehow find their way into that society through rare and highly valued interactions with other societies. Equally, one could argue a priori that such a society will shun everything unfamiliar. Similar arguments can be made, with similar validity, for communities occupying open environments. They might be receptive to novelty or they might not. In the absence of empirical data, it seems optimistic in the extreme to attempt to generalize.

Influential variables

In-breeders

Few in-breeders are absolutely self-pollinating. For most, a small amount of introgression, from wild and weedy relatives and from other varieties growing nearby, represents an injection of novel genes that can change the population's genetic make-up. This new genetic material will be of much greater consequence in a farmer-maintained variety than in a modern variety for which fresh seed is purchased in most seasons. Rice, for example, is mostly grown through in-breeding, but when two varieties, or even two closely related species, grow next to one another some crossing will occur. For farmers who buy fresh seed for every planting season, and even if they save their own seed from one or two harvests, this crossing is of no consequence. For farmers who routinely save their own seed and do not obtain fresh seed in bulk, it is an important source of adaptation and innovation. In Pahang, Malaysia, Pesagi swamp farmers grow 'sticky' and 'normal' (*japonica* and *indica*) varieties of rice together, which permits an exchange of genes between the two types (Lambert 1985). The farmers harvest ear by ear, which gives them an opportunity to select desired types for next season's seed and to banish undesirable types. Each household improves its varieties continuously, and although different households may share varieties with the same name, those varieties can be genetically very distinct. Farmers also experiment with newly received varieties and with some of the off-types that they reject as seed for an already recognized variety. The rice landraces of the Pesagi farmers are thus in a constant state of dynamic flux despite being an in-breeding species.

Even for in-breeders, seed exchange among communities through what is generally referred to as the informal seed system will often result in a reduced distinctiveness of farmers' varieties in the absence of selection. In Nepal, for example, groups of villages commonly exchange barley and wheat seed roughly every 3 years, which results in varieties being more similar across the district than might otherwise be expected (Iijima 1964).

Out-breeders

Out-breeders, as explained, will generally be more dynamic and change more rapidly than in-breeders precisely because individual plants must be fertilized by a genetically different individual, thus mixing genes. Farmers are therefore more likely to be actively concerned with maintaining the characteristics of a variety against ongoing introgression from other varieties and wild relatives. Perhaps the most detailed investigation of farmer selection in an out-breeder is

Louette's (2000) study of maize in Cuzalapa, Mexico. The farmers in the study grew several kinds of maize in the presence of one another and in the presence of teosinte, a wild relative of maize. Among the more interesting results from this fascinating study is the constant interplay between the local varieties, including the farmers' own stock and that of their neighbours and exotic seed from outside the area, governed by the active selection carried out by farmers. Louette studied the occurrence of purple kernels within white or yellow ears, which indicates pollination by a variety called Negro. Farmers made no attempt to isolate varieties, and in the outside rows of a plot of white or yellow varieties, 20–30 percent of the kernels might be purple. This number fell to 1 percent after moving two or three metres within the plot, but since Negro is itself not homozygous for kernel colour these figures probably underestimate the amount of out-breeding.[11]

Farmers maintain the characteristics of named varieties by selecting ears at harvest to use for seed lots. They generally choose well-filled kernels from healthy ears, and the mean weight of the ears selected for seed was approximately 30 percent higher than ears chosen at random from the harvest. Farmers did not deliberately select against ears from outside rows, and yet they somehow selected against cross-pollinated kernels. White and yellow kernels are recessive and, therefore, do not show up when they have crossed with Negro as the maternal parent. Negro normally yields about 7.5 percent nonpurple seeds when grown under controlled conditions with no crossing from other varieties. Unselected seeds from a deliberate cross of Negro with a white or yellow variety increased the proportion of nonpurple kernels to 16.5 percent, but when farmers selected seed lots from these deliberately crossed Negro plots, the proportion of nonpurple seeds remained constant relative to the parental generation at about 7.5 percent. Clearly, Cuzalapa farmers are able to maintain some aspects of variety distinctiveness in the face of considerable genetic introgression.

Louette (2000) concluded that the ways in which farmers actually managed their seed 'call into question the genetic definition of a landrace' (p. 110), although it is noteworthy that she did not herself offer such a genetic definition. She also concluded that 'the assumption that traditional systems are closed and isolated with respect to the flow of genetic material is clearly contradicted by the results of this study' (p. 133). Landraces of maize in Cuzalapa are distinct and diverse, but they are also subject to intense selection and rely on constant infusions from outside the area.

Clones

Clonally propagated species offer perhaps the clearest examples of identifiable varieties persisting unchanged over generations, even millennia (Robinson 1996). The Dottato fig, for example, is mentioned by Pliny (23–79 C.E.) and is still grown in Italy. The Ari lop fig has been grown in Turkey for 2,000 years, and the ancient variety Verdonne has been cultivated around the Adriatic for centuries. Aroids, date palms, olives, ginger, garlic, grapes, saffron, sisal, vanilla and black pepper are among the many crops in which several ancient clonal selections exist. These clonal selections have presumably changed little, and the very ease of their propagation means that many have spread widely around the globe. Many of the most ancient varieties are of long-lived perennial species, and any one of these selections would, if it were novel, almost certainly qualify for protection under existing laws.

A further distinction can be made between those varieties that people choose to maintain clonally, such as potato and cassava and most fruit trees, and those that lack all ability to reproduce sexually, such as garlic and many kinds of banana. Those species that can reproduce sexually often make important contributions to the diversity and identity of landraces. Brush, Carney and Huaman (1981) argue that potato cultivation in Peru favours diversity and change because the people sow mixtures of different species and genotypes in a single field, and the results of chance pollination are often harvested as 'rogue' tubers, which may then be saved for planting the following year. The same is true in the Altiplano of western Bolivia (Johns and Keen 1986). In addition, communities throughout the Andes frequently exchange tubers, further promoting a dynamic system. One might thus consider potatoes to be rather like common beans, in which the genotypes in a field at any one time might represent either a landrace themselves or a mixture of several landraces. For a community, its landraces might be characteristic of the particular locale in the medium term, but they are also components of a large, open and interlinked genetic system.

Similar considerations apply to other crops that are capable of sexual reproduction but that farmers normally choose to propagate clonally. Cassava, for example, is a perennial out-breeder in the wild. Boster (1984) has shown that among the Jivaro people of the Peruvian tropical rainforest, there is a constant turnover as old varieties are lost and new ones that appear as volunteers, possibly hybrids with wild relatives, are nurtured and join the farmer's arsenal of landraces. However, the cassava-farming communities of the Amazon are relatively isolated, and so local populations of cassava remain distinct from one another (Salick, Cellinese and Knapp 1997). Sweet potato is subject to a very similar regime in New Guinea, where considerable local knowledge is associated with volunteer seedlings and their incorporation into existing varieties (Schneider 1995).

Finally, it should be noted that Berg's (2009) distinction between landraces and folk varieties adds another dimension to the matrix, namely, how people interact with their crop. Such activity can also depend on the circumstances. Berg draws attention to Harlan's record of great diversity among barleys in Ethiopia and contrasts that with his conclusion that in Europe there were a limited number of types. Ethiopian farmers created large numbers of folk varieties from their barley, while European farmers maintained a few landraces.

Preliminary evaluation of policy implications

This necessarily brief summary of the interacting influences of the reproductive system, culture and environment is enough to suggest that a blanket approach to the identification of farmers' varieties for the purposes of protection is unlikely to be fruitful. Two conclusions about landraces emerge. Landraces are generally components of large, interconnected and dynamic networks of exchange (among communities and among gene pools) that defy strict definition. And without continuous intervention by farmers, their varieties would cease to exist. This understanding applies even to those long-lived, clonally propagated species, which would inevitably die out if not replanted or regrafted.

Thus, it might be useful to consider individual cases along the three dimensions that have been examined: breeding system, environment and human activities. At one extreme (closed, closed, closed) are cases such as cassava among the Jivaro people of the Peruvian Amazon. Individual selections are maintained clonally, and there is little trade or commerce among the different communities. Hence, farmers' varieties are DUS, at least in the short term. However, because of the use that farmers make of introgression from the wild and from other varieties, stability cannot be guaranteed in the long term. At the other extreme (open, open, open) are cases such as the maize farmers of Cuzalapa in Mexico. Seed is brought in from neighbours and from outside, crossing takes place readily, and farmers take no great pains during the growing season to isolate their varieties. One might expect farmers' varieties in such cases to be indistinct, nonuniform and unstable, and, yet, as a result of the selection by farmers of seed lots that are a nonrandom subset of the harvest, varieties do maintain a certain identity that is DUS. Among these two extremes will lie other clusters of conditions that make farmers' varieties more or less identifiable, both in themselves and with a specific community. The improved lines of Jethobudho rice in Nepal (discussed in Chapter 4 of this volume) represent a concerted and successful effort to develop a farmers' variety that was both identifiable and protectable. Other similar cases may well arise, but the properties of the farmers' varieties concerned cannot be predicted in advance. (See Box 2.2 on page 40.)

Perverse incentives

A word of caution about perverse incentives – in general, farmers are managing systems whose strength may be based, in part, on high levels of informal exchange with neighbouring farmers and on the integration of new materials from manifold sources. They do so primarily to retain and enhance the adaptability and resilience of their varieties and systems. If the genetic make-up of a variety changes as a result, but it continues to deliver what they require, including organoleptic qualities, the genetic changes are of no concern to them. Legalistic definitions, however, often require the process of ongoing change to stop or slow considerably. Protection associated with definition requires a variety to be 'owned' by reasonably few individuals or communities. It would be counterproductive to create a system of protection that encourages farmers to stop sharing their materials with others in an attempt to maintain their privileged relationship with the variety in question. Equally, it could well be important not to reward farmers for preserving the identity or 'purity' of the landrace if that threatens the ability of the landrace to deliver the qualities that prompted its selection and maintenance in the first place. Essentially, one does not want any system to radically change farmers' behaviour. In a small community, a refusal to share could be a problem. In a large community, freeloaders, who make use of the variety but do nothing to maintain it, could be a problem. In both cases, vigilance against such perverse incentives will be needed.

Conclusions

Other contributors to this volume have treated in detail the various reasons for which one might want to be able to identify a landrace and the consequences that might flow from this identification. As far as the biological issues are concerned, there will indeed be occasions in which the confluence of factors is such that a variety can both be identified and associated with individuals or a community. In this sense, one could define and protect a farmers' variety. It is, however, impossible to predict with any degree of certainty how often the constellations of factors will be aligned, and, in general, pessimism seems more appropriate than optimism.

An analogy from the world of music may be helpful. While 'music' has connotations of 'art' and 'aesthetic content,' organized sound is a universal element in all human cultures. In every human culture, 'organized sound' (music *sensu latu*) serves a purpose and is not necessarily supposed to be beautiful or aesthetic. Different cultures, and different members of one society, may disagree as to what specifically constitutes music, but all would agree that there is something that can be called music. Much of this music can be called folk music. It is not written down but is transmitted orally and aurally. Themes, melodies, storylines and all other elements are often shared among neighbouring communities. Sometimes a folk song will be carried far afield when its 'owners' emigrate. Words change, emphasis shifts, melodies mutate and remembered histories adapt, until only an expert academic musicologist can demonstrate that a tune such as 'Cumberland Gap' is the mutated descendant of 'Bonnie George Campbell.' Folk tunes are landraces.

A minor part of music can be considered 'high art' or 'classical.' It is written down or codified in some other way, and each composition is essentially fixed, apart from such fleeting aspects as interpretation or expressiveness. Many composers, however, find inspiration in folk tunes, using them as a basis on which to build works that, while they have much in common with the folk tunes on which they are based, are nevertheless not folk tunes. They are formal compositions – registered varieties, if you will.

There are opportunities for confusion in music and in plant varieties – compositions mistaken for folk tunes and scientifically bred varieties passing into lore as 'heirlooms,' but seldom in the other direction.[12] The question then arises: why would you want to capture a folk tune? I do not presume to know why a composer would want to do so, but collectors do it for much the same reason that farmers exchange varieties and gene banks save landraces – as a foundation for scholarship, as the basis for future developments and, in many cases, as a means of specifically strengthening the identification of a society with its geographical antecedents.[13] Collectors do indeed manage to overcome the technical challenges of identifying folk tunes, but the continuing evolution of those tunes means that the collection represents one instance that 'belongs' to the nexus between singer and collector alone.

The various definitions of cultivar in the ICNCP all include 'attribute[s] or combination of attributes' that are 'clearly distinct, uniform and stable when propagated by appropriate means.' As we have shown, landraces or farmers' varieties certainly can meet these conditions. If they could not, farmers would not be able to identify them from place to place or from time to time, which they can. (One way in which they might be encouraged to do so in a way that would help researchers to make use of their knowledge would be through the widespread adoption of a standard set of descriptors, such as the List of Descriptors for Farmers' Knowledge of Plants prepared by Bioversity International [2009].) Whether a particular landrace maintains its unique characteristics for a long enough period or can be unequivocally associated with one farmer or small community is much more debatable. Certainly, when farmers are propagating their varieties 'by appropriate means,' the expectation is that they will not retain their characteristics because one reason farmers keep landraces, and one goal of their management, is to allow the crop to adapt to fluctuating conditions. If the farmers were to be persuaded to fix the variety, it might no longer meet their needs and might no longer be considered a landrace.

The question has been asked: are the traits selected for by farmers consistent enough over space and time to be able to say that a landrace is distinct or identifiable enough for protective systems such as *sui generis* intellectual property rights, registration, defensive publication or other schemes that require a definition? To this the only answer is, 'possibly.'

Box 2.1 Names are not varieties are not names

The naive view that equates distinct names with genetically distinct varieties has now been shown repeatedly to be just that – naive. There are two problems. The same name may refer to genetically different populations. Or the same genetic population may go under many different names. In fact, the name a farmer gives to a variety is almost inevitably an imperfect piece of information if our interest is actually in the genetic make-up of the variety.

Chakauya and his colleagues studied sorghum landrace diversity over a 20-year period in Zimbabwe and Mali. In the humid south of Mali, perhaps 70 percent of the variety names have been lost over the past 20 years. Nevertheless, far fewer agronomically important traits were lost over the same period. Genetic diversity has been largely maintained even though particular names have not. Bean farmers in Malawi and elsewhere often distinguish the different seed coat colours in a mixture with different names, but a single coat colour could be associated with several different underlying genotypes and an otherwise uniform genotype could possess two different coat-colour alleles (Martin and Adams 1987).

Busso and his colleagues (2000) used molecular markers to investigate the underlying genetic diversity of pearl millet landraces in two Nigerian villages. Differently named varieties growing on a single farm were more similar than identically named varieties grown by different farmers in the same village. Where clonally reproduced varieties are incorporating new material from hybrids and volunteers, the genetic make-up is bound to change even though the names remain the same.

Box 2.2 Specific research questions

One characteristic of nonclonally propagated farmers' varieties is that they are genetically heterogeneous. The limits of this variation are very poorly understood. Over time, the composition of a landrace will trace a path through genetic space that may keep it within certain boundaries. Will the space occupied by two landraces that are accepted as different always be completely nonoverlapping? Or will there be seasons in which one landrace is, at least as far as its genetic make-up is concerned, 'the same' as a different landrace in a different season? Similarly, if the landrace is undergoing some sort of directional selection, at what point will it have moved far away enough from the previous population to merit a new name, at least in the eyes of researchers, if not farmers?

The UPOV Convention permits the registration of population varieties that show a certain permissible level of variation in their distinguishing characteristics. In general, farmers' varieties will be considerably more variable than this level. However, is there some level of variability that would permit farmers' varieties to be registered? Or would this possibility undermine the standards for conventional cultivars? Indeed, information on the frequency distributions and variability of traits important to farmers in their landraces is still lacking, and better information on customary levels of variation in farmers' varieties might help to answer questions of definition and identification.

Notes

1 This chapter revisits and expands upon issues that I addressed as part of a research team in a previous co-authored publication: Halewood, M., J.J. Cherfas, J.M.M. Engels, T.H. Hazekamp, T. Hodgkin and J. Robinson (2006). 'Farmers, Landraces, and Property Rights: Challenges to Allocating Sui

Generis Intellectual Property Rights to Communities over their Varieties,' in S. Biber-Klemm and T. Cottier (eds.), *Rights to Plant Genetic Resources and Traditional Knowledge: Basic Issues and Perspectives.* CAB International Publishing, Wallingford, UK, pp. 173–202. This earlier publication was supported by the Swiss Development Corporation through a project coordinated by the World Trade Institute.

2 International Committee on Systematics of Prokaryotes, online: <www.the-icsp.org> (last accessed 15 June 2012).

3 International Code of Botanical Nomenclature, online: <http://ibot.sav.sk/icbn/main.htm> (last accessed 15 June 2012).

4 International Code of Zoological Nomenclature, online: <www.iczn.org/iczn/index.jsp> (last accessed 15 June 2012).

5 International Committee on Taxonomy of Viruses, online: <www.ictvonline.org/virusTaxonomy.asp?version=2008&bhcp=1> (last accessed 15 June 2012).

6 International Code of Nomenclature for Cultivated Plants, online: <www.ishs.org/sci/icracpco.htm> (last accessed 15 June 2012).

7 UPOV Convention International Convention for the Protection of New Varieties of Plants, adopted on 2 December 1961, online: <www.upov.int/en/publications/conventions/index.html> (last accessed 15 June 2012).

8 Here, in the absence of suitable definitions (see later), being used interchangeably.

9 But see also the discussion under in-breeders of heterogeneous landrace versus mixture of homogeneous landraces.

10 Note that a farmer variety of an out-breeding species will consist of individuals that are heterozygous at most alleles, while a farmer variety of an in-breeding species will be made up of several different, but largely homozygous, types of individual. This in itself is a source of confusion. Some authors (Martin and Adams (1971) cited by Zeven 1998) describe the mix of diverse genotypes of common bean (*Phaseolus vulgaris*, an in-breeder) as a landrace. Others (Voss 1992) also cited by Zeven (1998) would say that the same farmer is growing a mixture of several landraces.

11 The allele for purple kernel colour is dominant. So any given distinctively purple seed of Negro might be homozygous, with two *purple* alleles, or heterozygous, with one *purple* allele and one other colour, perhaps *yellow* or *white*. These heterozygotes will produce two kinds of pollen grain, *purple* and *yellow*. An ovum of the white or yellow variety in the test plot that is fertilized by a *purple* pollen grain will be visible and counted. One that looks yellow might nevertheless have been pollinated by a *yellow* pollen grain from the adjacent Negro plants, rather than by a *yellow* pollen grain from the test plot. The percentage of purple kernels thus underestimates the amount of gene flow from the Negro plants, and underestimates the amount of out-breeding affecting the plants in the test plots.

12 The 'enola bean' case may be one example, although in the end this attempt to 'recognize' a landrace within a legal system of intellectual property protection was exposed as unjustified.

13 The parallels are considerable, as a reading of the life of Cecil Sharp, the collector of English folk tunes, and some modern criticism reveals. Critics of Sharp 'reflect an idiosyncratic Trotskyist Marxist framework that views any and all folk song collecting, scholarship, and attempts at revival as malign forms of appropriation and exploitation by the bourgeoisie of the working class.' <http://en.wikipedia.org/wiki/Cecil_Sharp> (last accessed 15 June 2012). Similar criticisms can be found of efforts to identify, collect and study farmers' varieties and landraces.

References

Berg, T. (2009). 'Landraces and Folk Varieties: A Conceptual Reappraisal of Terminology,' *Euphytica: Netherlands Journal of Plant Breeding* 166(3): 423–30.

Bioversity International (2009). *Bioversity and the Christensen Fund Descriptors for Farmer's Knowledge of Plants,* Bioversity International and the Christensen Fund, Rome, Italy and Palo Alto, CA.

Boster, J.S. (1984). 'Classification, Cultivation and Selection of Aguaruna Cultivars of Manihot Esculenta (Euphorbiaceae),' *Advances in Economic Botany* 1: 34–47.

Brush, S.B. (1995). 'In Situ Conservation of Landraces in Centers of Crop Diversity,' *Crop Science* 35(2): 346–54.

Brush, S.B., H.J. Carney and Z. Huaman (1981). 'Dynamics of Andean Potato Agriculture,' *Economic Botany* 35(1): 70–88.

Busso, C.S., K.M. Devos, G. Ross, M. Mortimore, W.M. Adams and M.J. Ambrose (2000). 'Genetic Diversity within and among Landraces of Pearl Millet (*Pennisetum glaucum*) under Farmer Management in West Africa,' *Genetic Resources and Crop Evolution* 47(5): 561–8.

Chakauya, E., P. Tongoona, E.A. Matibiri and M. Grum (2006). 'Genetic Diversity Assessment of Sorghum Landraces in Zimbabwe Using Microsatellites and Indigenous Local Names,' *International Journal of Botany* 2(1): 29–35.

Halewood, M., J.J. Cherfas, J.M.M. Engels, T.H. Hazekamp, T. Hodgkin and J. Robinson (2006). 'Farmers, Landraces, and Property Rights: Challenges to Allocating Sui Generis Intellectual Property Rights to Communities over their Varieties,' in S. Biber-Klemm and T. Cottier (eds.), *Rights to Plant Genetic Resources and Traditional Knowledge: Basic Issues and Perspectives*, CAB International Publishing, Wallingford, UK, pp. 173–202.

Harlan, J.R. (1975). 'Our Vanishing Genetic Resources,' *Science* 188(4188): 617–21. Available at: http://science.sciencemag.org/content/188/4188/617.short (accessed 11 February 2016).

Hodgkin, T., R.V. Ramanatha and K. Riley (1993). *Current Issues in Conserving Crop Landraces in Situ*, Proceedings from Workshop on On-Farm Conservation, 6–8 December 1993, Bogor, Indonesia.

Iijima, S. (1964). 'Ecology, Economy and Social Systems in the Nepal Himalaya,' *Development Economics* 2: 91–105.

Johns, T., and S.L. Keen (1986). 'Ongoing Evolution of the Potato on the Altiplano of Western Bolivia,' *Economic Botany* 40(4): 409–24.

Lambert, D.H. (1985). *Swamp Rice Farming: The Indigenous Pahang Malay Agriculture System*, Westview Press, Boulder, CO.

Louette, D. (2000). 'Traditional Management of Seed and Genetic Diversity: What Is a Landrace?' in S.B. Brush (ed.), *Genes in the Field: On-farm Conservation of Crop Diversity*, Lewis Publishers, Boca Raton, pp. 109–42.

Martin, G., and M.W. Adams (1987). 'Landraces of Phaseolus Vulgaris (Fabaceae) in Northern Malawi,' *Economic Botany* 41: 190–203.

Prain, G., and D. Campilan (1997). *Farmer Maintenance of Sweet Potato Diversity in Asia: Dominant Cultivars and Implications for In Situ Conservation*, International Potato Center Program Report, International Potato Center, Lima, Peru.

Robinson, R.A. (1996). *Return to Resistance: Breeding Crops to Reduce Pesticide Dependence*, in association with the International Development Research Centre, Davis, CA.

Salick, J., N. Cellinese and S. Knapp (1997). 'Indigenous Diversity of Cassava: Generation, Maintenance, Use and Loss among the Amuesha, Peruvian Upper Amazon,' *Economic Botany* 51(1): 6–19.

Schneider, J. (1995). 'Farmer Practices and Sweet Potato Diversity in Highland New Guinea,' in J. Schneider (ed.), *Indigenous Knowledge in Conservation of Crop Genetic Resources*, Doc. CIP-ESEAP/CRFIC, Bogor, Indonesia.

Teshome, A., L. Fahrig, J.K. Torrance, J.D. Lambert, T.J. Arnason and B.R. Baum (1999). 'Maintenance of Sorghum (Sorghum bicolor, Poaceae) Landrace Diversity by Farmers' Selection in Ethiopia,' *Economic Botany* 53(1): 79–88.

Zeven, A.C. (1998). 'Landraces: A Review of Definitions and Classifications,' *Euphytica: Netherlands Journal of Plant Breeding* 104(2): 127–39.

3 The farmer's role in creating new genetic diversity

Carlo Fadda

Introduction

The heritage of locally adapted major and minor crops was created by female and male farmers using their own knowledge of the surrounding environment, their taste preferences, their cultural legacy and knowledge system and their ingenuity to find new solutions to new problems. Smallhold farmers in marginal areas – farmers that rely on a diverse portfolio of seeds as one of the few resources they have to meet their livelihood needs – are the custodians of genetic diversity and have the skills to adapt their resources to changing conditions. Under normal conditions, these farmers are touched marginally by new varieties and external inputs. On the one hand, they cannot always find the seeds and inputs that are needed, and when they are available, they can be too expensive for them or, more simply, they are not competitive with their existing seeds (Jarvis et al., 2011). On the other hand, their seeds are well adapted to the marginal environment in which they live (Barry et al., 2007), and they are useful for buffering against biotic and abiotic stresses (Smale, 2006; Bhandari, 2009), climate change and changing market pressures (Smale, 2006) and for satisfying cultural and religious needs (Rana, Garforth and Sthapit, 2008).

The amount of diversity available at any given moment in farmers' fields depends largely on the decisions made by farmers – in particular, on which seeds to keep for the next season and which seeds to acquire and from where (Jarvis et al., 2011). This choice has evolutionary consequences, as it affects the genetic structure and the distribution of different alleles in the following year. The selected seeds, furthermore, need to be adapted to the climatic condition and to the biotic and abiotic pressures they face when they are in the field (e.g. drought, pests and diseases). If they are well adapted, they will survive and they will be available for the next planting season. If they are not well adapted, they will not have an opportunity to be selected by farmers for the next cropping season and, therefore, they will be lost from the production system of that specific area.

Farmers keep changing the genetic diversity they deploy in their production system to keep pace with various economic, cultural and environmental changes (Gauchan and Smale, 2007). Farmers' selection combined with natural selection (in combination with the breeding system of the crops concerned) is what determines the genetic diversity we find today in the production system (Morimoto et al., 2006). To ensure that useful and adapted genetic diversity continues to be available on farm, it is essential that farmers continually engage in the process of selecting the varieties that better fit their environment as well as their cultural and personal preferences. This is essential to ensure that the evolutionary processes that evolve from the interaction between human and natural selection can keep producing well-adapted varieties. The traits developed through this mechanism are also very important for breeders to develop improved varieties and, therefore, it is very important that the process is maintained.

Ex situ conservation efforts (and some *in situ* conservation efforts) aim at conserving the genetic integrity of specific varieties. Clearly, while such conservation is critically important, it does not respond to the need to maintain systems of farmer innovation that continue to create new diversity and conserve existing diversity through its use. This chapter presents four case studies, selected from existing literature, which highlight the processes through which female and male farmers manage the diversity in their production systems, conserving some traditional varieties, contributing to the evolution of others and discontinuing the use of other varieties. I have selected case studies that involve different crops, on different continents and from different socioeconomic contexts, to highlight how the farmer's role in variety conservation and development is a critically important, common feature around the world. This focus will enlighten the difficulties of developing policies that can respond to the complexities in the farmers' communities that conserve and develop these varieties.

The first case study emphasizes the role of women as custodians of genetic diversity. The second and third case studies will explain the role of farmers in domesticating crops and developing new varieties from wild materials. In particular, the second study highlights how home gardens in Cuba are important 'laboratories' where crop varieties can interbreed either among themselves or with their crop wild relatives. The third case study will emphasize the importance of traditional knowledge in the farmers of Benin who manage wild relatives and cultivated varieties in a dynamic way that creates new varieties and new combinations of genes (Scarcelli et al., 2008). The fourth case study will highlight how a network of farmers in France has contributed to developing, managing and creating their own varieties. Before proceeding further, it is important to note that in this chapter I use the term variety to refer to traditional or farmers' varieties and landraces. When modern varieties are relevant to illustrate the case study, the term cultivar will be used to indicate varieties created through formal breeding.

The role of women in creating agricultural biodiversity

The broadest review published so far on the role of women in biodiversity management and conservation can be found in the book *Women and Plants: Gender Relations in Biodiversity Management and Conservation* (Howard, 2003).

In most societies, women are heavily involved in agriculture. Other than the work they do in the community's agricultural plots, they take care of their home gardens, store the food and cook it. As a result of their role in the household, they know what their family's needs are, and therefore they use their ingenuity to make sure that those needs are met. In other words, they shape the diversity to meet the needs of themselves and their family. Alongside cultivated crops, they use wild plants harvested from the surrounding agricultural landscape or grown in the home garden for food preparation, meaning that these women have a deep knowledge about their surrounding environment. Such knowledge is extremely important since it is the unifying factor linking the use of agricultural biodiversity to cultural aspects – loss of knowledge and culture leads to genetic erosion (Brush, 1999). Hence, it is extremely important to emphasize the role of women in making sure that the biodiversity they manage is constantly adapted to changing conditions.

One particular case study is relevant in shedding light on the role that women play in creating new varieties from the existing genetic material, namely the development of new maize varieties in southwest China (Song and Jiggins, 2003). This case is particularly interesting because maize is one of the most important crops in the world and is the staple food for many communities. In addition, it is an open-pollinated crop and, therefore, controlling crosses among varieties is very challenging. The Chinese case study shows how women's role in this process has become even more relevant since socioeconomic forces have pushed men to migrate to cities to look for waged work, leaving women to care for the farms. It also demonstrates the influence of socioeconomic and policy factors.

As has happened in several rural areas around the world, women are more and more in charge of agricultural activities, while men tend to migrate to larger cities to seek waged employment outside the agricultural field. In this particular case, a survey has shown that women represent more than 85 per cent of the agricultural labour force in southwest China (Song, 1998). In addition, they cannot receive much help from their children due to the 'one-child policy,' the result of which is the fact that most families only have one child at best. The agricultural policy in China tends to support the adoption of high-yielding varieties developed by the formal system. In this context, women face two major challenges: (1) accessing the varieties released by the formal system and (2) maintaining a range of varieties that suit their preferences and that are adapted to the particular environment of southwest China. The high-yielding varieties do not perform well under these farmers' conditions and they are difficult to access. Therefore, the main source of seeds has become the informal seed system – that is, the farmer-to-farmer exchange of seeds. In other words, the extension service has failed to ensure that farmers are adopting the high-yielding cultivars developed by breeders.

There are several advantages for farmers to use their own seeds in an open-pollinated crop such as maize, including the fact that farmers' seeds do not lose vigour after one generation, as do hybrid seeds. They can be manipulated to produce varieties that meet farmers' needs in terms of resistance to biotic and abiotic stress, taste, storage, cooking properties and so on, and they can be used in complex agricultural systems where they can be crossed with material from the formal system to improve their quality. However, such efforts require particular skills.

Women are very active in the selection of seeds. They choose the best plants from their seeds, the best cobs from the plants and the best seeds from the cobs. In addition, they collect the pollen from the plants for artificial pollination and, when needed, they remove the tassels from the seed plant before they shed pollen in order to avoid as much as possible unintended crosses.

By using their own seeds and capacity, therefore, women farmers can shape the diversity that meets their needs. This case study demonstrates that these farmers have been making informed decisions about what to do in their production systems, despite strong pressure to act differently and to adopt improved varieties. And it is not an isolated case. Several studies in different parts of the world reveal that projects aiming at promoting improved varieties by extension services have failed because the proposed varieties did not meet the farmers' needs. Such results were reported for maize in Ethiopia (Alemo et al., 2008), for beans and maize in Ecuador (Jose Ochoa, 2010, personal communication) and for beans in Uganda (Mulumba Wasswa, personal communication).

In this case study, a solution was adopted that made sure that the formal system was able to operate in partnership with the farmers, particularly poor women farmers, to develop varieties that serve multiple tasks:

1. to increase the yield in order to satisfy the government's request to increase production to meet food security goals;
2. to ensure that the conservation goal is also being met by increasing the variety of genetic material available to breeders and including farmers' varieties as well as increasing the amount of material available to farmers;
3. to ensure that poor women farmers have the varieties they need to satisfy their needs;
4. to enhance the leadership capacity of women farmers to ensure that their opinions are taken into consideration by decision makers.

The partnership involved collaboration at all levels and throughout the farming season, from the selection of varieties to be used in the breeding program to the actual implementation of the breeding activities.

Home gardens: a natural laboratory for new diversity

Home gardens can be considered a repository of genetic diversity. Unlike agricultural plots, in which only one or a few intercropped crops are normally grown, home gardens contain a much higher number of different crops as well as a significant amount of intraspecific diversity. They can be considered highly diversified ecological niches (Galluzzi, Eyzaguirre and Negri, 2010). Home gardens are found almost everywhere in the world, in rural as well as in urban areas.

Very often, they are actively managed by farmers to create and improve crops (Hughes et al., 2007). Such improvement is possible because these gardens have a very high intraspecific diversity of different varieties that can be crossed with each other either naturally or artificially. This diversity of landraces is an important source of genes adapted to a specific environment, and therefore the home gardens have a high resilience against biotic and abiotic stress (Galluzzi et al., 2010).

Households keep different species in their home gardens, with different life cycles and different domestication statuses, including wild and semidomesticated species. Plant species range from medicinal plants to fodder species and from food to ornamental species. Many species and crop varieties that are not found anymore in productive systems are still found in home gardens (Galluzzi et al., 2010). Home gardens, therefore, serve multiple purposes within the household. Among other purposes, they can be considered as natural laboratories where ingenious farmers can test new varieties, cross landraces with their wild relatives and ultimately create new varieties and exchange information about the newly developed material (Engels, 2002). Such efforts are possible because home gardens are largely used for the subsistence of the household with minimum commercial purposes, and are therefore an ideal place to test new solutions for new problems. The home garden is certainly a particularly dynamic production system that changes every year as a result of its management strategies (Hodgkin, 2002). In addition to farmers' maintenance of diversity, Colin Hughes and his colleagues (2007) showed that in Mexico, by bringing different varieties together that would otherwise grow in isolation, home gardens have created an opportunity for spontaneous hybridization that would not normally be present.

The selected case study is a description of home gardens in Cuba and how they are used by households to maintain and increase genetic diversity (Castineiras et al., 2000). The size of the home garden in Cuba is variable. Normally, the garden surrounds the house, but sometimes there is a shift in the garden when a part is left fallow for about three years. One characteristic of the garden is that it mimics an agroforestry system – that is, it has a multilayer arrangement that brings different species together. Everywhere in Cuba, the different strata are maintained in home gardens (subterranean, herbaceous, bushes and trees). Normally, more than 30 species are kept in a single home garden, including vegetables, fruit trees and medicinal plants. Overall, more than 300 species are kept in a given region as a result, most of which are domesticated crops (approximately 80 percent); the remaining species are wild. One of the most valuable species kept in the home garden is the sapote (*Pouteria sapote*). Sapote or mamey is a very important tropical fruit tree in America, which grows from Mexico to the northern part of South America. The tree is found in home gardens of Cuba, either wild when it is a remnant of previously existing forest vegetation, or in managed form when it is the result of human selection over time (Shagaradodksy et al., 2004). Despite the fact that sapote is not a fully domesticated tree, it is possible to observe differences between trees according to the geographic location of the home garden and the farmer's preferences. As a matter of fact, there are clear differences between fruits based on the geographic location of the home gardens. Generally, and despite its importance in tropical America, sapote is a poorly studied tree and knowledge has mostly been derived from farmers' knowledge.

Farmers have several uses for the sapote, which they highly value, considering its presence in their garden to be a sign of wealth. It is valued for the taste of the fruit, for making drinks, ice cream and jams. In addition, the green fruit has medicinal properties to control diarrhoea, and the nut is used to treat skin discolouration and to protect children's hair from lice. Seeds are also used to make handicrafts and jewellery, and the wood is used for construction, although farmers generally do not cut sapote trees. It is important to note that sapote has a high market value and can also be used as a cash crop during the harvesting season.

Despite the fact that sapote is a perennial tree and, therefore, that the selection process for diversity takes a long time, farmers show preferences for some of the characteristics of the fruits. Thus, if new trees are to be planted, they are selected from those that have the preferred traits, such as the size of the fruits, the number of seeds per fruit (lower is better) and the thickness of the pericarp. These preferred traits are chosen through human selection, but because farmers are also fully aware of the importance of the interaction between the surrounding environment and the tree, they need to adapt their choice in accordance with the surrounding environment. This example shows clearly how evolutionary forces driven by farmers' choices and environmental pressures operate together to determine an increase in diversity. Such selection is only possible if farmers actively manage this important resource in their home gardens.

Farmers' management of crop wild relatives

One way to create new diversity for the use of farmers and to ensure that the varieties available to farmers possess the diversity required to cope with an unpredictable environment is to keep bringing wild relatives into the production systems. Since wild relatives possess adaptive genes that are lost during breeding or domestication, they are generally considered a very important resource to be conserved even *ex situ*, and a strong emphasis is placed on their conservation. Farmers throughout the world are also aware of the importance of crop wild relatives. There are several case studies that have reported the use of wild relatives that have been grown around farmers' production systems as a way of enhancing the diversity on their farm. The ability of farmers to bring genes from wild relatives into their cultivated populations is a management practice that keeps their varieties adapted to changing conditions.

One very significant and interesting case study is the use of wild yam (*Dioscorea* sp.) in Benin (Scarcelli et al., 2008). Yam is a vegetatively propagated tuber crop mainly grown in West Africa by subsistence farmers in traditional agroecosystems (Baco et al., 2004). Yam plays a very important role in these traditional agroecosystems, being the second most important tuber after cassava. Farmers normally use about 30 percent of the production to sow a new field, through vegetative propagation. Despite the fact that many varieties are still able to produce flowers and seeds, there is no evidence that farmers use seeds to sow new plots (Scarcelli et al., 2008). Varieties cultivated by farmers belong to two species: the *D. cayenensis* and *D. rotundata* complex (*D. cayenensis Lam.*, *D. rotundata* Poir) (Scarcelli et al., 2006). In addition, two wild relatives exist in Benin: one has its distribution in the savannah environment of north and central Benin (*D. abyssinica*), and the second one grows in the forests of central and south Benin (*D. praehensilis*). Despite the fact that wild relatives and cultivated varieties can potentially produce hybrids, no evidence exists that hybridization actually occurs naturally (Akoroda, 1985), although wild relatives are found in sympatry with the cultivated varieties.

Farmers of West Africa and particularly of Benin have developed a farming practice that makes use of the wild relatives of yam. This practice is very complex and is called 'ennoblement' (Mignouna and Dansi, 2003). Farmers collect a piece of a tuber from the wild plant according to specific morphological criteria – that is, it has to be similar to a cultivated tuber with large green stems, large tubers, white flush and lacking spines (Dumont, Vernier and Zoundjihèkpon 2006). These tubers are then subjected to different types of stress to change their morphological characteristics – that is, a barrier is put in the soil in order to avoid the excessive growth of wild tubers and to attain a cultivated-like shape. This process lasts 3–6 years and is called pre-ennoblement (Scarcelli et al., 2008). What happens exactly from a biological point of view during pre-ennoblement is far from understood (Scarcelli et al., 2006), but as a result of this process farmers, if they are happy with the characteristics of the newly developed yam, will bring it into production system, either to merge it with already existing varieties with similar characteristics or as a new variety. By using ennoblement, therefore, farmers can increase the diversity in their production systems. However, what exactly happens from a genetic diversity point of view during the ennoblement process? In theory, somatic mutations are the basic evolutionary forces operating for vegetatively reproduced crops such as yam; however, we also know that yam, both cultivated and wild, produces flowers and therefore cannot be considered a strictly vegetatively propagated crop.[1] A genetic analysis at different levels (wild, hybrid and landraces), therefore, is needed to evaluate how much diversity is actually generated through this practice. A first step to better understanding how much diversity is generated through the ennoblement process lies in understanding the existing diversity found on farm – do varieties belong to either the wild species or to the cultivated one? Are they hybrid? Results have shown that of the five genotypes represented on farm, there are two wild species (which are the ennobled varieties), a cultivated one and two hybrids, *D. abyssinica* × *D. rotundata* and *D. praehensilis* × *D. rotundata* (Scarcelli et al., 2006; Scarcelli et al., 2008). The presence of hybrids shows that, through this practice, farmers have created new varieties with new genetic combinations via the sexual reproduction of wild and cultivated yams. This system, whereby a sexual cycle and asexual propagation are mixed, ensures potential large-scale cultivation of the best genotypes while preserving the potential for future adaptation. The potential for future adaptation is ensured by the sexual reproduction that occurs during ennoblement, while the best yam genotypes are preserved from recombination by using asexual vegetative reproduction.

Farmers' management of mixtures

French wheat producers in southern France have been cultivating a group of wheat varieties called Touselle since 1042, when the first record for this group was found in the annals. Touselle mixtures are characterized by being soft wheat varieties, which are appreciated for their baking qualities, their early maturation and their relatively high yield under marginal conditions. The grains can be either white or red, and the ears can have awns or be awnless. Touselle wheat is well adapted to the south of France, but after the First World War these varieties were almost completely replaced by modern varieties (Osman and Chable, 2009). A group of farmers interested in recovering Touselle organized themselves into a network and requested the seeds of a mixture of varieties from the National Gene Bank at the Institut National de la Recherche Agronomique (INRA). For the first 2 years, Touselle was cultivated in a farmer's garden and then was tried out in his fields. Gradually, the farmer learned more about it – how densely the seed has to be planted, how long it takes to ripen, how resistant it is to heavy rain and so on – and his experiments became well known in the region (Zaharia, 2007).

Other farmers began to copy him, and by 2004 Touselle was being grown experimentally on a fairly large number of peasant farms in the south of France. In 2005, the Syndicat de Promotion de la Touselle was founded with the idea of promoting the production of bread made from Touselle, supported also by consumer groups and organizations, because it is healthier and produces nutritious and tasty bread. Farmers growing Touselle became real scientists, devoting an area of their farms to experiments with other varieties of Touselle brought in by other farmers. Even if it was not easy to recover varieties that had been abandoned many years before, enough seeds were now available. Together, farmers started crossing varieties and developing new strains. In other words, they were developing new varieties, even if some of the varieties they were using had not been cultivated for many decades and required special treatment.

While some of the farmers simply multiplied the seeds, others began developing the varieties so that they would be appropriate not only for different regions, climates and types of farming but also for the type of use that would be made of the wheat after it was harvested. It is precisely because one is able to choose and develop old varieties for a wide range of applications that they are so innovative. To give an idea of the amount of diversity that has been created by farmers, consider the case of Jean-François, who spent 3 years testing other old varieties, getting seeds from the INRA and from old peasant farmers, and becoming more and more convinced of their excellent qualities. Finally, he decided to put his energy into building his 'living collection' of old seeds. Today he has 80 varieties in his back garden, sown in 4- to 10-metre lines (which he himself has multiplied from a few seeds); 200 mini-plots in a half-hectare plot (each 7 square metres), with 160 varieties, and 15 varieties on a 1-hectare plot, multiplied separately, which he uses for experimental bread making and to supply five or six organic farmers in the region. Some of the varieties, such as Bon Fermier, Richelle, Rallet, Blé du Lot and Bladette, date from the nineteenth century.

Jean-François's idea was to multiply the use of these old varieties throughout the region by supplying seeds to organic farmers who were willing to test them on their own farms and then supply them to other farmers. He believed that the seeds could be improved by the farmers themselves, who would then be able to furnish a regular supply of good wheat to the peasant bakers. The farmers he chose had a strong commitment to conserving and creating diversity that has been adapted to different environmental conditions and to different uses, and they were determined to achieve this goal despite many serious obstacles, including the fact that the exchange of seeds of unregistered varieties is prohibited and that, as a result, these varieties cannot be registered because they do not conform to the DUS criteria.

Synthesis

While far from being an exhaustive description of farmers' ability to adapt diversity to their needs and to specific environmental conditions, these case studies illustrate the ability of farmers to develop new varieties and to manage the available seeds to adapt to, and buffer against, environmental changes. Interestingly, in all of the cases, the farmers have faced obstacles in continuing to manage their seeds in ways that will allow for the continuing evolution of their varieties. These obstacles have also been embedded in government policies and in the erosion of traditional knowledge. In the first case study, one of the major constraints women faced in managing their varieties was the deployment of modern varieties that do not meet local needs and are not adapted to the particular environment of their communities. This situation is possible because of the seed law in China. In the second case study, Cuban farmers might have been discouraged to continue the cultivation of the sapote tree because of regulations that make it difficult for farmers to access the market. In the third case study, the erosion of traditional leadership was also eroding the traditional knowledge that recognized the need to conserve the sacred forests where wild relatives of yam are found. Such ignorance threatened the existence of wild yams and, therefore, the possibility of using them to create new diversity in the production system. In the fourth case study, one major obstacle for farmers is the strict French seed law that does not allow for the commercialization of varieties that do not conform to DUS criteria. Indeed, cumulatively, the case studies highlight the widespread disincentives that farmers face in continuing their traditional practices.

These practices are essential to maintain the evolutionary potential embedded in traditional varieties and offer farmers an opportunity to continue to cultivate the adapted versions of their original varieties (Dawson et al., 2012). There are many reasons why it might be desirable to conserve genetic diversity on farm (Jarvis et al., 2011), and climate change is just one of those reasons. Landraces are the product of human selection to satisfy a broad range of needs and natural selection that allows for specific traits to be adapted to the surrounding environment. The linkages between climate change and genetic diversity are therefore strongly associated with this adaptive potential.

As a matter of fact, the adaptive potential of landraces and the way they are harnessed by smallhold farmers is particularly important under the various conditions of climate change. While it is not possible to draw a sound conclusion on the impact of climate change on global-scale agriculture productivity (Gornall et al., 2010), many authors agree that this impact will be negative in areas in developing countries that are already food insecure (Lobell et al., 2008). There is a general consensus that genetic diversity is very important in adapting to climate change since it can provide many of the traits that can adapt to climate change. Yet genetic diversity is usually seen as a source of the traits that can be used by breeders to develop varieties that are better adapted to changing environmental conditions (e.g. Ceccarelli et al., 2010; Reynolds, Hayes and Chapman, 2010). However, if landraces have traits that can be used by breeders to develop varieties that are better adapted to changing climatic conditions, they can also be used directly by farmers to adapt their production systems to climate change and to continue adapting their varieties to the surrounding environment in ways that will also satisfy cultural needs (Esquinas-Alcázar, 2005).

As shown in the preceding case studies, farmers have the skills, knowledge and capacity to create new diversity and to continue to manage their existing varieties. In the context of climate change, farmers can face an additional problem. The seeds they are currently using might not be adapted for future agriculture, and they may not currently have well-adapted seeds in their seed systems (Bellon, Hodson and Hellin, 2011). In order to address this problem and to ensure that farmers have the required genetic diversity to be resilient in times of climate change, one possible solution is to increase the portfolio of varieties from which they can choose. Gene banks are conserving millions of accessions collected from farmers' fields, along with information on each variety's potential suitability. Returning part of this diversity to farmers so that they can use it to adapt to climate change and reestablish the evolutionary process that was stopped when the landraces were collected from farmers' fields could be an option. The same evolutionary process that allowed farmers to develop the accessions now conserved in gene banks could now be used to adapt to changing climatic conditions.[2] Farmers have demonstrated their ability to adapt their systems to climate change. Considering the speed at which climate change is occurring, it might be necessary to provide them with landraces collected from places with similar environmental conditions that the farmers are now facing.

Conclusion

Farmers have been creating and managing genetic diversity all over the world and for many different production systems, and this genetic diversity is now under threat and has been eroded (Brush, 1999). Far from being an exhaustive collection of case studies, this chapter has emphasized the importance of the role played by farmers in creating new genetic diversity. This type of evolutionary management of genetic diversity is based on two critical elements: the cultural element that shapes diversity in such a way that satisfies farmers' needs in terms of taste, size, shape and colour of seeds, cooking capacity and so on; and the interaction of the genetic diversity with the surrounding biophysical environment. A farmers' ability to select for these traits is very important since it serves to increase the amount of diversity within a crop in a context-dependent way. Thus, farmers in different areas will select differently for certain traits, and, by doing so, they will select for specific alleles regulating those traits. At the same time, a farmer's selection process will interact with the environment to select for traits that make the variety more adapted to the surrounding environment, and this process will strongly affect the productivity of their landraces. Traits such as pest and disease resistance, time to maturity and plant height are all affected by the environment, and farmers can make such traits their goals to improve productivity. This selection process is what creates new diversity, which is almost always optimally adapted to the local climatic conditions (Enjalbert et al., 2011). Cultural diversity, interaction with the environment and population genetics are what makes farmers' varieties function in a specific context, with farmers playing a key and unique role as evolutionary forces. This role can be even more important in a climate change scenario.

Notes

1 Somatic mutations occur in cells outside the dedicated reproductive group and therefore are not usually transmitted to descendants. In clonally propagated crops all mutations are in principle somatic as they do not reproduce sexually.
2 Bioversity International, online: <www.bioversityinternational.org/announcements/seeds_for_needs.html>.

References

Akoroda, M.O. (1985). 'Pollination Management for Controlled Hybridization of White Yam,' *Tropical Agriculture* 60: 242–8.

Alemo, D., W. Mwangi, N. Mandefro and D.J. Spielman (2008). 'The Maize Seed System in Ethiopia: Challenges and Opportunities in Drought Prone Areas,' *African Journal of Agricultural Research* 3(4): 305–14.

Baco, M.N., S. Tostain, R.L. Mongbo, O. Dainou and C. Agbangla (2004). 'Gestion Dynamique de la Diversite varietale des ignames cultivees (Dioscorea cayanensi-D. rotundata),' *Plant Genetic Resources* 139: 18–24.

Barry, M.B., J.L., Pham, J.L. Noyer, B. Courtois, C. Billot and N. Ahmadi (2007). 'Implications for In Situ Genetic Resource Conservation from the Ecogeographical Distribution of Rice Genetic Diversity in Maritime Guinea,' *Plant Genetic Resources* 5: 45–54.

Bellon, M.R., D. Hodson and J. Hellin (2011). 'Assessing the Vulnerability of Traditional Maize Seed Systems in Mexico to Climate Change,' *Proceedings of the National Academy of Sciences of the United States of America*. http://dx.doi.org/10.1073/pnas.1103373108.

Bhandari, B. (2009). *Summer Rainfall Variability and the Use of Rice (Oryza sativa L.) Varietal Diversity for Adaptation: Farmers' Perceptions and Responses in Nepal*, Master's thesis, CBM Swedish Biodiversity Centre, Uppsala, Sweden.

Brush, S.B. (1999). *Genetic Erosion of Crop Populations in Centers of Diversity: A Revision*, Proceedings of the Technical Meeting on the Methodology of the Food and Agriculture Organization, Prague, Czech Republic.

Castineiras, L., Z.F. Mayor, S. Pico and E. Salinas (2000). 'The Use of Home Gardens as a Component of the National Strategy for the In Situ Conservation of Plant Genetic Resources in Cuba,' *Plant Genetic Resources* 123: 9–18.

Ceccarelli, S., S. Grando, M. Maatougui, M. Michael, M. Slash, R. Haghparast and M. Rahmanian (2010). 'Plant Breeding and Climate Changes,' *Journal of Agricultural Science* 148(6): 627–37. http://dx.doi.org/10.1017/S0021859610000651.

Dawson, J.C., E. Serpolay, S. Giuliano, N. Schermann, N. Galic, V. Chable and I. Goldringer (2012). 'Multi-Trait Evolution of Farmer Varieties of Bread Wheat after Cultivation in Contrasting Organic Farming Systems in Europe,' *Genetica*. http://dx.doi.org/10.1007/s10709-012-9646-9.

Dumont, R., A. Dansi, P. Vernier and J. Zoundjihèkpon (2006). *Bioversity and Domestication of Yams in West Africa: Traditional Practices Leading to Dioscorea rotundata Poir*, CIRAD-IPGRI Collection Repere, Rome, Italy.

Engels, J. (2002). 'Home Gardens: A Genetic Resources Perspective,' in J.W. Watson and P.B. Eyzaguirre (eds.), *Home Gardens and In Situ Conservation of Plant Genetic Resources in Farming Systems*, Witzenhausen, Germany / IPGRI, Rome.

Enjalbert, J., J.C. Dawsona, S. Paillard, B. Rhoné, Y. Rousselle, M. Thomas and I. Goldringer (2011). 'Dynamic Management of Crop Diversity: From an Experimental Approach to On-Farm Conservation,' *Comptes Rendus Biologies* 334(5–6): 458–68.

Esquinas-Alcázar, J. (2005). 'Science and Society: Protecting Crop Genetic Diversity for Food Security: Political, Ethical and Technical Challenges,' *Genetics* 6(12): 946–53. http://dx.doi.org/10.1038/nrg1729.

Galluzzi, G., P. Eyzaguirre and V. Negri (2010). 'Home Gardens: Neglected Hotspots of Agro-Biodiversity and Cultural Diversity,' *Biodiversity Conservation*. http://dx.doi.org/10.1007/s10531-010-9919-5.

Gauchan, D., and M. Smale (2007). 'Comparing the Choices of Farmers and Breeders: The Value of Rice Land Races in Nepal,' in D.I. Jarvis, C. Padoch and H.D. Cooper (eds.), *Managing Biodiversity in Agricultural Ecosystems*, Columbia University Press, New York, pp. 407–25.

Gornall, J., R. Betts, E. Burke, R. Clark, J. Camp, K. Willett and A. Wiltshire (2010). 'Implications of Climate Change for Agricultural Productivity in the Early Twenty-First Century,' *Philosophical Transactions of the Royal Society: Biological Sciences* 365(1554): 2973–89. http://dx.doi.org/10.1098/rstb.2010.0158.

Hodgkin, T. (2002). 'Home Gardens and the Maintenance of Genetic Diversity,' in J.W. Watson and P.B. Eyzaguirre (eds.), *Home Gardens and In Situ Conservation of Plant Genetic Resources in Farming Systems*, Witzenhausen, Germany / IPGRI, Rome.

Howard, P. (ed.) (2003). *Women and Plants: Gender Relations in Biodiversity Management and Conservation*, Zed Books, New York.

Hughes, C.E., R. Govindarajulu, A. Robertson, D.L. Filer, S.A. Harris and C.D. Bailey (2007). 'Serendipitous Backyard Hybridization and the Origin of Crops,' *Proceedings of the National Academy of Sciences of the United States of America*, 104(36): 14389–94.

Jarvis, D.I, T. Hodgkin, B.R. Stahpit, C. Fadda and I. Lopez-Noriega (2011). 'A Heuristic Framework for Identifying Multiple Ways of Supporting the Conservation and Use of Traditional Crop Varieties within the Agricultural Production System,' *Critical Reviews in Plant Science* 30(1): 125–76.

Lobell, D.B., M.B. Burke, C. Tebaldi, M.D. Mastrandrea, W.P. Falcon and R.L. Naylor (2008). 'Prioritizing Climate Change Adaptation Needs for Food Security in 2030,' *Science* 319(5863): 607–10. http://dx.doi.org/10.1126/science.1152339.

Mignouna, H.D., and A. Dansi (2003). 'Yam (*Dioscorea spp.*) Domestication by the Nago and Fon Ethnic Groups in Benin,' *Genetic Resources and Crop Evolution* 50: 519–28.

Morimoto, Y., P. Maundu, M. Kawase, H. Fujimaki and H. Morishima (2006). 'RAPD Polymorphism of the White-Flowered Gourd (*Lagenaria siceraria*) (Molina) Standl. Landraces and Its Wild Relatives in Kenya,' *Genetic Resources and Crop Evolution* 53(5): 963–74.

Osman, A., and V. Chable (2009). 'Inventory of Initiatives on Seeds of Landraces in Europe,' *Journal of Agriculture and Environment for International Development* 103(1–2): 95–130.

Rana, R.B, C. Garforth and B.R. Sthapit (2008). 'Farmers' Management of Rice Varietal Diversity in the Mid Hills of Nepal: Implications for On-Farm Conservation and Crop Improvement,' *Plant Genetic Resources* 28: 1–14.

Reynolds, MP, D. Hayes and S.C. Chapman (2010). 'Breeding for Adaptation to Heat and Drought Stress,' in M.P. Reynolds (ed.), *Climate Change and Crop Production*, CAB International, Wallingford, UK.

Scarcelli, N., S. Tostain, M.N. Baco, C. Agbangla, O. Dainou and Y. Vigouroux (2008). 'Does Agriculture Conflict with In Situ Conservation? A Case Study on the SUs of Wild Relatives by Yam Farmers in Benin,' in N. Maxted et al., *Crop Wild Relative Conservation and Use,* CAB International, Wallingford, UK.

Scarcelli, N., S. Tostain, C. Mariac, C. Agbangla, O. Da and J. Berthaud (2006). 'Genetic Nature of Yams (*Dioscorea sp.*) Domesticated by Farmers in Benin (West Africa),' *Genetic Resources and Crop Evolution* 53: 121–30.

Shagaradodksy, T., L. Castineiras, V. Fuentes and R. Cristobal (2004). 'Characterization in Situ of the Variability of Sapote or Mamey in Cuban Home Gardens,' in P.B. Eyzaguirre and O.F. Linares (eds.), *Home Gardens and Agrobiodiversity,* Smithsonian Institution, Washington, DC.

Smale, M. (2006). *Valuing Crop Biodiversity: On-Farm Genetic Resources and Economic Change,* CAB International, Wallingford, UK.

Song, Y. (1998). *'New' Seed in 'Old' China: Impact Study of CIMMYT Collaborative Programme on Maize Breeding in Southwest China,* Ph.D. dissertation, Wageningen Agricultural University, Wageningen, The Netherlands.

Song, Y., and J. Jiggins (2003). 'Women and Maize Breeding: The Development of New Seed Systems in a Marginal Area of Southwest China,' in P.L. Howard (ed.), *Women and Plants,* Zed Books, New York, pp. 273–88.

Zaharia, H. (2007). 'Bread of Life,' *Seedlings* 4: 12–15.

Part II

Case studies at the interface of farmer variety enhancement efforts and national policies

4 Leveraging the successful participatory improvement of Pokhareli Jethobudho for national policy development in Nepal

Pratap Kumar Shrestha

Introduction to the enhancement project

Rice is a major crop in Nepal, both in terms of the area of cultivation and production, and it is the most preferred food in the country. The Pokhara Valley in the Kaski district is rich in rice diversity. D.K. Rijal and his colleagues (1998) reported 69 rice varieties under farmers' cultivation in this district, out of which 62 were landraces. Jethobudho is a popular rice landrace widely grown in the Pokhara Valley (grown by approximately 15 percent of rice growers) due to the high market demand for its quality traits such as aroma, taste, softness and other culinary qualities (Anonymous, 2006). Production of this variety has been fairly low, however, as it is susceptible to lodging as well as numerous diseases and has a low yield compared to modern varieties. This chapter is about a highly participatory project that was designed to enhance the qualities and performance of Jethobudho, to make it more attractive for use in the Pokhara Valley and to share associated benefits with local farmers.

The project, Strengthening the Scientific Basis of *In Situ* Conservation of Agricultural Biodiversity on Farm (*In Situ* Project), was jointly implemented by Local Initiatives for Biodiversity, Research and Development (LI-BIRD), which is a Nepal-based nongovernmental organization; the Nepal Agricultural Research Council (NARC); and Bioversity International, in collaboration with the Kaski District Agricultural Development Office (DADO), various farmers' groups and community-based organizations in operation at the project sites. It ran from 1997 to 2006. This chapter also draws on experiences with respect to the cultivation and use of Pokhareli Jethobudho after the *In Situ* Project came to a close.

One of the primary objectives of the enhancement program was to get the improved Jethobudho variety registered in order to facilitate the formal and legal production and marketing of its seeds as well as to provide some form of ownership right to the farmers who contributed to its development and conservation. The *In Situ* Project initiated a participatory landrace enhancement program for Jethobudho rice using methods of participatory plant breeding (PPB). PPB is a method of plant breeding that involves farmers in various stages of the breeding process – in parent selection, in the identification of preferred traits, in the selection of lines from the segregating or heterogeneous population and in the provision of feedback on the performance of the advanced lines (Witcombe et al., 2006). According to M.L. Morris and M.R. Bellon (2004, p. 32), PPB

> refers to a set of breeding methods characterised by many different forms of interaction between farmers and breeders . . . all designed to shift the locus to the local level by directly involving the end user in the breeding process.

The main objective of the PPB program for Jethobudho rice was to improve its production traits – reduce lodging, and increase disease tolerance and yield while maintaining the quality traits inherent to the variety – in order to maintain or increase its cultivation. Another objective of the enhancement program has been to develop a better, and relatively uniform and stable, variety compared to the local population but with the true traits of the Jethobudho landrace, making it eligible for official variety registration. This would then allow farmers to claim ownership rights over the new Jethobudho rice variety as per the existing Seeds Regulation, and commercially produce and market its seed. LI-BIRD, the Agricultural Botany Division of the NARC and Bioversity International, jointly developed the PPB program for the enhancement of the Jethobudho landrace in the Pokhara Valley, in collaboration with the DADO and the Fewa Seed-Producing Farmers' Group of Pame village in the district of Kaski, which represented farming communities of the area. It is interesting to note that while LI-BIRD, the NARC, the DADO and Bioversity International recognized that only by registering the enhanced variety and applying separately for a right to 'ownership' (equivalent to a breeder's right) under the existing regulations would the farmers be able to enjoy formal exclusive control over the crop and to benefit from its production, the farmers themselves were not aware of the need to register the enhanced variety. Without the assistance of the project partners, the farmers would not have been able to take advantage of this opportunity.

Two farming communities in the Pokhara Valley were selected as primary sites for the program, namely Fewa Phant (at 900 metres above sea level) and Malmul Phant (at 600 metres above sea level). Jethobudho rice is indigenous to the Pokhara Valley, and these two communities were well known for growing it. In fact, this particular variety has not been identified in any other part of Nepal. As mentioned earlier, the very qualities that make Jethobudho rice so popular, such as its aroma, taste and texture, have a great deal to do with the physical conditions of where it is grown. The soil condition, the water quality and the air temperature of the Pokhara Valley all serve to influence the quality of this cultivar. It is interesting to note the fact that this rice variety is so easily affected by its physical location is possibly a factor that limits its range of cultivation. While the project partners have started to assess its production in other hill districts of Nepal, its growth is clearly quite limited to date. Another reason for the selection of these two communities was the fact that, of all of the communities in Nepal, these two places were very interested in participating in the enhancement program in order to improve the production and the quality traits of their traditional Jethobudho variety. The farmers in these communities expected to benefit personally from the increased production and the higher market price that the enhanced Jethbudho rice variety would demand.

Description of the specific enhancement efforts

The initial survey of the Jethobudho rice production areas in the Pokhara Valley was carried out in 1998, and it revealed that there was considerable variation in the characteristics of the cultivar, in terms of its morphology (grain type and colour), its agronomy (plant height, tilling ability, panicle length, culm strength, yield, blast occurrence and lodging intensity) and postharvest quality traits (softness, flakiness, aroma and other cooking qualities). This information provided breeders and the project team with an adequate foundation from which to use a mass selection method for the enhancement of the Jethobudho rice landrace. The project team designed detailed procedures and methods for such a selection and started implementation of these activities in 1999. In 2000, two sets of 338 lines were cultivated in Malmul Phant and in Fewa Phant. Following the harvest, 183 lines were selected according to various traits, such as field tolerance to leaf and neck blast, a strong culm, plant height below 150 centimetres (in order to avoid lodging) and high productivity. From 2001 to 2003, these 183 lines were repeatedly assessed and tested, according to the desired traits outlined earlier, until the number of selections was reduced first to 46 varieties and then to the final six. The details of the procedures and methods adopted throughout the project are presented in Table 4.1.

From the evaluation of 338 lines of the Jethobudho landrace, six lines were selected that consisted of traits that were consistent with the true Jethobudho landrace, as described by farmers and consumers with reliable traditional knowledge. All of these six lines were better than the existing Jethobudho population in terms of their agronomic traits (tolerance to blast disease and lodging, yield components and fodder quality) as well as their postharvest traits (milling recovery, softness, taste, flakiness, aroma and other cooking qualities) (Gyawali et al., 2006; Gyawali et al., 2010). The traits of these six lines were adequately uniform, were distinctly separable from the normal Jethobudho population and were adequately reproduced in new plants in the subsequent years. These factors qualified the bulk of these six lines to be registered as a new rice variety.

The combined yield of the six lines was also higher than the normal population, ranging from 2.61–2.98 tonnes per hectare compared to 2.54–2.66 tonnes per hectare for the normal Jethobudho population (Gyawali et al., 2006; Gyawali et al., 2010). The milling percentage of the selected lines was also very high (71 percent), and local traders were willing to pay a higher premium price for it (approximately 100 Nepalese rupees per 0.1 tonne). The local traders were largely mill owners who were selling the milled rice to local consumers. The price for

Table 4.1 Procedures and methods adopted for the enhancement of Jethobudho rice in Nepal

Year	*Goals*	*Methods*	*Outcomes*
1998	• assessment of field diversity of Jethobudho for morphological, agronomic and quality traits	• participatory rural appraisal, baseline surveys and four-cell analysis[a] in farming communities throughout the Pokhara Valley	• considerable variation was found in the Jethobudho population throughout the seven major Jethobudho production areas
1999	• breeding goals were set for Jethobudho enhancement • analysis of the preferred traits • collection of accessions/lines for evaluation and enhancement	• breeding goals were set jointly by the project team and the farming communities • trait analysis was done with the help of expert farmers, traders/millers and hoteliers • farmers' fields were visited during the harvest, and five panicles were selected from each field	• breeding goals were aimed at improving the Jethobudho population by selecting preferred traits • landrace enhancement procedure was designed
2000	• diversity assessment • selection of lines for the identified preferred traits	• two sets of the 338 lines were assessed in farmers' fields: one at Malmul Phant and another at Fewa Phant • one set was assessed for blast tolerance at the NARC research station in Malepatan in the Pokhara Valley	• 338 lines of Jethobudho rice were collected from seven production areas
2001	• 183 selected lines of Jethobudho were evaluated for blast and lodging tolerance as well as yield components • consumers and market traders identified numerous quality traits	• the performance of 183 lines was assessed at the NARC research station in Malepatan and in farmers' fields in Fewa Phant • season-long data was collected for further screening • market surveys and consumer feedback were collected for the selection of quality traits • selection of lines was made on the basis of postharvest micromilling and organoleptic tests	• 183 lines were selected for awnless or short awn, field tolerance to leaf and neck blast, a strong culm, a height of no more than 150 centimetres and good productivity • 46 lines of Jethobudho landrace that confirmed most of the preferred tests were selected for further evaluation
2002	• evaluation of 46 lines of Jethobudho landrace for agronomic and postharvest traits • assessment of these cultivars for the aroma trait • assessment of these cultivars for wider adoption in the Pokhara Valley	• agronomic trials[b] were conducted in farmers' fields in Malmul Phant and Fewa Phant as well as at the NARC research station in Malepatan • micromilling and organoleptic tests[c] were conducted for postharvest traits	• six lines of Jethobudho landrace confirming most of the preferred test were selected for further evaluation

2003	• evaluation of six lines of Jethobudho landrace for agronomic and postharvest traits	• agronomic trials[b] were conducted in farmers' fields in Malmul Phant and Fewa Phant as well as at the NARC research station in Malepatan • micromilling and organoleptic tests for postharvest traits • simple sequence repeats assessment for aroma traits[d]	• all six lines performed better than the normal population • these lines were bulked for further evaluation
2004	• the bulk of six lines were evaluated for wider adoption • seed production was initiated	• participatory variety selection (PVS)[e] was used to test wider adoption • farmers began seed production in five farming communities in the Pokhara Valley	• bulk population performed better than the normal population
2005	• evaluation of bulk population and seed production • field visit by member of the Variety Release Committee (VRC)	• farmers participated in PVS and seed production throughout many locations in the Pokhara Valley • members of the VRC visited farmers' fields in three farming communities	• breeder and foundation seeds were produced for further multiplication • members of the VRC were satisfied with their performance
2006	• new Jethobudho population proposed for release and registration • seed production of new population	• meeting of the Variety Approval, Release and Registration Committee held on 4 July 2006 • seed production was initiated through the Fewa Seed-Producing Farmers' Group in the village of Pame	• the bulk of six lines was officially released and registered as Pokhareli Jethobudho • seed production and marketing continued by farmers' group

Source: Adapted from the original (Anonymous, 2006; Gyawali et al., 2006; Gyawali et al., 2010).

[a] Four cell analysis groups plant varieties into four quadrants with plant varieties grown by (1) many farmers in a large area (acreage); (2) many farmers in a small area; (3) a few farmers in a large area; and (4) a few farmers in a small area (Rana et al., 2005).

[b] Agronomic trials consisted of assessing agronomic traits, such as plant height, number of tillers, panicle length, grain yield, straw yield, occurrence of neck blast disease, sterility and culm strength for lodging resistance.

[c] Organoleptic test is done to assess cooking and taste qualities of the given plant variety. The milled rice of the enhanced Jethobudho was cooked and its cooking and taste qualities were compared to the traditional Jethobudho and other rice varieties.

[d] Simple sequence repeat (SSR) assessment uses a molecular marker technique to establish the genetic relationship among plant varieties. The SSR assessment showed that the selected six lines of enhanced Jethobudho consisted of traits for aroma and that these aroma traits were quite similar to those found in the well-known *indica* and *japonica* aromatic rice varieties (Bajracharya et al., 2004).

[e] Participatory variety selection is a participatory method of selecting new plant varieties, whereby farmers are provided with seeds of new plant varieties for their independent testing/experimentation using their own management practices and selection criteria. It provides an effective method of testing the suitability of new plant varieties, thus confirming farmers' diverse socioeconomic and production environment (Witcombe et al., 1996).

raw unprocessed rice was established through negotiation between the farmers and the traders, and it varied from year to year and even within the same season. One of the traders, a company called Karmacharya Traders, was a large enterprise based in Pokhara that operated nationally. It was selling Jethobudho rice with attractive packaging throughout Nepal. With the expanding market and increasing consumer preference, the demand and the price for the enhanced Jethobudho with better postharvest traits is higher than the existing Jethobudho variety.

After the development of the enhanced Jethobudho rice variety, the project partners began to look into the benefits of registering it. They began by discussing the advantages of registration with the Fewa Seed-Producing Farmers' Group, and the farmers showed interest and enthusiasm in following this course. Hence, the project partners approached the Variety Approval, Release and Registration Sub-Committee of the National Seed Board and invited them to make a field visit to monitor and assess the enhanced variety of Jethobudho in 2005. Such a visit is an established practice done with the objective of assessing the varietal characteristics (distinctness and uniformity criteria) of the potential new plant variety. The sub-committee members accepted the invitation and made a field visit to the Pokhara Valley in September 2005 (rice season). The team assessed the crop's performance and interacted with the performance of the project team and the farmers' group. The team was impressed with the performance of the enhanced Jethobudho variety and provided a positive report for its release and registration.

After receiving the favourable report, the project partners prepared and submitted the application form. It was assumed that the variety would be registered under the name of the farming community that was involved in its cultivation and conservation so that the farmers' right to ownership would be ensured. However, it was learned through the application process that the registration could not be made solely under the farming community's name due to the additional conditions set forth by the National Seed Board as per their interpretation of the related policy provisions in the National Seed Policy 1999. The application form required applicants to have a level of education and professional training for them to be able to maintain the breeder and foundation seeds that the farmers were unable to meet. Thus, the application was finally submitted under the names of the project partners and the Fewa Seed-Producing Farmers' Group as co-registrees. The farmers were in agreement with this decision.

However, this was not the only hurdle facing the project team. The partners were aware from the very beginning that registration would be very difficult due to the kind of scientific data that was traditionally required. There are currently four legal documents in Nepal that support the registration of new plant varieties: the National Seed Policy 1999, the Agricultural Biodiversity Policy 2007 (first amendment, 2014), the Seeds Act 1988 (first amendment, 2012), and the Seeds Regulation 2013 (after first amendment of Seeds Regulation 1997). These provisions, however, were suitable only for varieties that were developed through a formal breeding system. They require the adoption of more formal breeding procedures as well as the presentation of multiseason scientific data with statistical analysis, and skilled technical staff and infrastructure that were far beyond the common farmer's reach. Several years earlier, LI-BIRD and CAZS (Centre for Arid Zone Studies) Natural Resources of the United Kingdom had begun lobbying efforts with the National Seed Board to have these provisions changed in order to recognize new plant varieties developed from participatory plant breeding. These lobbying efforts eventually succeeded, and in 2004 the National Seed Board formed a committee to review the plant variety release and registration process. The review committee included members from the National Seed Board, the NARC, the Ministry of Agriculture, the Department of Agriculture, LI-BIRD, the CAZS Natural Resources and the Seed Entrepreneurs of Association of Nepal. While LI-BIRD did present the perspectives of farmers, farmers were not directly involved in the committee's decision making. In the course of the debate, LI-BIRD and CAZS Natural Resources provided evidence that:

1 participatory plant breeding did adopt scientific breeding principles and procedures;
2 the participation of farmers and the use of their knowledge and feedback in the breeding process increased breeding efficiency and the success of new plant varieties;
3 farmers' perception data was systematically collected, was subject to statistical analysis and was very useful in assessing the performance of the new plant varieties.

The evidence also demonstrated that the farmers were making rational decisions in selecting lines from the segregating population (Witcombe et al., 2006).

After lengthy discussions, the government finally agreed to recognize PPB as a complementary scientific method for developing new plant varieties, and to accept data from farmer participatory varietal research trials for the registration of PPB varieties. The proposal for new provisions for the national listing of local plant varieties, which are traditionally under cultivation, was also accepted, enabling formal registration of such varieties. These changes were made to the Seeds Regulation through a ministerial decision published in the Nepal Gazette on 6 June 2005. An industry breakthrough, these changes necessitated the inclusion of organoleptic data (cooking and taste qualities) and permitted the use of farmers' perception data in the process of registering new plant varieties. By accepting farmers' perception data, it was easier for farmers to meet the data requirements for variety release and registration. Ultimately, the use of farmers' perception data and the requirement for farmers' participation served (1) to reduce the experimental lag time as well as the time taken to register the variety and (2) to recognize the contribution of farmers' knowledge and breeding material in the cultivation process, thereby making it much easier for them to become co-registrees. This latter achievement, in turn, has paved the way to establish the farmers' ownership right to the enhanced Jethobudho variety along with the other (formal) partners.

With these various legal changes in place, it was finally possible to follow through with the application to register. On 4 July 2006, the enhanced variety of Jethobudho rice was officially released and registered under the name Pokhareli Jethobudho. Once the registration was complete, it was learned that it was also necessary to submit a separate application for ownership, as the right to ownership was not included with the registration. According to section 14 of the Seeds Regulation 2013, the application for ownership is optional and is left to the decision of the party registering the variety. The registration of Pokhareli Jethobudho has, for the first time in Nepal, recognized farmers as breeders and made them eligible to claim breeders' rights in the form of ownership right over the new variety. The farmers and the concerned organizations are now eligible to produce and market the seeds of this enhanced variety anywhere in the country, which was not possible for them prior to registration. However, as the scope of the ownership right is not defined in the existing Seed Law and Regulations, it is theoretically possible for anybody to produce and sell seeds of this variety. Because of this nobody has ever applied for the ownership rights for any registered plant variety. The Fewa Seed-Producing Farmers' Group, which is a co-registrant of the new variety, is responsible for the production and marketing of the seeds and is beginning to benefit from this business. The group works in close conjunction with LI-BIRD, which is providing technical support and marketing assistance. The majority of sales have been made to the DADO, LI-BIRD and private seed traders, as well as farmers within and outside the farming community. The monetary benefit to the farmer from the sale of seed was relatively small until 2010 due to low seed sold per year. This scenario, however, changed in 2011 when a local private seed entrepreneur, allowed to trade the seeds, started paying a higher premium price to the seed-producing farmers. The amount of profit that each farmer received for his or her rice depended on the amount of seed that was sold. From 2006 to 2014, each farmer has received a gross revenue of 54,494 Nepalese rupees per year (see Table 4.2).

The future of the Pokhareli Jethobudho landrace is still uncertain, but there are strong efforts being made to create adequate incentives for on-farm conservation and use. The current rate of seed production is low compared to the overall quantity of rice production in the Pokhara Valley. LI-BIRD, in collaboration with the DADO, has also been striving to popularize the variety in areas outside the Pokhara Valley. A total of 1,202 kilograms of Pokhareli Jethobudho rice seed was distributed to farmers for cultivation in eight neighbouring districts in 2010. LI-BIRD has been monitoring the performance of Pokhareli Jethobudho, and hopefully their success will increase the demand for this variety in the following seasons. Furthermore, eight farmers in the Thulakhet Agricultural Group in the Pokhara Valley also began to produce and market Pokhareli Jethobudho rice seeds. They produced a total of 1,121 kilograms of seed and earned a net revenue of 4,484 Nepalese rupees. Such success stories have encouraged further research on and development of landrace enhancement programs. The National Rice Research Programme of the NARC initiated the enhancement of Lalka basmati rice – a type of red husk basmati rice that is found in the southern plains of Nepal – and formally registered and released it with the National Seed Board in 2010. In addition, LI-BIRD is planning to initiate landrace enhancement programs for other endangered rice varieties that have potential commercial value, such as Jhinuwa, Anadi, Kariya

Table 4.2 Details of revenue generated from the sale of seeds of Pokhareli Jethobudho for seed-producing farmers in Kaski, Nepal

Year	*Number of farmers*	*Seed quantity (kg)*	*Seed price (NPR/kg)*	*Gross revenue (NPR)*	*Seed premium (NPR/kg)*	*Revenue from seed premium (NPR)*[a]
2006	7	3,425.0	35	119,875	4	13,700
2007	7	2,931.5	35	102,603	4	11,726
2008	7	3,240.5	44	142,582	5	16,203
2009	10	3,943.5	65	256,328	5	19,718
2010	10	4,788.0	75	359,100	5	23,940
2011[b]	10	4,850.0	120	582,000	70	339,500
2012	10	5,250.0	130	682,500	70	367,500
2013	10	6,500.0	150	975,000	80	520,000
2014	10	7,500.0	170	1,275,000	90	675,000
Total		42,428.5		4,494,988		1,987,287

Source: LI-BIRD for data from 2006–2010; Unnati Agro-vet for data from 2011–2014.

[a] The net revenue from premium prices was calculated by multiplying the total seed sold each year with the additional amount received as a premium for the improved seed.

[b] Until 2010 LI-BIRD facilitated selling seeds and regulated the price. Both the seed price and seed premium increased drastically after 2011 when Unnati Agro-vet, a local seed trader, took over buying seeds directly from seed-producing farmers, responding to actual market price.

Kamud, Tilki and others. It is also planning to start PPB for the enhancement of some neglected and underutilized species, such as amaranth, buckwheat, barley, beans and aromatic sponge gourd. There is now increasing attention to and research initiatives for the enhancement of threatened landraces in Nepal.

Analysis of the enhancement project: strengths, weaknesses, opportunities and threats (SWOT)

The genetic enhancement of the Jethobudho rice landrace has successfully led to the identification of a particular variety, registered as Pokhareli Jethobudho, which is better than the existing population and consists of all of the traits of the original variety, as confirmed by those with reliable traditional knowledge. In addition, it is a multiline variety and has maintained the original traits and genetic diversity of the Jethobudho landrace. The improvements that were made to the preferred traits of the variety have also created a greater demand for the product by all involved in the value chain (i.e. by farmers, traders and consumers). Initial observations show that farmers are now more willing to cultivate this variety, and this willingness will certainly create an incentive for farmers to conserve Jethobudho rice on their farms. The official registration of this variety has provided recognition domestically and internationally for the landrace genetic enhancement program as well as for the need for intellectual property rights for farmers.

The project has also demonstrated that farmers are clearly able to enhance and maintain a good quality landrace by adopting innovative selection procedures guided by their local knowledge about the traits of the variety. The six innovative farmers of the Pokhara Valley were responsible for maintaining the six lines that ended up being identified as Pokhareli Jethobudho. These farmers have now been deemed the custodians of the Pokhareli Jethobudho variety, and they are benefiting directly from their efforts and knowledge of this landrace. The experience shows that the participation of farmers and farming communities is extremely important for the success of any landrace enhancement program, and that external assistance in the form of training and resource development can only strengthen their capacity to undertake such initiatives on their own.

The case of the Jethobudho landrace has created opportunities for the enhancement, protection, utilization and conservation of local varieties and landraces of all other crops. Training and orientation programs could further empower farmers and farming communities to enhance local plant varieties and landraces as well as to officially register and establish farmers' rights over these varieties and landraces. The policy precedent set in the case of Pokhareli Jethobudho could be extended to other plants and plant varieties.

The Jethobudho landrace enhancement program was initiated and driven by the *In Situ* Project team, while farmers participated in the process and provided their input as it was invited. One of the challenges to the program was the fact that, aside from the six innovative farmers that were cultivating the six lines of the enhanced variety and who knew how to maintain the desired qualities of the Jethobudho landrace, many of the farmers in the region did not have the same know-how or the same desire to improve their landraces. The economic incentive to maintain the enhanced population of Jethobudho rice was probably not high enough for some farmers to invest their time and effort in such a process. This lack of incentive can be explained in three ways. First, a majority of the farmers engaged in the production of Jethobudho rice were tenant farmers who were cultivating the land for absentee landowners on a sharecropping basis. They were cultivating Jethobudho rice largely in the interest of the landowners and were not taking much of it for their own consumption. Moreover, since they were selling the majority of their share of the rice to local traders – who were willing to buy the rice no matter what its quality since the demand for the variety was always higher than the supply – the quality of the rice did not matter so much to them. Second, farmers in this district have always traditionally sold paddy rice (unmilled) and, therefore, were not accustomed to paying much attention to the quality of the milled rice. Third, farmers in the area had a tradition of selling their rice by volume and, therefore, were more interested in increasing the volume of their product than in improving its qualities, such as purity, uniform grain size, appearance and milling recovery. The six farmers who participated in the project, on the other hand, were producing the variety largely for their own consumption and, therefore, were giving careful attention to the selection of seed in order to produce high-quality Jethobudho rice with a high milling recovery.

With respect to threats to the genetic enhancement of farmers' varieties, it should be noted that the process of registration, while intrinsic to the ultimate goal of protecting traditional landraces, is not an easy task and requires continuous support of the project staff, especially in collecting, analyzing, presenting and defending the data as well as in the registration application. Registering the Pokhareli Jethobudho variety was one of the primary mechanisms used to ensure that farmers would benefit from the commercial production and sale of the seeds. However, it is not possible for farmers alone to fulfil all of the requirements of the variety registration process. The success of the landrace enhancement program, therefore, ultimately depends on the capacity of farmers and farming communities to undertake the enhancement process, to protect their varieties and to benefit from the production and marketing of the seeds.

Focus on relevant policies and laws

There are four existing legal documents in Nepal that support the participatory enhancement of the Jethobudho rice landrace as well as the protection of the enhanced variety, such as Pokhareli Jethobudho. The National Seed Policy 1999 emphasizes the need for the conservation of agricultural biodiversity and the establishment of rights over plant variety. It also encourages the development of new plant varieties. The Agricultural Biodiversity Policy 2007 (first amendment, 2014) proposes that participatory plant breeding be used as a working strategy to enhance the genetic performance of local plant varieties and landraces (MoAC, 2014). In addition to these official policies, the Seeds Act 1988 (first amendment, 2012) and the Seeds Regulation 2013 have made legal provisions for the registration of new plant varieties. These provisions, however, were previously suitable only for varieties developed through a formal breeding system and did not recognize enhanced varieties developed through participatory plant breeding. After considerable lobbying by organizations involved in the system, the government agreed to make changes to these regulations that would recognize participatory plant breeding as a scientific method of developing new plant varieties. These changes were made to the Seeds Regulation of 1997 through a ministerial decision published in the Nepal Gazette on 6 June 2005. Two major changes were made to the regulation. The first change enabled farmers to use qualitative data based on their traditional knowledge of the landrace in their application for registration. Such perception data included the major traits of the variety obtained through a preference ranking, consisting of aroma, flavour, texture and cooking qualities. The second change removed the requirement for cultivation data from several different locations in the country and for multiple years of production (at least three years per unit of land) and allowed for cultivation data that came from only one location and represented only one year's yield performance. In addition, the law's uniformity criteria were relaxed for the registration of local plant varieties and landraces.

The uniformity of the variety is a necessary condition for the registration of new plant varieties. This condition was removed in the case of farmers growing local plant varieties since it was acknowledged that farmers usually maintain a population with a certain degree of heterogeneity. These changes had three obvious advantages:

1 they accepted farmers' participation in the breeding process – that is, the collaboration of farmers and breeders;
2 they allowed for the straightforward use of farmers' perception data, which did not require a complicated statistical analysis that would be difficult for farmers to implement;
3 by not having to work in many different locations, they made it easier and less costly for breeders and farmers to accomplish the genetic enhancement of a species.

The registration of enhanced local plant varieties has enabled the commercial production and marketing of such varieties. Since farmers are eligible to apply for ownership over a particular variety, they may be interested in investing time and money into its development without fearing that it will be lost to other cultivators, and they can market the product for financial benefit. There are two obvious benefits to the process of registration and ownership. First, it serves to promote the conservation and utilization of local plant varieties and landraces. Second, it establishes the rights of farmers and communities over such genetic resources.

It should be noted, however, that the legal provisions made in the Nepalese Seeds Act 1988 (first amendment, 2012) and in the Seeds Regulation 2013 for the protection of farmers' rights over their plant varieties are very weak. Farmers can apply for a right of ownership over their new plant variety if they wish. However, the scope of this right is not defined by these legal instruments – that is, these do not mention the kind of rights the owner can exercise over the right-protected variety. There are also practical constraints in registering farmer-developed and/or farmers' local plant varieties/landraces. The technical and logistical requirements that have to be fulfilled for the registration of plant varieties make it very difficult for the farmers and farming communities to register new and/or local plant varieties/landraces on their own. Farmers are not used to collecting and processing this type of data, and many are not actually capable of doing it. Therefore, they are often incapable of completing the registration application. A related problem occurs with respect to the definition of a breeder. According to the Seeds Act, anyone, including the farmer who develops the new variety, should be able to qualify to be a breeder. However, in actual practice, this is not the case. Based on the policy guidelines set out in the National Seed Policy 1999, the National Seed Board issued a set of infrastructure requirements and conditions on 5 September 2003 for private and nongovernment organizations engaged in the development of new plant varieties and the production of foundation seeds. According to these guidelines, such private or nongovernment organizations must have:

- their own land or land rented or leased for the purpose of research;
- a plant breeder with at least a master of science degree in agriculture;

- a seed technologist with at least a master of science degree;
- individuals with bachelor of science degrees for each crop for which plant breeding and/or seed production is to be done;
- other human resources as required for the multidisciplinary team;
- a seed godown (store) of the required capacity;
- other equipment, as necessary, to carry out plant breeding and seed production.

These conditions indirectly make farmers and farmers' groups ineligible to qualify as breeders in the registration of new varieties. For ordinary farmers and farmers' groups in Nepal, meeting these conditions is neither possible nor viable. The registration of the Pokhareli Jethobudho variety was possible because the application for registration by the Fewa Seed-Producing Farmers' Group was made jointly with formally established partner organizations. So far, there are no cases where farmers have independently registered their varieties in Nepal. This discrepancy between Nepalese seed policy and the actual law was highlighted at the fourth National Seed Seminar on 19–20 June 2008 in Kathmandu, and a strong recommendation was made to remove the conditions set out for the production of breeder seeds (Shrestha et al., 2008). The draft Plant Variety Protection and Farmers' Right Bill, which is under preparation, has recognized these weakness and has made explicit provisions for the registration of farmers' varieties without imposing such conditions (ibid.).

Conclusion

This case study shows that value addition through genetic enhancement is effective in making local plant varieties and landraces competitive for continued production and conservation on farm. As is evident from the example of Pokhareli Jethobudho rice, farmers and farming communities have been making enormous contributions, through their local knowledge and innovative practices, to the continuous enhancement of genetic plant resources. However, the contribution of these farmers is not adequately recognized by the Nepalese government's policies and law. As a result, farmers' rights over their plant varieties are almost nonexistent in Nepal. Creating favourable administrative and legal procedures and conditions that would enable farmers to register and obtain ownership rights over their varieties is critical to establishing farmers' rights over their genetic resources. There is an urgent need to review the Nepalese seed policy and law with an aim to making changes that will guarantee the rights of farmers and create incentives for the conservation of genetic resources.

References

Anonymous (2006). *Proposal for the Release of Pokhareli Jethobudho Dhan: A Rice (Self-Pollinated Crop) Variety Developed from Landrace Selection in Nepal.* LI-BIRD, NARC, DADO Kaski, IPGRI and Fewa Seed Producer Group, Kaski, Nepal.

Bajracharya, J., S. Gyawali, B.R. Sthapit and D.I. Jarvis (2004). 'Jethobudho Landrace Enhancement V: Molecular Evaluation of Enhanced Populations of Jethobudho for Aroma My SSR,' in B.R. Sthapit, M.P. Upadhyay, P.K. Shrestha and D.I. Jarvis (eds.), *On-Farm Conservation of Agriculture Biodiversity in Nepal*, Volume 2, *Managing Diversity and Promoting Its Benefits*, Proceedings of the Second National Workshop, 25–27 August 2004, Nagarkot, Nepal, International Plant Genetic Resources Institute, Rome, Italy.

Gyawali, S., B.R. Sthapit, B. Bhandari, J. Bajracharya, P.K. Shrestha, M.P. Upadhyay and D. Jarvis (2006). 'Jethobudho Landrace Enhancement for In Situ Conservation of Rice in Nepal,' in B. Sthapit and D. Gauchan (eds.), *On-farm Management of Agricultural Biodiversity in Nepal: Lessons Learned*, Proceedings of a National Symposium, 18–19 July 2006, Budhanilkantha, Nepal, International Plant Genetic Resources Institute, Rome, Italy.

Gyawali, S., B.R. Sthapit, B. Bhandari, J. Bajracharya, P.K. Shrestha, M.P. Upadhyay and D. Jarvis (2010). 'Participatory Crop Improvement and Formal Release of Jethobudho Rice Landrace in Nepal,' *Euphytica* 176(1): 59–78.

Ministry of Agriculture and Cooperatives (MoAC) (2014). *Agricultural Biodiversity Policy*, MoAC, Kathmandu, Nepal.

Morris, M.L., and M.R. Bellon (2004). 'Participatory Plant Breeding Research: Opportunities and Challenges for the International Crop Improvement System,' *Euphytica* 136: 21–35.

Rana, R.B., B.R. Sthapit, C. Garforth, A. Subedi and D. Jarvis (2005). 'Four-Cell Analysis as a Decision-Making Tool for Conservation of Agro-Biodiversity on-Farm,' in B.R. Sthapit, M.P. Upadhyay and P.K. Shrestha (eds.), *On-Farm Conservation of Agricultural Biodiversity in Nepal*, Volume 1, *Assessing the Amount and Distribution of Genetic Diversity on-Farm*, Proceedings of the Second National Workshop, 25–27 August 2004, Nagarkot, Nepal, International Plant Genetic Resources Institute, Rome, Italy.

Rijal, D.K., R.B. Rana, K.K. Sherchand, B.R. Sthapit, Y.R. Pandey, N. Adhikari, K.B. Kadayat, Y.P. Gautam, P. Chaudhary, C.L. Poudyal, S.R. Gupta and P.R. Tiwari (1998). *Strengthening the Scientific Basis for In Situ Conservation of Agrobiodiversity: Findings of Site Selection in Kaski, Nepal.* Working Paper no. 1/98, NARC/LI-BIRD, Nepal/IPGRI, Italy, Rome.

Shrestha, P., K. Adhikari and M.N. Shrestha (2008). *Implications of Plant Variety Protection and Farmers' Right Bill on Seed Production and Marketing in Nepal.* Paper Presented at the Fourth National Seed Seminar organized jointly by the National Seed Board, the Ministry of Agriculture and Cooperatives, the NARC, the Seed Entrepreneurs Association Nepal, the National Salt Trading Corporation and LI-BIRD, 19–20 June 2008, Kathmandu, Nepal.

Witcombe, J.R., S. Gyawali, S. Sunwar, B.R. Sthapit, K.D. Joshi (2006). 'Participatory Plant Breeding Is Better Described as Highly Client-Oriented Plant Breeding: Optional Farmer Collaboration in the Segregating Generations,' *Experimental Agriculture* 42: 79–90.

Witcombe, J.R., A. Joshi, K.D. Joshi and B.R. Sthapit (1996). 'Farmer Participatory Crop Improvement: Varietal Selection and Breeding Methods and Their Impact on Biodiversity,' *Experimental Agriculture* 32: 445–60.

5 Promoting policy support for the enhancement and marketing of farmers' varieties in Vietnam

Nguyen Thi Ngoc Hue, Devra Jarvis and Michael Halewood

Tamxoan is a local rice variety cultivated by the Kinh people in the coastal lowlands of the Red River delta in northern Vietnam (Haihau district, Nam Dinh province).[1] The size of the area planted with Tamxoan in the district has decreased rapidly since the early 1990s due to the subsidized introduction of hybrid rice varieties (and chemical fertilizers) in the area, which has been part of the government's efforts to increase rice production to guarantee local food security and to supply Vietnam's burgeoning rice export market. The quality of Tamxoan rice had also been slowly worsening as a side effect of its neglect. In recent years, people had started to notice negative aspects associated with growing hybrids in the area, for example, environmental impacts, rising costs of fertilizers and susceptibility to some diseases. As a result, there has been a renewed interest in local varieties, at least in some areas. Since 1999, local farmers, the Vietnamese government, universities and an international agricultural research organization have been working together to enhance Tamxoan in order to improve its already favourable qualities and to make it a more competitive alternative, or complement, to hybrid rice. This chapter will provide an account of these efforts. It will also describe the way in which Vietnamese laws concerning intellectual property, seed quality and, most recently, farmers' innovation have supported, or created challenges for, the enhancement, protection and marketing of the variety.

The replacement of local varieties with hybrid rice

Haihau district is flat and just a half a metre above sea level. Most of the farms are small – approximately 800 square metres. There are a few small villages, but most people live in the countryside. Farming accounts for three-quarters of the local economy. The primary crop is rice, but farmers also grow maize, sweet potatoes, vegetables, legumes, litchis, oranges and pomelos. Farmers in the district currently plant 12 different rice varieties, including several hybrids and four local varieties: Tamxoan, Nep cai hoa vang, Nep Thau dau and Nep Ba Lao. Of these four local varieties, Tamxoan has historically been the most popular, partly based on the fact that the other three varieties are glutinous (i.e. 'sticky' rice) – which are generally less widely used in Vietnam than nonglutinous rice – and also because of its flavour, soft texture and length.

In 1991, the Vietnamese government launched a program to boost rice production. Among other things, the program involved boosting Vietnam's own rice-breeding capacity and the import of hybrid rice varieties from China. By 2006, 600,000 hectares of hybrid rice was being grown in the country, with the main concentration to be found in the north of the county in the Red River delta. The reason for initially concentrating hybrid production in the north is that the north was more food insecure than the south of the country, where the rich Mekong delta provides excellent conditions for high levels of rice production (Vien and Nga, 2008).

Many of the high-yielding Chinese hybrid seeds took far less time to reach maturity and produced a high-quality rice. Their yield was also considerably higher, averaging 5–6 tonnes per hectare, whereas Tamxoan (before it was enhanced through the project's efforts) yielded approximately 2.4–2.6 tonnes per hectare (Binh et al., 2004). Another advantage for farmers was the fact that such varieties could be grown both in the spring and summer, while the traditional varieties could only be planted in the summer and harvested once in the fall. Having two harvesting seasons made it easier for farmers to spread the work out over a longer period of time. In addition, the hybrid plants were much smaller and, therefore, not so susceptible to lodging. Lodging, which occurs when the plant grows tall and is blown over by the wind, is a common problem for Tamxoan. It often results in a damaged plant and a reduced yield per hectare. The improved varieties assured farmers in the community an increased production and, therefore, provided them with a secure livelihood. Farmers with average or good quality plots were even able to sell a part of their harvest since they often produced more than they needed for their own consumption.

Subsidies provided by the government made the expensive hybrid seeds available to farmers at half of the market price. In addition, falling prices for chemical fertilizers and pesticides made the cultivation of these hybrid varieties less costly. The government also advised farmers on the techniques for fertilizer and pesticide application. Within a few years, high-yielding hybrid varieties were being grown more extensively than the traditional varieties. By the end of the 1990s, only about 10 percent of the total cultivated area in Nam Dinh province was dedicated to local rice varieties (Trinh et al., 2003). From the perspective of many farmers, the cultivation of traditional landraces implied a return to the past. Ultimately, this nationwide production of improved Chinese and Vietnamese hybrid rice boosted agricultural production throughout Vietnam, allowing the country to become self-sufficient and reducing poverty and starvation. It also served to raise export levels, bringing in much-needed currency.

The pendulum swings back: rekindled interest in local varieties

However, this remarkable phenomenon in rice production also had some negative side effects, one of which was the erosion of traditional rice varieties, especially in the Red River delta and in Nam Dinh province, where the reliance on hybrids was growing most rapidly. Hybrid seeds cannot be saved and replanted by farmers each year. Instead, farmers have to return to commercial suppliers for seed each year. While on the one hand, the increased yield may justify this form of dependence on commercial seed producers, it left farmers vulnerable to fluctuating seed prices on the other hand, and prices generally increased as a result of the fact that government subsidies were eventually cut back. Finally, the improved varieties required far more chemical fertilizers and pesticides, which was an additional expense for farmers and also was beginning to be recognized as a hazard for human health and the environment. Over the past 5–10 years, the Chinese hybrids turned out to be particularly vulnerable to some pests, so increased use of pesticides was necessary. As the pests grew more resistant, still more pesticide was required, with increasing environmental consequences and costs to farmers. As a result, some farmers in Nam Dinh province started to become dissatisfied with the hybrid varieties and were interested in investigating alternatives.

Tamxoan clearly merited reconsideration. It is culturally important, used in local rituals such as food in the village festivals, local conferences and meetings, wedding parties, Tet holidays and ceremonial offerings, funeral repasts and banquets. Its taste and smell is considered to be superior to that of the Chinese hybrids, and it consistently received a higher price at the local markets. Farmers could save and replant its seed, and it required less expensive inputs (i.e. less fertilizer and pesticides). However, the price differential was not enough to overcome, on a significantly wide scale, the economic incentive to cultivate the Chinese hybrids. To make Tamxoan more competitive, it needed to be improved to increase its yield, to ensure it was consistently high quality and to decrease its tendency to lodging.

At roughly the same time, the Vietnam Agricultural Science Institute (VASI) and Cantho University, in partnership with the International Plant Genetic Research Institute,[2] initiated the Vietnamese component of an internationally coordinated project called Strengthening the Scientific Basis of *In Situ* Conservation of Agricultural Biodiversity on Farm (*In Situ* Project) (Chuong et al., 2003). The project sought to:

1. support the development of a framework of knowledge on farmer decision-making processes that would influence the *in situ* conservation of agricultural biodiversity;
2. strengthen national institutions for the planning and implementation of conservation programs for agricultural biodiversity;
3. broaden the use of agricultural biodiversity and participation in its conservation by farming communities and other groups.

The partners of the *In Situ* Project were surveying the country for crops and communities with which to work, giving preferences to sites where relatively high levels of crop diversity currently were or had recently been deployed as well as to local varieties with market potential.

Ultimately, it was agreed that the *In Situ* Project would support a participatory Tamxoan rice enhancement program in Haihau district, using rehabilitation methods with the participation of farmers and breeders. Similarly to the Nepalese case study in Chapter 4 of this volume (which was also supported partially through the *In Situ* Project), the objectives of this program were to reduce the occurrence of lodging, increase disease tolerance and increase the yield of the selected variety, Tamxoan. At the same time, however, it was important to maintain the traits of Tamxoan for which farmers and consumers had an evident preference. The national partners involved in the program were the Plant Genetic Resources Centre (PGRC) at the VASI, the Centre for Agrarian System Research and Development, the Agricultural Development Office of Haihau district and the Tamxoan-producing farmers' group of Haitoan village.

The government was behind the enhancement project because it wanted to support a local landrace with market potential that could be grown in Nam Dinh province. It approached the farmers and other stakeholders and asked them to invest in the project in order to help to make Haihau the seed production centre of the project and to make Tamxoan a major source for consumption. The farmers responded favourably. They believed in the potential for improving Tamxoan and were convinced that they could benefit as a result.

Participatory enhancement of Tamxoan

The first step was to take a survey of the places Tamxoan was being grown in the area, identifying how such traits as propensity for lodging, height, length of panicle, aroma, taste and yield varied across the populations studied. Based on the survey results, it was possible to identify (through participatory exercises involving farmers) which traits were most appropriate to enhance. Approximately 200 lines related to the Tamxoan landrace were eventually identified and collected within the area. The project team surveyed the availability and quality of the existing seed and investigated the potential areas for increased seed production in the future (assuming that the project was successful and that there would be increased demand for the improved variety's seed).

In 2000, a group of farmers, representatives from the agricultural department of Nghia Hung, agronomists and cultivators of the PGRC at VASI began the lengthy selection of Tamxoan cultivars through a participatory rural appraisal process. The first step was to identify the typical traits of Tamxoan. Farmers were asked what they considered to be typical Tamxoan characteristics, and these characteristics were put into a ranking matrix, which was the basis of the selection process. Using this ranking matrix, different lines representing the typical Tamxoan characteristics were taken from 20 different farmers' plots. This group of farmers was known as the key farmer group. After the 2001 harvest, the first selection took place, which focused on morphological traits such as the height of the plants, the colour of the husks (brown-yellow), the colour, size and cooking properties of the rice, and the susceptibility of the plant to disease. This process led to the selection of 40 different lines, which were planted by the 20 key farmers in the 2002 season. After the 2002 harvest, seven lines were identified as best representing the most important combined traits of Tamxoan, according to farmers and consumers. In 2003, the group evaluated the selected seven lines of Tamxoan and found that each of them had outperformed the reference Tamxoan population for the prioritized traits. The average yield of the seven lines was 3.06–3.20 tonnes per hectare compared to 2.43–2.66 tonnes per hectare for the reference populations (Suu, Trinh and Loan, 2007).

In 2004 and 2005, the new Tamxoan cultivars were evaluated for the first and second time for uniformity and stability as well as for agronomic and postharvest traits. Varieties that did not pass these evaluations were removed. Before planting in 2006, the Haihau farmers and other stakeholders set the seed quantity target for that year's harvest at 3,000 kilograms. The key farmers cultivated a total of 2 hectares of Tamxoan in 2006 and then raised it to 30 hectares in 2007 since they were beginning to realize that increasing the cultivation of Tamxoan varieties was beneficial to their financial well-being. The government encouraged farmers to cultivate Tamxoan on at least 100 hectares by 2010. The farmers signed a contract with the district government that year that obliged the government to buy Tamxoan seeds from them for a certain price per kilogram. This seed was stored away for the next season's cultivation, and in this way the government secured a supply of good-quality seed. In addition, the farmers received support in the form of financial resources and education during the selection process and throughout the season. They also received money for organic and chemical fertilizer. At the beginning of the 2007 season, farmers were provided with Tamxoan seed by the district government at a subsidized price that was 75 percent below market value, which was meant to persuade farmers to increase the cultivation of Tamxoan in Haihau district. The subsidy was gradually reduced over the following 4 years. More details about the methods that were followed to enhance and promote Tamxoan are set out in Table 5.1.

As of 2010, the area of land cultivated with Tamxoan began to be reduced. As of 2011, Tamxoan is being cultivated on only 40 hectares of land. The reduction is largely due to the competition from, and increased use of, a new high-yielding variety called Bacthom. Bacthom is a pure line (not a hybrid) introduced originally from China to Vietnam in 2003, and it has been gaining popularity among farmers in North Vietnam. It has a short growth duration and is very similar to Tamxoan in aroma and softness, and it has a higher yield and can grow in two season. The Department of Agriculture encourages farmers to grow Bacthom through field visits and the propagation of Bacthom by extension staff as well as by making seeds available to buy.

Despite these pressures, it is likely that at least 40 hectares of Tamxoan will continue to be grown in the coming years. When Tamxoan was given geographical indication protection, on the basis of its useful traits, the local authorities (including the People's Committee of commune district agriculture division, Haitoan commune cooperative) included plans to grow Tamxoan in the Economic and Social Development Plan of Hai Hau District (Haihau District

Table 5.1 Procedures and methods adopted for the enhancement of Tamxoan rice in Vietnam

Year	*Goals*	*Methods*	*Outcomes*
1999	• assessment of field diversity for morphological, agronomic and quality traits	• participatory rural appraisal • baseline survey • an analysis of the cultivation of Tamxoan in four groups of farming communities in Nam Dinh province (large households with a large farming area; large households with a small farming area; small households with a large farming area; small households with a small farming area)	• considerable variation in Tamxoan population was found • three major Tamxoan production areas were identified in Nam Dinh province
2000	• development of selection tasks for Tamxoan enhancement • analysis of preferred traits • collection of accessions for evaluation and enhancement	• selection work was undertaken jointly by project team and farming communities • trait analysis done with the help of expert farmers and traders • project team visited primary farmers' field at harvesting stage and selected 10 panicles from each household plot	• selection task set to improve the Tamxoan population through the evaluation of preferred traits • landrace enhancement procedure designed • 200 lines of Tamxoan collected from three production areas
2001	• diversity assessment and selection of lines for the identified preferred traits	• two sets of 200 lines were assessed in farmers' field – one in Haihau and another at Nghia Hung	• 40 lines were selected according to grain size, field tolerance to leaf blight, strong culm, plant height lower than 150 centimetres, aromatic flavour and productive plants
2002	• evaluation of 40 selected lines of Tamxoan for leaf blight and lodging tolerance, yield components and the identification of quality traits preferred by market and consumers	• performance assessment trial set up in Haihau • data collected for further screening • simple sequence repeats were used to test the genetic relationship of 40 lines • market survey and consumer feedback used to select for quality traits and the selection of lines on the basis of postharvest micromilling and organoleptic tests	• seven lines of Tamxoan landrace confirming most of the preferred traits and high genetic similarity coefficient were selected for further evaluation

2003	• evaluation of seven lines of Tamxoan for agronomic and postharvest traits • assessment for aroma traits • assessment for wider adoption in Haihau	• agronomic trials set up in farmer's field at Haitoan in Haihau district for postharvest traits (aroma, stickiness, amylose content and so on)	• bulk lines of Tamxoan to create the multiline Tamxoan variety
2004	• the first evaluation of the new Tamxoan variety for agronomic and postharvest traits was completed on a large scale • improved cultivation technique test for Tamxoan was given	• first trial for uniformity and agronomic traits was conducted at Haitoan in Haihau district • tests were conducted for aromatic levels • lines that did not pass were removed	• the new Tamxoan performed better than the normal cultivars • surveys were completed to identify the Tamxoan market
2005	• second evaluation of the new Tamxoan for agronomic and postharvest traits was completed on a large scale • improved cultivation technique test for Tamxoan was given	• second trial for sustainability and agronomic traits was conducted at Haitoan in Haihau district • tests were conducted for aromatic levels • lines that did not pass were removed	• the new Tamxoan performed better than the normal population • new seed multiplication • the cultivation technique for Tamxoan was improved
2006	• trial of large scale (production pilot) and cultivation technique was developed for new Tamxoan variety • evaluation and seed production	• evaluation for yields, uniformity and quality of grain • evaluation by Ministry for Agriculture and Rural Development and the Department for Agriculture of Nam Dinh	• 2 hectares were set aside for the production of seed for next season's production • production pilot was developed for Tamxoan rice market • management protocol for Tamxoan was created

(*Continued*)

Table 5.1 (Continued)

Year	*Goals*	*Methods*	*Outcomes*
2007	• seed production of new cultivar • field visit by members of the Variety Test and Trail Centre and the Ministry for Agriculture and Rural Development	• participatory varietal selection and seed production through farmers' participation in many locations in Haihau district • application submitted to Department of Intellectual Property Rights	• 30 hectares of the new Tamxoan were cultivated • marketing was continued by farmers' group • production pilot initiated for Tamxoan rice market
2008	• new Tamxoan population was proposed for release • it was registered for protected geographical indication • seed is produced for new population	• management of Tamxoan production and market • application submitted to the Department of Intellectual Property Rights • 432 farmers of Haihau Tamxoan Association cultivated 60 hectares of the new Tamxoan variety	• 60 hectares of the new Tamxoan were cultivated • Tamxoan rice registered for protected geographical indication as 'Haihau Tamxoan' variety
2009	• Tamxoan market is increased	• 432 farmers of Haihau Tamxoan Association cultivated 60 hectares of new Tamxoan variety	• 60 hectares were cultivated with Tamxoan rice for production and marketing
2010–11	• the sustainable production of Tamxoan was maintained	• 432 farmers of Haihau Tamxoan Association cultivated their land with the new Tamxoan variety	• 35–40 hectares were cultivated with Tamxoan rice for production and marketing

Source: Centre for Agrarian Systems Research and Development (2009). The format for this table was reproduced from P. Shrestha's case study in Chapter 4 of this volume concerning the enhancement of Pokhareli Jethobudho in Nepal.

People Committee, 2004). The local authority still works to raise awareness among farmers about the value of Haihau Tamxoan and encourages them to grow it. Agricultural scientists with the Vietnam Academy of Agricultural Sciences also have plans to further increase the quality and yield of Haihau Tamxoan and to encourage the marketing and spread of Tamxoan in the future.

Policy and legal support for cultivation and marketing of Tamxoan: Vietnamese intellectual property and seed marketing laws

In Vietnam, at the time of the project, there was no possibility for farmer or farming communities to obtain intellectual property protection over plant varieties in their own names. Nor was it possible for farmers or farming communities to be acknowledged under Vietnamese variety release and seed quality regulations. To address this situation, the stakeholders involved in the Tamxoan project in Nghia Hung and Haihau (i.e. the key farmers, the community leader, the Centre for Agrarian Systems Research and Development, the VASI and the district government) established an association named the Haihau Tamxoan Production and Marketing Association. The association has been operational since January 2006. Under the Plant Variety Ordinance, which was issued by the Vietnam National Assembly in 2004, this association has registered the new Tamxoan variety under the name Tamxoan Haihau rice.[3] To be registered under the Plant Variety Ordinance, a variety must satisfy the conditions of being distinct, uniform and stable (DUS) and represent enhanced value for cultivation and use (VCU). Once a variety is registered, it may be sold commercially within Vietnam.

The association also applied, in its own name, for geographical indication protection for Tamxoan Haihau rice, pursuant to the Intellectual Property Law of Vietnam.[4] Geographic indication protection was granted, and the new cultivar was officially registered (and made subject to geographic protection) by the Department of Intellectual Property Rights in July 2008. The Ministry of Agriculture and Rural Development supported the decision of the local authorities to register geographic indications for agricultural products. The protection afforded by the law extends across the entire country.

For the first time in Vietnam, the registration of Tamxoan Haihau rice under both of these laws enabled farmers to be recognized as the co-developers of the new variety since they were the members of the association. The farmers, as well as the concerned organizations, were thereby exclusively eligible to produce and market the seed of this enhanced Haihau Tamxoan cultivar anywhere in Vietnam. The Haihau community considers itself the original source of Tamxoan, and the association has been responsible for designing the labels, creating the packaging and arranging the transport of the cultivar to external markets. Some of the rice, with its new packaging, has been sold in the Haihau district and also in Hanoi, which is relatively easy to arrange since the road from Haihau to Hanoi is in excellent condition.

Some of the experiences in this project have confirmed the need for further policy reform to support farmers as conservers and developers of crop diversity and new crop varieties. Some of the people and organizations involved in the project have continued to be involved (once the project ended) in policy development activities that culminated in another project entitled the Regulation on Production Management of Farm Households' Plant Varieties (Household Plant Varieties Regulation), which was adopted by the Ministry of Agriculture in 2008 (see Chapter 23 of this volume for more details about this regulation).[5]

Among other things, the Household Plant Varieties Regulation allows farm households to register varieties under the national seed law and to apply for intellectual property protection for varieties under the national intellectual property law in their own name. It also validates processes such as participatory plant breeding and participatory variety selection. The regulation sets the legal and administrative basis for the government to provide financial support for farmers' activities to use, improve and conserve local crop diversity. However, while the Household Plant Varieties Regulation is still relatively new, the procedures that farmers must follow are still very complicated, and it seems likely that most small farmers will not have the capacity or funds necessary to satisfy the requirements for the DUS and VCU tests.

The project's experiences in looking for ways to recognize farmers' contributions to the development of Haihau Tamxoan under the laws that existed at that time, and through the creation of the Haihau Tamxoan Production and Marketing Association, were partially responsible for the decision to draft and adopt the Household Plant Varieties Regulation.

Notes

1. The Kinh people are the major ethnic group in Vietnam, comprising approximately 85% of the countries population (Wikipedia, online: <http://en.wikipedia.org/wiki/Vietnamese_people> (last accessed 12 November 2012).
2. Now named Bioversity International, one of the fifteen International Agricultural Research Centres of the Consultative Group on International Agricultural Research.
3. Plant Variety Ordinance (*Phap lenh giong cay trong*), Vietnam National Assembly (March 2004).
4. Intellectual Property Law of Vietnam (*Luat So huu tri tue*), Vietnam National Assembly (29 November 2005).
5. Regulation on Production Management of Farm Households' Plant Varieties, Decision no. 35/2008/QĐ-BNN (15 February 2008).

References

Binh, V.T., L.D. Thinh, B.T. Thai and D.D. Huan (2004). 'Studying and Developing the Appellation of Origin for Hai Hau Fragrant Rice', *Magazine of Agriculture and Rural Development*, 10, Ministry of Agriculture and Rural Development, Vietnam Ministry of Agriculture and Rural Development.

Centre for Agrarian Systems Research and Development (2009). 'Improving Quality and Commercialising the Product of Tam Xoan Hai Hau at Hai Hau, Nam Dinh', in final report of the project *Conservation and Commercialising the Product of Tam Xoan Hai Hau*, Casrad, Vietnam Academy of Agricultural Science, pp. 17–23.

Chuong, P.V., H.D. Tuan, N.T.N. Hue, L.N. Trinh, D. Jarvis and B. Sthapit (2003). 'The Historical Development of In Situ Conservation in Vietnam: Institutional Arrangements for Project Implementation with Multipartners,' in *On-Farm Management of Agricultural Biodiversity in Vietnam*, Proceedings of Symposium, 6–12 December 2001, Hanoi, Vietnam, International Plant Genetic Resources Institute, Rome, Italy.

Haihau District People Committee (2004). 'Plan on Rice Production of District in Period 2005–2009', in *Report on the Orientation of Socio-Economic Development of Haihau District Period 2005–2009*, pp. 5–6 (*Báo cáo định hướng phát triển kinh tế xã hội của huyện Hải Hậu năm 2005–2009, phần về kế hoạch sản xuất lúa gạo của huyện trong giai đoạn 2005–2009*).

Suu, T.D., L.N. Trinh and H.M. Loan (2007). 'Evaluation Results of Yield and Quality of Promising Tam Rice Varieties,' *Journal of Vietnamese Agricultural Science and Technology* 3(4): 10–15.

Trinh, L.N., P.H. Cuong, N.T.N. Hue, N.P. Ha and D.V. Nien (2003). 'Genetic Diversity within Farmers' Rice Varieties in Three Eco-regions of North Vietnam over Time,' in *On-Farm Management of Agricultural Biodiversity in Vietnam*, Proceedings of Symposium, 6–12 December 2001, Hanoi, Vietnam, International Plant Genetic Resources Institute, Rome, Italy.

Vien, T.D. and N.D. Nga (2008). *Economic Impact of Hybrid Rice in Vietnam: An Initial Assessment*. Available at: www.cares.org.vn./webplus/Article/ECONOMIC%20IMPACT%20OF%20HYBRID%20RICE%20IN%20VIETNAM.pdf (last accessed 4 April 2014).

6 Participatory barley breeding in Syria

Policy bottlenecks and responses

Salvatore Ceccarelli

Introduction to the enhancement project

This case study is based on the development, evolution and achievements of a participatory plant-breeding project on barley that was implemented in Syria in 1996. Barley is the major rain-fed crop in Syria and is the main animal feed in the country. In the first phase of the project, which lasted 3 years, those involved in the project experimented with farmers in nine villages on a number of technical and methodological options, while in the second phase, which started in 2000, these individuals continued cultivation until the 2010–11 season. Although it had to be put on hold because of security issues in the countryside, it is hoped that a full-fledged participatory plant-breeding program will eventually be implemented.

In both phases of the project, the emphasis has been the improvement of the two main landraces grown in the country, which follows an explicit request by the farmers. In the first phase, the project was initiated by the International Centre for Agricultural Research in the Dry Areas (ICARDA) with minor involvement from the Ministry of Agriculture and Agrarian Reform (MAAR) of Syria. In the second phase, the collaboration with MAAR was more intense and allowed the expansion of the project in 24 villages covering 90 percent of the barley-growing areas. Gradually, the varieties produced by the project, which were based on crosses with or selections from landraces, started to be grown over large areas of land, and the seed being exchanged between farmers and/or sold from participating to nonparticipating farmers began to increase. At this point, the collaboration with MAAR came to an end, and in 2008, the research staff of MAAR were instructed in writing to refrain from collaborating with the project. The reason was that varieties not officially released cannot be legally cultivated and, thus, the commercialization of their seed was not legal.

Barley is a major crop in Syria and is grown over an area of between 1.0 and 1.5 million hectares annually. Its production, both as grain and straw, is used exclusively as livestock feed for small ruminants. A very small area is planted with malting barley varieties. The crop is strictly rain-fed and is predominant in parts of the country that receive at least 350 millimetres of annual rainfall. In these areas, barley is replaced by wheat only if and where farmers have access to water for supplementary irrigation. In the drier parts of the country, it is the only possible rain-fed crop, making these farmers very vulnerable to climatic changes.

Two main landraces are cultivated: Arabi Abiad (*Abiad* means white in Arabic), which has white seeds and is grown in the wetter areas; and Arabi Aswad (*Aswad* means black in Arabic), which has black seeds and is grown in drier areas. There is very little (if any) overlapping of the two types, and farmers have a strong preference for either one or the other. Farmers in dry areas consider that the grain and straw quality of the black-seeded landrace is better that the white-seeded one. Considerable phenotypic and genotypic heterogeneity for many plant characteristics that are of agronomic importance exists among the landraces collected in different farmers' fields (even if they are designated by the same name) as well as among individual plants within the same farmer's field (Ceccarelli, Grando and Van Leur, 1987; Ceccarelli and Grando, 2000; Newton et al., 2010; Ceccarelli, 2012). Considerable phenotypic and genotypic heterogeneity for disease reaction also exists (Van Leur, Ceccarelli and Grando, 1989). Syrian landraces of barley are also polymorphic in both their chloroplast and nuclear genomes (Russell et al., 2003).

The project area

The project area lies in the northern part of the Fertile Crescent within Syria. It receives between 350 and 200 millimetres of average annual precipitation and stretches in a wide arc from Dara'a and Suweida provinces in the south to Hassakeh in the northeast towards the border with Iraq (see Figure 6.1). The area encompasses a range of agroecological conditions, from high to low potential environments for cereal production. Barley is the main winter cereal. It is planted in the fall, usually after the first rainfall (mid-October to mid-December), and harvested from May to June. At the wettest end of the area (with 350 millimetres of annual rainfall) and on fertile soils, farmers can obtain up to 5 tonnes per hectare of grain in a good season by using fertilizer. In contrast, at the driest end of the area (with 200 millimetres of annual rainfall), soils are generally poor, input levels are low, and grain yields vary from nothing to around 1.5 tonnes per hectare.

The first phase

A number of preparatory meetings with farmers set the scene for the first phase of the project, with the majority of the farmers expressing interest in being able to compare other types of barley with their own landraces. The experiment, which was designed in consultation with farmers in regard to plot size, number of lines, agronomic management systems and so on, included 200 barley lines and populations: 50 fixed lines that were unrelated to Syrian landraces; 50 segregating populations (F3 bulks) from crosses between fixed lines that were unrelated to Syrian landraces; 50 lines derived from Syrian landraces; and

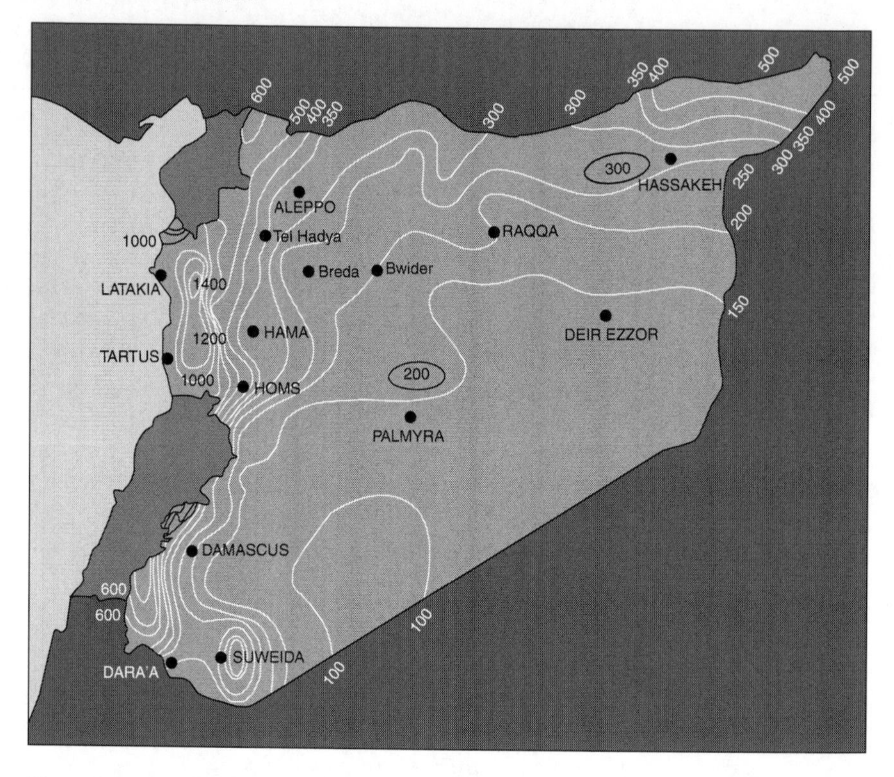

Figure 6.1 Map of Syria

50 segregating populations (F3 bulks) from crosses in which at least one parent was a Syrian landrace. In addition, there were eight farmers' cultivars (from seed purchased from eight of the nine host farmers, as one of them was growing an improved cultivar already included in the trial) (Ceccarelli et al., 2000). The 208 entries were deliberately chosen to test farmers' and breeders' preferences for different attributes and/or characteristics. The entries could be classified according to four distinctions:

1 modern germplasm (100) versus landraces (108)
2 fixed lines (100) versus segregating populations (108)
3 two rows (158) versus six rows (50)
4 white seed (161) versus black seed (28) or segregating (mixed) seed colour (19).

All of the material was grown under rain-fed conditions.

Each farmer was given a field book in which he recorded daily precipitation (measured through a rain gauge) and observations of the evaluative plot. The quantitative scoring method that most farmers preferred was a numeric scale (highest = best, lowest = worst). Some farmers used qualitative scoring such as ticking or classifying the plots as 'bad,' 'medium,' 'good,' 'very good' and 'excellent.' Eventually, farmers used a mixture of quantitative scores for some traits and qualitative descriptors for others. Farmers used the earlier observations at the time of the final selection to assign the final score.

There were large differences in average grain yield, biomass, harvest index and plant height between the nine farmers' fields and the two research stations (see Table 6.1). The highest yield was nearly 3.7 t ha^{-1} of grain and over 7.7 t ha^{-1} of total biomass. The two most stressed sites were farmers' fields in drier

Table 6.1 Rainfall, average grain yield, total biomass, harvest index and plant height of 208 barley entries in nine farmers' fields and two research stations (Breda and Tel Hadya)

Location (code)		*Rainfall (mm)*	*Grain yield (kg/ha)*	*Biomass (kg/ha)*	*Harvest index*	*Plant height[a] (cm)*
Ibbin (1)	Mean	436	3,248	8,600	0.37	102
	standard error		81	147	0.005	0.6
Ebla (2)	Mean	460	2,857	8,000	0.36	98
	standard error		58	113	0.005	0.7
Tel Brak (3)	Mean	278	3,685	7,661	0.48	88
	standard error		69	101	0.006	0.9
Jurn El-Aswad (4)	Mean	284	1,415	7,259	0.20	45
	standard error		51	228	0.005	0.7
Baylonan (5)	Mean	193	280	2,599	0.11	46
	standard error		13	60	0.004	0.7
Al Bab (6)	Mean	350	376	1,514	0.24	33
	standard error		15	39	0.009	0.5
Melabya (7)	Mean	241	713	2,733	0.26	–
	standard error		29	103	0.005	–
Bari Sharki (8)	Mean	248	1,017	4,534	0.22	52
	standard error		36	163	0.006	0.7
Sauran (9)	Mean	303	2,515	7,117	0.36	69
	standard error		46	101	0.006	0.9
Breda (BR)	Mean	233	811	2,689	0.31	44
	standard error		18	51	0.005	0.6
Tel Hadya (TH)	Mean	434	4,495	12,336	0.36	96
	standard error		63	110	0.003	0.7

Source: Ceccarelli et al. (2000).

[a] Melabya was not sampled.

areas, with less than 0.5 t ha^{-1} of grain, a very short crop and a very low harvest index.

Three of the most interesting conclusions to be drawn from this first phase of work were as follows. First, for some broad categories – in this case, modern germplasm versus landraces – the farmers' selection was mostly influenced by the environment, with a large preference for landraces in drier locations. Second, the entries selected by the farmers, regardless of whether they were landraces or not, yielded as much, and in one case significantly more, than those selected by the breeder. Third, farmers can handle a large number of lines (often scientists believe that farmers can only handle a small number of lines, implying that it is only possible to do participatory variety selection [Ceccarelli, 2009], and this finding makes it possible to transfer a plant breeding program from research stations to farmers' fields). One additional output was the increased interest of farmers towards their landraces, since it was the first opportunity for most of them to systematically compare their own varieties with exotic germplasm under local conditions.

Second phase: the participatory plant-breeding program

Based on the experience of the first phase and the enhanced skills of the farmers, a proper cyclical participatory plant-breeding (PPB) program was started in 2000. One of the specific requests of the farmers was that the material used in the second phase should be predominantly based on landraces. The methodology used in this second phase is described in detail by Ceccarelli, Grando and Baum (2007), Ceccarelli and Grando (2007) and Ceccarelli (2009). Like any breeding program, it consists of four stages of evaluation and selection, with entries selected from stage 1 being tested for a second year in stage 2, those selected in stage 2 being tested for a third year in stage 3, and those selected in stage 3 being tested for a fourth year in stage 4. All of the trials involved row and column designs, partially replicated in stage 1 and fully replicated (two replications) in stages 2, 3 and 4. In all of the trials, the researchers collected data on agronomic traits and grain yield, and the farmers expressed their opinion on every single plot in a numerical form. All of the data were subjected to spatial analysis, and the results were tabulated in Arabic and used by the farmers to decide which entries to discard and which to promote to the next stage.

In the second phase, as PPB was gaining popularity in Syria, the number of villages increased from nine to eleven. In 2003, following a workshop in which the then minister of agriculture expressed his personal support for PPB, the number increased to 24 villages. This increase was made possible by the collaboration of the MAAR, which had research stations in every province, and of the extension service, which had staff located in most villages.

Both phases of the project were expected to lead to an institutionalization of PPB in Syria. In fact, the MAAR in Syria is in charge of plant breeding (there is no private breeding) through the General Commission for Scientific Agricultural Research (GCSAR). The General Organization for Seed Multiplication

(GOSM) is in charge of seed production, multiplication and the supply of strategically important crops, which includes wheat and barley. However, the amount of seed produced annually is well below what is needed (about 10 percent of barley and 40 percent of wheat). The GCSAR is also the organization responsible for the evaluation of the variety for release. These lines are tested in on-farm field trials for 3 years before the respective breeder proposes one or more for release to the national Variety Release Committee (Bishaw, 2004). To fully understand the institutional issues described later in this section, it is important to mention that the on-farm trials used in Syria suffered all of the problems identified by Robert Tripp and his colleagues (1997) in the system of variety testing in relation to variety release. With particular reference to the Syrian situation, these problems included:

- inappropriate site selection – in some cases, the sites were actually within research stations and not in farmers' fields;
- unrepresentative trial management – the level of inputs, particularly of fertilizers, were higher than those used by the majority of the farmers, and the same applied to crop rotation, which, in the best of cases, was only one of the rotations used by the farmers;
- trial analysis was biased against poor environments – usually sites with low or variable yields and those with entries failing to give a measurable yield were discarded from the analysis;
- use of suboptimal experimental designs and statistical analysis – for example, little or no use of spatial analysis and use of unweighted means across sites, which because of the scale effect leads to the selection of the highest yielding entries in the highest yielding sites – the GCSAR has used the same experimental designs and statistical analysis during the last 35 years;
- lack of farmer participation and lack of attention to farmer-preferred variety traits – farmers are only involved in providing the land for the trials.

The period from 2003 to 2007 was very fruitful. The GCSAR and the extension staff collaborated fully in running the trials, in note-taking, in interacting with farmers and in analyzing data. Farmers, after each full cycle of breeding (four cropping seasons), started identifying and naming superior lines (by 2007, more than 30 new varieties and populations were been named by farmers in this period). It was hoped that because the PPB trials did not suffer from any of the problems listed earlier, the data generated by the PPB trials – conducted, as mentioned earlier, for 4 years across a number of locations as row and column design and analyzed with spatial analysis in GenStat with the estimation of best linear unbiased predictors – could be accepted by the GCSAR as the best PPB entries for release.

The first policy-related tensions arose with respect to the project's promotion of two pure lines – selected from Arabi Aswad, Tadmor and Zanbaka – that were developed before the PPB program was developed and that had been rejected by the national variety release committee. These lines were introduced into the PPB program and almost immediately adopted by farmers. In their peak period of adoption, Tadmor and Zanbaka were estimated to be grown on 5,000 and 20,000 hectares of land after they were tested in the PPB trials at the beginning of the second phase. When the results of the on-farm trials were compared with those of the PPB trials, a considerable discrepancy was found in the case of both Tadmor (Figure 6.2) and Zanbaka (Figure 6.3).

In the on-farm trials, Tadmor and Zanbaka outyielded Arabi Aswad by 3.6 percent and 1.5 percent, respectively (on average, in 26 trials in 4 years). At the end of the 4 years of testing, they were withdrawn without being submitted for release because the GCSAR breeder did not consider these varieties to demonstrate a significant yield improvement. When tested in the PPB trials, Tadmor and Zanbaka outyielded Arabi Aswad by nearly 20 percent and 17 percent, respectively (on average, in 55 trials in 3 years).

Zanbaka, and to a lesser degree Tadmor, have also been successfully tested in the Gezira region of Iraq, an area where farmers have a strong preference for black-seeded barley, much like the Raqqa and Hassakeh provinces in Syria. The reasons for the different results between the on-farm trials and the PPB trials are not clear. They could actually be caused by a combined effect of the locations of the on-farm trials (which were usually planted on the best fields in the area and were often in natural depressions that favoured water harvesting), their

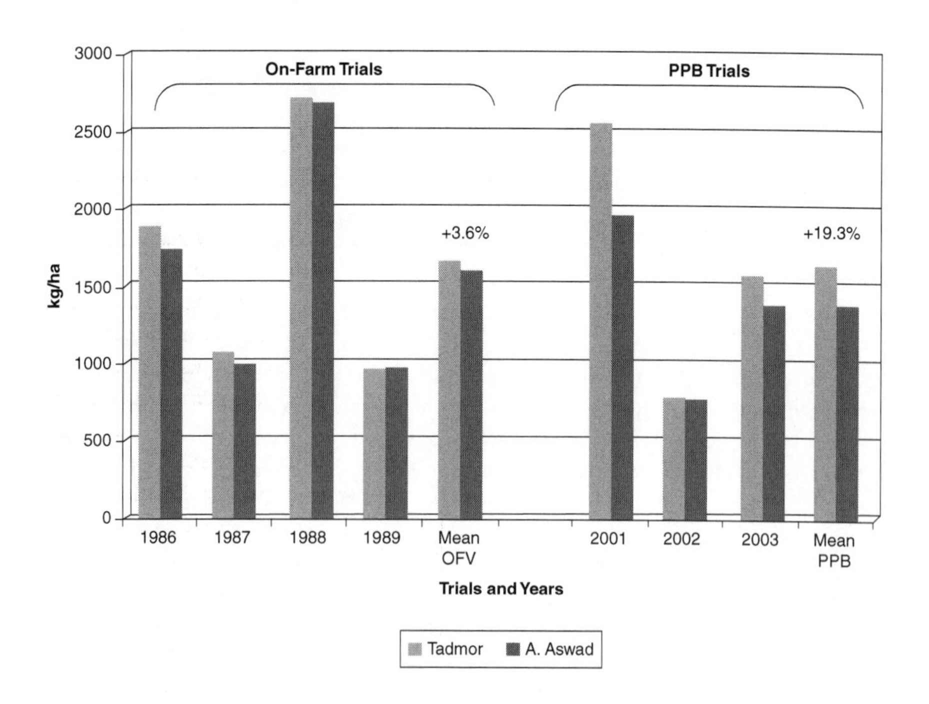

Figure 6.2 Grain yield of Tadmor and Arabi Aswad in the on-farm trials conducted between 1986 and 1989 and in the PPB trials conducted between 2001 and 2003

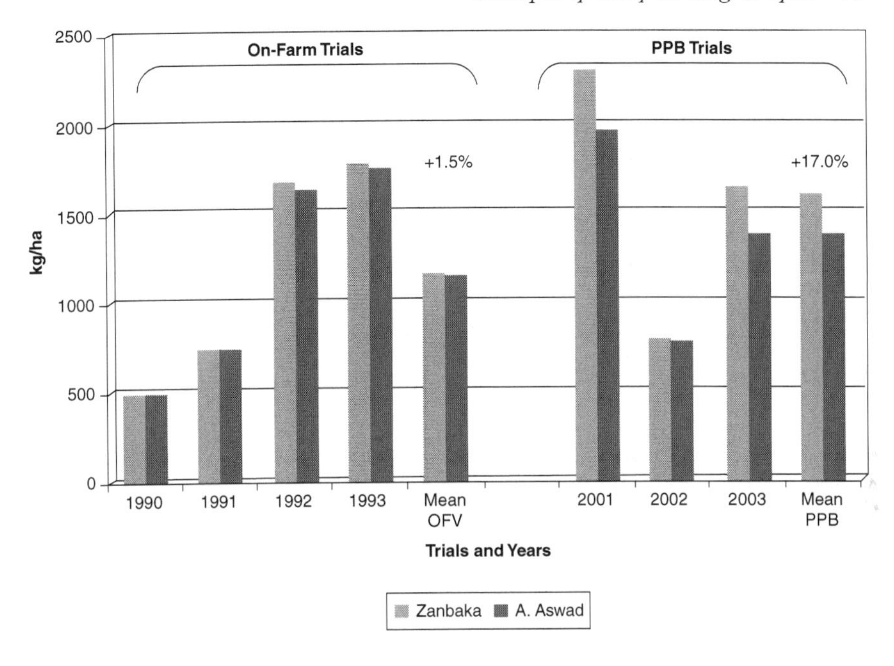

Figure 6.3 Grain yield of Zanbaka and Arabi Aswad in the on-farm trials conducted between 1990 and 1993 and in the PPB trials conducted between 2001 and 2003

management (optimum rotation and agronomic practices) and the statistical analysis, which did not take into account the spatial variability.

The results shown in Figures 6.2 and 6.3 were shared with the staff of the GCSAR, including the director general of the time, in order to stimulate discussion on the possible revision of the variety release process. It was decided that changes in experimental design and statistical analysis as well as the opinion of the farmers in the initial adoption, which usually takes place in a PPB program, should play a key role in the process. If successful, it was hoped that these changes would help to facilitate the release of numerous PPB varieties, which have been being continuously produced (at the moment of writing, the number of entries named by farmers has reached more than 70).

However, this approach did not lead to the expected outcome. Despite providing the GCSAR with the software and the training needed to modernize the trial's design and analysis, the on-farm trials continued to be conducted in the usual way, and the data from the PPB trials that took place between 2003 and 2008, which were collected by the same GCSAR staff, were not even considered for a possible release. The closest attempt to integrate PPB into the National Breeding Program was when the GCSAR proposed that, after 4 years of testing in farmers' fields and after consolidating the farmers' opinions, the selected lines, which were named and adopted by the farmers, should be tested again in on-farm trials for three cropping seasons and that *only* the data from these 3 years should be used for eventually recommending one or more of these lines for release. This enormous delay was a result of a law that had to be respected. This proposal was discussed with the farmers, and it was agreed that it was entirely unacceptable as it would cause unnecessary delay the entire process by at least four years (three years for the trials and one for writing the report), during which time the additional information collected on the breeding material would be of dubious scientific value. An effort was also made to adjust the law that was always being quoted in these discussions but without success.

Are policies and laws relevant?

As more and more varieties were identified and selected by farmers, it became clear that availability of seed was a major bottleneck in adopting these varieties, and on the effectiveness of PPB projects in the longer term (Moustafa, Grando and Ceccarelli, 2006). Since there had been no progress in the stalemate with the GCSAR concerning the release of these varieties – releases that could have been implemented using the GOSM facilities to produce and distribute seed – farmers were provided with locally made equipment to clean and treat the seed from PPB varieties. This seed began to be commercialized, and several farmers started seeing results, despite the limitation of the informal seed systems, either from the higher yield of the varieties or from both the higher yield and profits from the sale of the seed (Moustafa et al., 2006).

Most of the work described earlier was supported by the International Development Research Centre through various projects. One of these projects ended in 2007 with a workshop held in Jordan (the project included Syria and Jordan) with the participation of about 200 farmers and high officials of the two Ministries of Agriculture (the Jordanian minister opened the workshop). During the workshop, several farmers shared their experiences with PPB and with its outcome and exchanged experiences on the issue of seed multiplication and distribution.

In the closing ceremony, the highest representative of the Syrian Ministry of Agriculture accused the farmers, and indirectly ICARDA, of engaging in illegal activities since the cultivation of varieties has not been officially released and the commercialization of seed from these projects is prohibited by law. A few weeks later, the minister of agriculture sent a letter to the director general of ICARDA complaining that these activities were a threat to the national food security. The letter was followed by instructions to all of the staff of the GCSAR to refrain from collaborating in PPB trials. Some of the GCSAR researchers actually continued to collaborate on the projects on weekends. As of today and despite several requests, no copy of the law has been received.

At the time of writing, we are not aware that there is a national seed law that restricts the exchange of seed – the only existing law is a ministerial decree from 1975 stating that there are no restrictions about the movement of seed. A law was drafted in 2002 with the Food and Agriculture

Organization (FAO) to regulate the exchange of plant genetic resources based on the International Treaty on Plant Genetic Resources for Food and Agriculture in conformity with the treaty's provisions.[1] In this draft law, national sovereignty remains the basic principle regulating access to Syrian genetic resources. The draft law further recognizes the right of farmers and local communities to participate in national decision making about conservation and the use of plant genetic resources and related benefit sharing. Farmers and local communities are also to be consulted before access is granted for collecting *in situ* plant genetic resources. No further progress has been made, and the draft law remains in limbo (personal communication with an FAO representative, 2009).

The gender dimension

One distinctive aspect of the PPB program in Syria is the lack of female participation, which has not been a problem in other countries where PPB programs have been implemented with the same methodology. One of these countries was Jordan, and due to the relatively short distance between Jordan and Syria, it was decided at the time of field selection to invite a group of 10 Jordanian women farmers to visit three PPB villages in Syria (two in the South and one in the centre of the country). This decision had the expected effect of bringing Syrian women out of their houses and into the field. It was thereby discovered that women farmers in Syria are interested in PPB, but they are not being informed about the possibility of collaborating or they are assuming that they cannot participate. Since that experience, a female researcher has been supporting the integration of Syrian women farmers into the PPB efforts by combining gender analysis with action research. This work has revealed gender-based differences in agronomic management, crop preferences and project needs (Galié et al., 2009).

Conclusions

The PPB Syrian experience has shown that it is entirely possible to organize a plant-breeding program with the full participation of farmers while maintaining the science of plant breeding. Thus, there is no scientific justification for avoiding the use of PPB. The Syrian experience has also shown that PPB can make a contribution to three of the most frequently debated global problems (i.e. biodiversity, climate change and world famine). The continuous decline of biodiversity has been widely documented (World Conservation Monitoring Centre, 1992; Butchart, Walpole and Collen, 2010; Frison, Cherfas and Hodgkin, 2011) as well as the entity and the effects of climate change. In addition, famine, as well as hidden hunger among various groups of people, is still widespread. A recent report to the United Nations establishes a relationship between agrobiodiversity, seed systems, hunger and participatory plant breeding. It underlines the fact that hunger is not only a problem of production but also a problem of accession and availability, and it recommends that donors and international institutions, including the Consultative Group on International Agricultural Research and the FAO, should fund, among other initiatives, breeding projects on a large diversity of crops. These crops should include orphan crops and varieties for complex agro-environments, such as dry regions and breadbasket regions, in order to address the needs of the most vulnerable groups and put farmers at the centre of research through participatory research schemes such as participatory plant breeding. Eventually, participatory plant breeding will be able to combine the maintenance and enhancement of agro-biodiversity with the need to feed everyone by making more food available and accessible and being able to cope with a continuous and gradual change in climatic conditions.

Note

1 International Treaty on Plant Genetic Resources for Food and Agriculture, 29 June 2004, online: <www.planttreaty.org/texts_en.htm> (last accessed 12 November 2012).

References

Bishaw, Z. (2004). *Wheat and Barley Seed Systems in Ethiopia and Syria*, Wageningen University, Wageningen. Available at: library.wur.nl/WebQuery/wda/lang/1717865 (last accessed 19 February 2016).

Butchart, S.H.M., M. Walpole and B. Collen (2010). 'Global Biodiversity: Indicators of Recent Declines,' *Science* 328: 1164–8.

Ceccarelli, S. (2009). 'Selection Methods. Part 1: Organizational Aspects of a Plant Breeding Programme,' in S. Ceccarelli, E.P. Guimarães and E. Weltizien (eds.), *Plant Breeding and Farmer Participation*, FAO, Rome, pp. 63–74.

Ceccarelli, S. (2012). 'Landraces: Importance and Use in Breeding and Environmentally Friendly Agronomic Systems,' in N. Maxted et al. (eds.), *Agrobiodiversity Conservation: Securing the Diversity of Crop Wild Relatives and Landraces,* CAB International, Wallingford, UK, pp. 103–17.

Ceccarelli, S., and S. Grando (2000). 'Barley Landraces from the Fertile Crescent: A Lesson for Plant Breeders,' in S.B. Brush (ed.), *Genes in the Field: On-Farm Conservation of Crop Diversity*, International Development Research Centre, Boca Raton, Florida, pp. 51–76.

Ceccarelli, S., and S. Grando (2007). 'Decentralized-Participatory Plant Breeding: An Example of Demand Driven Research,' *Euphytica* 155: 349–60.

Ceccarelli, S.S. Grando and M. Baum (2007). 'Participatory Plant Breeding in Water-Limited Environment,' *Experimental Agriculture* 43: 411–35.

Ceccarelli, S., S. Grando, R. Tutwiler, J. Baha, A.M. Martini, H. Salahieh, A. Goodchild and M. Michael (2000). 'A Methodological Study on Participatory Barley Breeding in the Selection Phase,' *Euphytica* 111: 91–104.

Ceccarelli, S., S. Grando and J.A.G. Van Leur (1987). 'Genetic Diversity in Barley Landraces from Syria and Jordan,' *Euphytica* 36: 389–405.

Frison, E.A., J. Cherfas and T. Hodgkin (2011). 'Agricultural Biodiversity Is Essential for a Sustainable Improvement in Food and Nutrition Security,' *Sustainability* 3: 238–53.

Galié, A., B. Hack, N. Manning-Thomas, A. Pape-Christiansen, S. Grando and S. Ceccarelli (2009). 'Evaluating Knowledge Sharing in Research: The International Farmers' Conference Organized at ICARDA,' *Knowledge Management for Development Journal* 5: 107–25.

Moustafa, Y., S. Grando and S. Ceccarelli (2006). Assessing the Benefits and Costs of Participatory and Conventional Barley Breeding Programs in Syria. Available at: http://impact.cgiar.org/assessing-benefits-and-costs-participatory-and-conventional-barley-breeding-programs-syria (last accessed 29 September 2011).

Newton, A.C., T. Akar, J.P. Baresel, P.J. Bebeli, E. Bettencourt, K.V. Bladenopoulos, J.H. Czembor, D.A. Fasoula, A. Katsiotis, K. Koutis, M. Koutsika-Sotiriou, G. Kovacs, H. Larsson, M.A.A. Pinheiro de Carvalho, D. Rubiales, J. Russell, T.M.M. Dos Santos and M.C. Vaz Patto (2010). 'Cereal Landraces for Sustainable Agriculture: A Review,' *Agronomy for Sustainable Development* 30: 237–69.

Russell, J.R., A. Booth, J.D. Fuller, M. Baum, S. Ceccarelli, S. Grando and W. Powell (2003). 'Patterns of Polymorphism Detected in the Chloroplast and Nuclear Genomes of Barley Landraces Sampled from Syria and Jordan,' *Theoretical and Applied Genetics* 107: 413–21.

Tripp, R., N. Louwaars, W.J. Van der Burg, D.S. Virk, J.R. Witcombe (1997). *Alternatives for Seed Regulatory Reform: An Analysis of Variety Testing, Variety Regulation and Seed Quality Control*, AGREN Network Paper no. 69. Available at: http://dlc.dlib.indiana.edu/dlc/bitstream/handle/10535/4581/Alternatives%20for%20seed%20regulatory%20reform%20an%20analysis%20of%20variety%20testing%20variety%20regulation%20and%20seed%20quality%20control.pdf?sequence=1 (last accessed 19 February 2016).

Van Leur, J.A.G., S. Ceccarelli and S. Grando (1989). 'Diversity for Disease Resistance in Barley Landraces from Syria and Jordan' *Plant Breeding* 103: 324–35.

World Conservation Monitoring Centre (1992). *Global Biodiversity: Status of the Earth's Living Resources*, in B. Groombridge (ed.), Chapman and Hall, London, p. 585.

Part III

The international policy context

Global systems of conservation and use and farmers' rights

7 Historical context

Evolving international cooperation on crop genetic resources

Regine Andersen

When the first hunters and gatherers became farmers some 10,000 years ago, they started out with only a few crops and crop varieties. Through careful selection of the best seeds and propagating material, and exchange with other farmers, it became possible to develop these few varieties into many different varieties. In addition, new crops were found in the wild that could be cultivated. Through the continuous management and innovation by farmers over thousands of years, the few initial crops and varieties evolved into an inconceivable wealth of crop diversity. Some 7,000 crop species have been cultivated or collected by humans for food (Wilson, 1992, 275), and the estimated number of distinct varieties of some of these crops exceeds 100,000 (FAO, 1998, 18).[1]

In the last century, however, the development of crop diversity changed profoundly with the breakthrough of modern plant breeding and the introduction of genetically homogeneous modern varieties. Although this change has contributed to a substantial increase in food production, it has also decimated untold food crop varieties that were vital to small-scale farmers and reduced the diversity that is essential for the future of plant breeding (FAO, 1998, 30–40). With this accelerating genetic erosion, concern and awareness have arisen among stakeholders with regard to the emerging threat to plant breeding and food production. At the international level, this concern was first taken up by the Food and Agriculture Organization (FAO) and has gradually become an issue within other international arenas as well.

In this chapter, we will see how current international cooperation on the management of crop genetic resources has taken form in the FAO – in response to this emerging awareness and, later, in response to the demands emerging from other international negotiation processes – and how farmers' rights have evolved through this process.

Emerging awareness in the FAO

The FAO's engagement in plant genetic resources for food and agriculture (PGRFA) began shortly after the organization was established in 1945. In 1948, the FAO conducted a conference for 44 member states on biological research in forestry, addressing the use and exchange of genetic material on the global scale (Pistorius, 1997, 11). In 1957, the organization started publishing a newsletter on plant genetic resources[2] and, from 1961 on, it convened a series of technical meetings and conferences on plant genetic resources. A FAO Panel of Experts on Plant Exploration was established in 1963 to advise the organization and set international guidelines for the collection, conservation and exchange of PGRFA. The conference on plant genetic resources that was held in 1967 formulated several resolutions that were subsequently adopted by the UN Conference on Environment and Development in Stockholm in 1972. In 1973, a new conference interpreted the resolutions of the Stockholm Conference in the context of PGRFA (FAO, 1998, 2).

Following these efforts within the FAO, the International Board for Plant Genetic Resources (IBPGR) was founded in 1974 under the auspices of the Consultative Group on International Agricultural Research (CGIAR).[3] The IBPGR was located at the FAO headquarters in Rome and drew on the staff designated for the FAO program on genetic resource conservation (Fowler, 1994, 184). Collecting missions were accelerated, and gene banks were constructed and expanded at the national, regional and international levels (FAO, 1998, 2). During these years, the erosion of PGRFA had increased at an unprecedented rate, and the efforts of the IBPGR were vital to saving plant genetic resources that were in danger of extinction.

As a result, the IBPGR had considerable power over the direction of conservation activities and was assumed to have the authority to designate certain gene banks for holding particular collections. Only 15 percent of the samples collected were designated for storage in developing countries, whereas 85 percent were stored in industrialized countries and in the gene banks of the International Agricultural Research Centres (IARCs) of the CGIAR (Fowler, 1994, 184), most of which are located in developing countries. There have been some obvious reasons for this choice. First, gene banks in the South – with the exception of those maintained by the IARCs – were rare at the time that most of the materials were collected. Second, they were often unable to ensure the safety of seed accessions as adequately as gene banks in the North could, due to shortage of electricity, the lack of institutional capacity and financial constraints. Unlike today, the majority of genetic resource samples provided by the IARCs in the 1970s went to countries in the North,[4] and the IARCs were, together with the IBPGR, heavily criticized by nongovernmental organizations (NGOs) for serving private interests in the North (see, inter alia, Fowler and Mooney, 1990, 182–93). The IBPGR and the IARCs did invaluable work in saving fast-eroding plant varieties from extinction, but at the same time, developing countries lost control over their own genetic resources.

The disparities between the North and the South regarding PGRFA were taken up in *Seeds of the Earth: A Private or Public Resource?* by Pat Roy Mooney in 1979.[5] This book provoked much concern and discussion, and is held to have given a decisive push to a multilateral process towards the international regulation of the management of PGRFA (see Fowler, 1994, 180). At the FAO Conference in 1981, the Mexican delegate successfully proposed that the elements of a legal convention on the feasibility of establishing an international gene bank should be drafted and reported back to the FAO Conference in 1983 (ibid., 187).

From the very beginning of the discussions in the FAO, there were major disagreements as to what the organization should be striving for in this regard. Most developing countries were concerned about the loss of diversity in PGRFA, combined with the disparity of the designated storage facilities between North and South, and related issues of access and control. They supported the Mexican proposal to draft a legal convention (Fowler, 1994, 186). Most industrialized countries, fronted by the United States, Great Britain, and Australia together with the seed industry, were also concerned about genetic erosion but sceptical of an international regime of this type, since they feared the politicization of the issue and the possible loss of control over PGRFA conserved in gene banks by the FAO (ibid., 188).

The adoption of the International Undertaking on Plant Genetic Resources

During the negotiations, the major conflict lay between those in favour of plant breeders' rights over improved varieties of plants and those in favour of unrestricted access to all plant varieties (Fowler, 1994, 187–91). The United States and representatives of the seed industry were the leading proponents of the former stance, and developing countries made up the latter position. This point is worth noting since most developing countries were later to change their position on access in order to provide for benefit sharing from the use of genetic resources, which required a stricter regulation of access. This regulation came a decade later under the Convention on Biological Diversity (CBD) and in response to the emerging Agreement on Trade-Related Aspects of Intellectual Property Rights (TRIPS Agreement), which was then being negotiated in the Uruguay Round that led to the establishment of the World Trade Organization (WTO).[6]

When the International Undertaking on Plant Genetic Resources was adopted in 1983, it attracted wide support, and, after some time, it was adhered to by 113 countries.[7] However, several countries that were important to its implementation remained outside the International Undertaking, including the United States, Canada, Switzerland, and Australia. The adoption of the International Undertaking can be seen as a partial victory for developing countries because it was achieved despite the opposition of major industrialized countries led by the United States, which also chaired the FAO Conference at the time of its adoption. It was only a partial victory, however, because the new agreement ended up as a legally nonbinding undertaking, without the adherence of industrialized countries that were important to the international management of PGRFA. This factor severely limited its prospects for implementation.

The objective of the International Undertaking was to ensure that PGRFA would be explored, preserved, evaluated, and made available for plant breeding and scientific purposes. The two-pronged goal was clear: conservation and access. The International Undertaking was based on 'the universally accepted principle that plant genetic resources are a heritage of mankind and consequently should be available without restriction' (Article 1). Its main tasks were:

- to explore a variety of crops, their wild relatives, and nondomesticated food plants in International Undertaking signatory countries;
- to preserve, evaluate, and document plant genetic resources in the areas of their natural habitat in the major centres of genetic diversity as well as outside, in gene banks, and in living collections of plants;
- to ensure open access to the plant genetic material preserved.

These tasks were to be carried out not only by the states adhering to the International Undertaking but also by multilateral actors such as the FAO and the institutions of the CGIAR. The implementation of the International Undertaking indicated close international cooperation in the form of international gene banks and information-sharing networks, steps to enhance the performance and numbers of gene banks, and efforts to encourage financial support and capacity building.

Along with the International Undertaking, the Commission on Plant Genetic Resources (CPGR) was established by the 22nd session of the FAO Conference in Rome in 1983.[8] The CPGR was an intergovernmental body charged with ensuring the implementation of the International Undertaking and monitoring it, especially the operation of international arrangements for the management of PGRFA.

Seeking wider adherence to the International Undertaking

A big problem with the International Undertaking was that a number of major industrialized countries had not adhered to it when it was adopted.[9] At the first session of the CPGR in 1985, proposals for modifying the text of the International Undertaking in order to attract greater adherence were already being considered (FAO, 1985, paras. 12–13). The commission recommended that the Secretariat prepare a paper for consideration at its next session, whereby it could analyze countries' reservations to the International Undertaking and delineate possible courses of action, including suggestions for possible interpretations of the text that could increase acceptance (ibid., para. 13). The CPGR also established a working group, led by its own chairman and consisting of 23 members from different country groups, to consider the progress made in implementing the commission's program of work and any other matters referred to it (ibid., paras. 78–80). Much of the negotiations pertaining to adherence took place in this working group.[10]

The main reason that countries did not adhere to the International Undertaking was because of its objective stating that genetic resources should be available without restrictions. This objective was seen to be in conflict with plant breeders' rights. Therefore, countries could only adhere to the International Undertaking if the text was modified in some way (Andersen, 2005a). It was in this context that the concept of farmers' rights was taken up for the first time in the FAO. The first documented use of the concept was at a meeting of the working group in 1986 (FAO, 1986a). The idea came up as a countermove in response to the increased demand for plant breeders' rights in order to draw attention towards the unremunerated innovations of farmers that were seen as the foundation of all modern plant breeding (ibid., para. 14):

> The Working Group emphasized that, in addition to the recognition of plant breeders' rights, specific mention should be made of the rights of the farmers of the countries where the materials used by the breeders originated. These materials were the result of the work of many generations and were a basic part of the national wealth. FAO should study this subject with a view to formulating a constructive solution.[11]

The working group produced a report on how to deal with the reservations by many countries to the International Undertaking and on how to attract greater adherence (FAO, 1986b, para. 8). The third chapter of the report is devoted to farmers' rights. It not only links the issue to the question of access to genetic resources but also reveals substantial uncertainties as to the understanding of the concept and calls for further elucidation.

At the second meeting of the working group in 1987, farmers' rights were addressed in greater detail:

> During the discussion of document CPGR/87/4, the Working Group agreed that the breeding of modern commercial plant varieties had been made possible first of all by the constant and joint efforts of the people/farmers (in the broad sense of the word) who had first domesticated wild plants and conserved and genetically improved the cultivated varieties over the millennia. Thanks were due in the second place to the scientists and professional people who, utilizing these varieties as their raw material, had applied modern techniques to achieve the giant strides made over the last 50 years in genetic improvements. In recent years some countries had incorporated the rights of the latter group into laws as 'Breeders' Rights,' i.e. the right of professional plant breeders or the commercial companies which employ them to participate in the financial benefits derived from the commercial exploitation of the new varieties. However, as document CPGR/87/4 pointed out, there was presently no explicit acknowledgement of the rights of the first group, in other words, no 'Farmers' Rights.' The Working Group considered such rights to be fair recognition for the spadework done by thousands of previous generations of farmers. And which had provided the basis for the material available today and to which the new technologies were in large measure applied. The Group agreed, that what was the issue here was not individual farmers or communities of

> farmers but the rights of entire peoples who, though having bred, maintained and improved cultivated plants, had still not achieved the benefits of development nor had they the capacity to produce their own varieties. Alternative names such as 'right of the countries of origin' or 'gene donors,' were proposed, but the conclusion was that the name 'farmers' rights' was the most expressive.
>
> (FAO, 1987, Appendix F)

As these quotations show, the main element of the farmers' rights concept concerned the need to reward farmers for their contribution to PGRFA. The rights holders were not to be single farmers or communities but, rather, entire peoples – that is, a form of a collective right. This concept can be seen to be the foundation for the stewardship approach to farmers' rights that is discussed in Chapter 8 of this volume.

The idea of developing farmers' and plant breeders' rights simultaneously in order to seek a balance between the two also emerged at this meeting:

> The Working Group concurred that Breeders' Rights and Farmers' Rights were parallel and complementary rather than opposed and that the simultaneous recognition and international legitimization of both these rights could help to boost and speed up the development of the people of the world.
>
> (FAO, 1987, para. 12)

At the second session of the CPGR in 1987, farmers' rights were brought up, and delegations expressed a wide range of opinion. The commission agreed to adopt practical measures to ensure wider adherence to the International Undertaking and established a contact group composed of 17 members to work out the various proposals (ibid., paras. 12 and 34). The contact group met during the second session of the commission and agreed that

> while the so-called 'farmers' rights' could not yet be given a precise definition, some sort of compensation for their most valuable contribution to the enrichment of the plant genetic resources of the world was well-founded and legitimate. It was pointed out that one way of giving practical recognition to this right could be in a form of multifaceted international cooperation including a freer exchange of plant genetic resources, information and research findings, and training. Another way could be through monetary contribution for financing a programme for the furtherance of the objectives of the International Undertaking on Plant Genetic Resources.
>
> (ibid., App. G)

Thus, the contact group did not arrive at a definition of farmers' rights but outlined some ways and means of according practical recognition within the framework of the International Understanding. Nevertheless, there were still deep controversies over these issues between the countries of the Organisation for Economic Co-operation and Development (OECD) on the one side, and the group of developing countries and their NGO supporters on the other.

Effects of the Uruguay Round on the process

These controversies were also fuelled by another process, the Uruguay Round of the General Agreement on Tariffs and Trade (GATT), which ultimately led to the WTO, where intellectual property rights (IPRs) were brought into the negotiations by the United States.[12] During the first years of the Uruguay Round, which started in 1986, an agreement on IPRs was strongly opposed by several developing countries. Indeed, by the 1988 midterm review of the round, it was determined that such an agreement was going to be impossible to achieve (Evans and Walsh, 1994, 39). Several developing countries insisted on the insertion of 'social clauses' in the agreement and were ambivalent as to whether such an agreement should be incorporated into the GATT at all.

In the course of 1989, however, those developing countries that were in opposition changed their positions and dropped their earlier resistance to an agreement on IPRs. This radical shift clearly resulted from their recognized need to make concessions within the negotiations, since a consensus on all of the agreements was going to have to be achieved before the package could be adopted (Yusuf, 1998, 9). By yielding to the agreement on IPRs, it was felt that gains could be achieved in other important areas of the Uruguay Round – such as in agriculture, textiles and tropical products. Hence, the agreement on IPRs was seen as one of the few bargaining cards that was available to developing countries.

Two years later, in a detailed proposal submitted in 1990, developing countries sought to include a range of social objectives as well as to exempt from patentability, inter alia, plant varieties and essentially biological processes for the production of animals and plants (Andersen, 2008, 150). This proposal paved the way for the negotiation of the TRIPS Agreement, which was adopted in 1994 and entered into force in 1996 (one year after the Marrakech Agreement Establishing the World Trade Organization [WTO Agreement]).[13] In the meantime, the social objectives had been deleted in the text that should have become the TRIPS Agreement (but were included in the WTO Agreement), and plant varieties had been made subject to IPRs. Thus, the resulting TRIPS Agreement excluded from patentability plants and animals (other than microorganisms) and essentially biological processes for the production of plants and animals (other than nonbiological and microbiological processes), but it did oblige members to provide for the protection of plant varieties either by patents or by an effective *sui generis* system (a system of its own kind) or a combination of the two (Article 27.3.b). An effective *sui generis* system for plant varieties was never defined, but it was understood to mean some sort of IPR for plant varieties (Andersen, 2008, 145–71). Even though there are several different *sui generis* systems in operation, the term has most often been related to the International Union for the Protection of New Varieties of Plant (UPOV) (ibid., 164–68).

Keystone dialogues

By 1987–88, most developing countries were still opposed to IPRs related to plants and plant varieties, and this opposition also affected negotiations in the FAO. With these deepening controversies, William Brown, who was then chair of the US National Board for Plant Genetic Resources (prior to which he had been president of Pioneer Hi-Bred), initiated contact with the Keystone Center in Colorado, seeking to establish a dialogue on plant genetic resources among international stakeholders (Fowler, 1994, 197). The Keystone approach was to invite stakeholders to participate as individuals in order to reduce conflict and seek dialogue, to keep the discussions off the record and to produce a report solely on the basis of consensus. These discussions took place from 1988 to 1991 and were chaired by M.S. Swaminathan, who also led an interim steering committee that gave direction to the dialogues. The facilitators were the staff of the Keystone Center.

During its three sessions, the Center gathered together 92 stakeholders from 30 countries, many of whom were central to the negotiations, and succeeded in leading international discussions on such issues as farmers' rights, the common heritage of mankind, international funding and, to some extent, IPRs (Keystone Center, 1991). Some of the proposals from the Keystone dialogues found their way into the FAO Conference resolutions on the International Undertaking in 1989 and 1991, which were initiated to deal with the reservations to the International Understanding expressed by the core industrial countries. The Keystone dialogues ended up being instrumental in achieving a breakthrough in this regard.

Agreed interpretations of the International Undertaking

The preparatory work of the working group and the contact group of the CPGR and, in particular, the Keystone dialogues paved the ground for two resolutions that were adopted on 29 November 1989: Resolution 4/89 on the Agreed Interpretation of the International Undertaking, and Resolution 5/89 on Farmers' Rights. Both resolutions were annexed to the International Undertaking. These two resolutions were adopted by consensus, but they came about only as a result of tense negotiations. There had been fierce resistance to the idea of plant breeders' rights among developing countries, and the interpretations that provided for the acceptance of these rights could only be adopted with the simultaneous recognition of farmers' rights (Andersen, 2005a). In this way, the opponents of plant breeders' rights gained recognition of farmers' rights in exchange for something that already existed: plant breeders' rights.

Resolution 4/89 on the Agreed Interpretation of the International Undertaking endorsed the idea that the interpretation was intended to provide an equitable and therefore solid and lasting global system; by doing so, it eased the concerns of many countries that had withheld their approval of the International Undertaking and secured the adherence of others. The resolution presented the rationale behind the agreed interpretation and then stated that 'Plant

Breeders' Rights as provided for under UPOV (International Union for the Protection of New Varieties of Plant) are not incompatible with the International Undertaking' (para. 1) and that

> states adhering to the Undertaking recognize the enormous contribution that farmers of all regions have made to the conservation and development of plant genetic resources, which constitute the basis of plant production throughout the world, and which form the basis for the concept of Farmers' Rights.
>
> (para. 3)

Furthermore, it set out that

> the adhering states consider that the best way to implement the concept of Farmers' Rights is to ensure the conservation, management and use of plant genetic resources, for the benefit of present and future generations of farmers. This could be achieved through appropriate means, monitored by the Commission on Plant Genetic Resources, including in particular the International Fund for Plant Genetic Resources, already established by FAO. To reflect the responsibility of those countries which have benefited most from the use of germplasm, the Fund would benefit from being supplemented by further contributions from adhering governments, on a basis to be agreed upon, in order to ensure for the Fund a sound and recurring basis. The International Fund should be used to support plant genetic conservation, management and utilization programmes, particularly within developing countries, and those which are important sources of plant genetic material.
>
> (para. 4)

Resolution 5/89 on Farmers' Rights represented a milestone since it was the first recognition by the FAO Conference of farmers' contributions to the global genetic pool, and it outlined the contents and implications of the concept itself:

> The FAO Conference ... [e]ndorses the concept of Farmers' Rights (Farmers' Rights mean rights arising from the past, present and future contributions of farmers in conserving, improving, and making available plant genetic resources, particularly those in the centres of origin/diversity. These rights are vested in the International Community, as trustee for present and future generations of farmers, for the purpose of ensuring full benefits to farmers, and supporting the continuation of their contributions, as well as the attainment of the overall purposes of the International Undertaking) in order to:
>
> a ensure that the need for conservation is globally recognized and that sufficient funds for these purposes will be available;
> b assist farmers and farming communities, in all regions of the world, but especially in the areas of origin/diversity of plant genetic

resources, in the protection and conservation of their plant genetic resources, and of the natural biosphere;

c allow farmers, their communities, and countries in all regions, to participate fully in the benefits derived, at present and in the future, from the improved use of plant genetic resources, through plant breeding and other scientific methods.

(FAO, 1989)

In 1991, a new annex to the International Undertaking was adopted as Resolution 3/91 (FAO, 1991). This time, the FAO Conference stated that the concept of genetic resources being the heritage of mankind, as applied in the International Undertaking, was subject to the sovereignty of states.[14] This interpretation might be seen to have been heavily influenced by the ongoing negotiations for a Convention on Biological Diversity (CBD), which was adopted only six months later and which also incorporated the principle of national sovereignty in Article 3. As a result of the CBD negotiations and in response to the emerging intellectual property regime, negotiators from developing countries demanded control over access to their genetic resources as well as the fair and equitable sharing of the benefits arising from their use. In many circles, this demand brought about a shift in thinking on genetic resources, from a perspective based on the common heritage of mankind to a bilateral approach to benefit sharing. This shift can be seen as the beginning of the ownership approach to farmers' rights, which is discussed in more detail in Chapter 8 of this volume.

After Resolution 3/91, FAO members stated that the conditions for access to plant genetic resources required further clarification (FAO, 1991, para. d). The original purpose of the International Undertaking – which was to ensure unrestricted access to genetic resources – was no longer clear, and the principles of 'the common heritage of mankind' that had controlled these resources were blurred. New factors had been introduced which complicated the follow-up and limited the prospects of implementation.

In the same annex to the International Undertaking, the FAO Conference also decided to set up an international fund to implement farmers' rights. This initiative was a direct result of the Keystone dialogues, where the need for such a fund had been noted and generally accepted by all. Participants from industry ensured that such a fund would be a way to recognize farmers' rights rather than simply compensating farmers for their efforts (Fowler, 1994, 201). Third world participants were so surprised by this consensus that they did not start a debate on this precondition. However, despite all best intentions, except for a few initial contributions, the fund never came to fruition.

Some achievements of the International Undertaking

The greatest achievement of the International Undertaking at the international level was the establishment of the International Network of *Ex Situ*

Collections under the auspices of the FAO. In 1994, the CGIAR centres with gene banks concluded agreements to place their germplasm collections under the auspices of the FAO.[15] These agreements constituted the cornerstones of the international network – some 600,000 accessions of plant varieties, including information about each of them, were made available by the FAO for plant breeding and direct use under formally agreed terms (FAO, 1998, 83). The greatest advantage of this system was not the fact that all of these varieties were now readily available (after all, much of the material had been available under similar terms for years) but, rather that the system was now being controlled: management had now been placed under the auspices of a multilateral organization, thereby putting an end to any insecurity about long-term access to these resources.

In 1998, these accessions made up between 20 and 50 percent of all genetic material conserved in gene banks (FAO, 1998, 280).[16] In fact, the CGIAR collections contain the highest genetic diversity of PGRFA among gene banks in the world, due to the heavy emphasis on landraces and wild relatives (Fowler, Smale and Gaiji, 2001; Fowler and Hodgkin, 2004).[17] They are also known for their well-maintained and documented collections. To ensure continued availability and to meet the demands of developing countries on this point, the centres agreed that any recipient of genetic resources should not claim ownership to them or seek IPRs over them or over any related information. This mandate was set out very clearly in the material transfer agreements that were used by the centres. However, this obligation did not extend to new varieties of plants that resulted from the use of genetic resources from the gene banks – a distinction that was later applied in the Multilateral System on Access and Benefit Sharing under the International Treaty on Plant Genetic Resources for Food and Agriculture (ITPGRFA) (discussed later in this chapter).[18]

The most comprehensive review of the situation at the national level was the *State of the Worlds' Plant Genetic Resources for Food and Agriculture*, which was finalized in 1996 and published in 1998 (FAO, 1998). It concluded that about 40 percent of all countries (industrialized as well as developing countries) had relevant national programs and coordination mechanisms (ibid., 223). However, these programs focused mainly on *ex situ* conservation, with poor institutional linkages to utilization efforts. Only 27 of 154 countries reported that *in situ* conservation had been included in their national programs, and only 26 countries reported that utilization was even an integral part of their programs (ibid., 202).

The 1996 *State of the World* report and the Global Plan of Action for the Conservation and Sustainable Utilization of Plant Genetic Resources for Food and Agriculture, which is discussed later in this chapter, were both milestones for the realization of the International Undertaking.[19] It should be noted that without the International Undertaking, the ITPGRFA would probably not exist today. However, the initiative to negotiate the treaty was to come from another angle: the CBD.

The CBD and new negotiations on PGRFA

The adoption of the CBD was a decisive event for the development of the International Undertaking regime. The CPGR had not had the opportunity to clarify the uncertainties that concerned the availability of PGRFA under the International Undertaking before the CBD was adopted in 1992. Thus, the International Undertaking was the first international agreement (albeit non-binding) to deal with the conservation and sustainable use of PGRFA. The CBD became the first legally binding international agreement to address the sustainable management of biological diversity worldwide.[20] The CBD was developed as a stand-alone convention as well as a framework convention (Andersen, 2008, 135). One important aspect of the framework convention approach was that the details regarding issue-specific areas were to be negotiated in separate protocols.

At the Conference for the Adoption of the Agreed Text of the Convention on Biological Diversity in May 1992, the Nairobi Final Act was adopted (UNEP, 1992). Under this Act, a resolution on the interrelationship between the CBD and the promotion of sustainable agriculture was adopted on 22 May (Resolution 3). This resolution dealt particularly with the importance of PGRFA and recommended that ways and means be explored to develop complementarity and cooperation between the CBD and the Global System for the Conservation and Sustainable Use of Plant Genetic Resources for Food and Sustainable Agriculture (ibid., para. 2), which had been established under the FAO with the International Undertaking acting as a central component. Finally, the resolution recognized the need to seek solutions to two outstanding matters concerning PGRFA:

- access to *ex situ* collections that had not been acquired in accordance with the CBD;
- the question of farmers' rights.

At its 27th session in November 1993, the FAO Conference accordingly requested the FAO Director-General to provide a forum for negotiations for harmonizing the International Undertaking with the CBD (Resolution 7/93). These negotiations were to consider the issue of access to PGRFA on mutually agreed terms, including *ex situ* collections not dealt with in the convention, as well as the issue of realizing farmers' rights. The CPGR followed up with a mandate and a proposed process (FAO, 1994). These efforts formed the point of departure for the lengthy negotiations that finally resulted in the adoption of the ITPGRFA in 2001.

Revising the International Undertaking in harmony with the CBD was a challenging task. The specific features, uses and management needs of PGRFA had to be taken into consideration.[21] PGRFA constitute the basis of farming and are essentially domesticated resources (except for their wild relatives). Since access to PGRFA is a condition for the further domestication and thus the continued existence of these resources, expeditious facilitation of access was a major concern to the negotiators. To ensure access, it was also important that transferred PGRFA should remain in the public domain and not be made subject to exclusive IPRs. A different solution to benefit sharing had to be found than the one envisaged under the CBD. It was necessary to focus on those who conserve and sustainably use these resources, rather than on the specific providers, for several reasons:

1 For most crops, it is difficult to identify the countries of origin (the countries entitled to provide access under the CBD) since the crops have been developed through the exchange of seeds across borders for centuries and even millennia (Andersen, 2001; Fowler, 2001).
2 All countries are interdependent on PGRFA so a complicated system of transfers between providers and recipients would hamper expeditious access to these resources (Palacios, 1998).
3 Rewarding only the current providers of genetic resources would not be fair to all of those farmers around the world who maintain and develop crop genetic diversity that will benefit future generations.

The CBD approach envisaged that each country would establish its own legislation and procedures for access and benefit sharing. However, for breeders, such an approach was problematic. Breeding often requires access to a wealth of plant varieties from different countries, and negotiating separate agreements with all of these countries would impose heavy burdens on the breeders, obstructing plant breeding for food and agriculture. In addition, many breeders regarded the first generation of laws on access and benefit sharing (ABS) as overly restrictive and bureaucratic (Ten Kate and Laird, 1999, 17–33, 293–312). Such a bilateral approach to the regulation of access to PGRFA appeared to many as an unattractive option.[22]

Thus, the parties sought to establish, inter alia, a multilateral system on ABS for PGRFA.[23] This move can be seen to some extent as a return to the principle of genetic resources as the common heritage of mankind. However, a list of crops to be included in the multilateral system was to be established, and it was not easy to decide which plants should be included in this list. Although the countries agreed that the list should cover crops that were important for food security and for which there was an interdependence between countries, these criteria were not precise enough to guide the selection of crops without controversies (Fowler, 2004). For instance, Latin American countries argued that pears were important for food security but that tomatoes and groundnuts were not; African countries refused to include forage grasses, which resulted in Latin American countries withholding 'their' forage grasses despite the fact that they had originally been inclined to include them in the list; China refused to include soybeans; and a range of other plants were kept outside the system.

Although all of the negotiators agreed on the need for accessibility, there was enormous suspicion among developing countries that was aimed specifically at developed countries, since the latter were seen as reaping the greater financial benefits from ABS and being able to impose IPRs on their products. Therefore, the majority of developing countries advocated for the prohibition of IPRs on PGRFA under the ITPGRFA and emphasized benefit-sharing arrangements. In contrast, the majority of OECD countries advocated open access to PGRFA and minimal restrictions, if any, on IPRs. Some of the small industrialized countries acted as bridge-builders in this context. For example, Norway supported the positions of developing countries on several occasions, mediating between the parties and contributing to developing a consensus text.

Commission on Genetic Resources for Food and Agriculture (CGRFA)

In 1995, shortly after the negotiations began, the CPGR was renamed the CGRFA, and its mandate was broadened to cover all components of biological diversity of relevance to food and agriculture, including animal genetic resources, forestry and fisheries. Its advisory role for the FAO was strengthened, as was its coordinating role with regard to other multilateral institutions.[24] According to its statutes, the CGRFA was to review all matters relating to FAO policy, programs and activities related to PGRFA and to recommend measures for developing a comprehensive global system on these resources.[25] Furthermore, the CGRFA was to provide an intergovernmental forum for negotiations (like the forums that led to the ITPGRFA), a body to oversee coordination with other relevant international agreements and additional international, governmental and nongovernmental bodies dealing with the conservation and sustainable use of genetic resources. The CGRFA was also to oversee the Global System for the Conservation and Sustainable Use of Plant Genetic Resources for Food and Sustainable Agriculture (FAO, 1998, 256), which was to cover a range of FAO initiatives to ensure safe conservation and to promote the availability and sustainable use of PGRFA by providing a flexible framework for sharing the benefits and burdens, including first the International Undertaking and later the ITPGRFA.[26]

At the 11th regular session of the CGRFA in 2007, the commission adopted a multiyear program of work covering plant, animal, forestry, aquatic, microbial and invertebrate genetic resources. That same year, the Global Plan of Action for Animal Genetic Resources, which was overseen by the commission, was adopted by the first International Technical Conference on Animal Genetic Resources. As of April 2014, 177 countries and the European Union were members of the CGRFA.

The 1996 Leipzig Conference and the Global Plan of Action

However, let us first return to the 1990s. No one might have imagined that it would take over a decade for the ITPGRFA to become fully operational – starting with the (re)negotiations of the new regime, its eventual adoption, another 3 years until it entered into force, and a further 2 years for the Standard Material Transfer Agreement (SMTA) to be adopted making the treaty operational. Action to halt genetic erosion and improve the management of PGRFA was urgently needed and could not await the conclusion of these drawn-out negotiations. It was during this time that the fourth International Technical Conference on Plant Genetic Resources was held in Leipzig in 1996, bringing together delegates from 150 countries. In a declaration from the meeting, the representatives stated that major gaps existed in countries' national and international capacities to conserve, characterize, evaluate and sustainably use plant genetic resources.[27] They also stated that access to and sharing of genetic resources and technologies were essential for ensuring world food security and meeting the needs of the growing world population. Based on this understanding, the representatives adopted the Global Plan of Action for the Conservation and Sustainable Utilisation of Plant Genetic Resources for Food and Agriculture (Global Plan of Action).[28]

The Global Plan of Action was prepared with the active participation of 154 countries. Each country prepared comprehensive reports on the state of PGRFA in its territories. These reports were compiled and analyzed in a comprehensive and detailed report covering biological, technical and institutional concerns. This first *State of the World* report not only represented the most important reference work on PGRFA since Jack Harlan's *Crops and Man* (1975), but it also represented a process that galvanized policy makers and practitioners for the management of these vital resources throughout the world (FAO, 1998).[29] As such, the report stands as a major achievement in itself.[30]

The Global Plan of Action provides a framework for identifying priority areas by the countries and support for capacity enhancement towards that end. Priority activities are to be identified within the areas of *in situ* conservation and development, *ex situ* conservation, the utilization of plant genetic resources and institutional development and capacity enhancement.[31]

International Treaty on Plant Genetic Resources for Food and Agriculture (ITPGRFA)

Throughout the negotiations, farmers' rights were one of the most contested issues leading to the ITPGRFA. Most developing countries, as well as some industrialized countries – among them Norway – had advocated comprehensive and internationally binding recognition of farmers' rights, whereas countries such as the United States and Australia were unwilling to support this stand. The controversies were complex and a breakthrough seemed unlikely when, in 1999, negotiators from the North decided to meet some of the demands from the South – this compromise led to the long-awaited breakthrough. What resulted was the final text of the ITPGRFA on farmers' rights as we know it today.[32] However, it still took almost two years to negotiate the final modalities of the treaty.

When the ITPGRFA was finally adopted in November 2001, many observers had almost given up on ever reaching a final consensus. Negotiations had proven extremely difficult, core provisions in the text were in brackets right up to the last hour, and it seemed impossible to unite all fronts on various joint solutions. Since full consensus proved impossible, the matter had to be put to a vote. At the conference, 116 countries voted in favour of the treaty and two countries abstained (Japan and the United States). The ITPGRFA was finally able to be adopted by the FAO Conference at its 31st session on 3 November 2001. It was the first legally binding agreement to deal exclusively with PGRFA, and it was also incidentally the first international treaty of the new millennium.[33] Since that time, the United States has revised its policy and has signed the treaty.[34] The ITPGRFA entered into force on 29 June 2004.[35] As of January 2016, 136 states are party to the treaty.

The objectives of the ITPGRFA are the conservation and sustainable use of PGRFA as well as the fair and equitable sharing of the benefits arising from their use – in harmony with the CBD – for sustainable agriculture and food security (Article 1). The provisions on conservation set out that the contracting parties shall be 'subject to national legislation, and in cooperation with other Contracting Parties where appropriate,' promote an integrated approach to the exploration, conservation and sustainable use of PGRFA (Article 5). Suggested measures include improving the *ex situ* conservation of plant varieties (and also wild crop species) and providing farmers with support for on-farm management and conservation of PGRFA – the latter being particularly relevant farmers' varieties and farmers' rights.

The ITPGRFA stipulates that contracting parties shall develop and maintain appropriate policies and legal measures that promote the sustainable use of PGRFA (Article 6). This provision is an obligation for all contracting parties and may include such measures as promoting diverse farming systems; encouraging research that enhances and conserves biological diversity; developing plant breeding with the participation of farmers in developing countries; broadening the genetic bases of crops; increasing the range of genetic diversity available to farmers; expanding the use of local and locally adopted crops and underutilized species; making wider use of a diversity of varieties and species in on-farm management; encouraging conservation and sustainable use; and adjusting the breeding strategies and regulations on variety release and seed distribution.

In the preamble to the ITPGRFA, the contracting parties affirm that the past, present and future contributions of farmers in all regions of the world – particularly those in the centres of origin and diversity – in conserving, improving and making available these resources constitute the basis of farmers' rights. They also affirm that the rights recognized in the ITPGRFA to save, use, exchange and sell farm-saved seed and other propagating material, to participate in relevant decision making and to encourage fair and equitable benefit sharing are fundamental to the realization of farmers' rights. Article 9 of the ITPGRFA recognizes the enormous contribution of farmers in the conservation and development of PGRFA and that this contribution constitutes the basis of food and agriculture production throughout the world. It explicitly states that responsibility for the implementation of farmers' rights, as they relate to the management of PGRFA, rests with national governments. Certain measures to protect and promote farmers' rights are suggested. These measures encompass the protection of relevant traditional knowledge, equitable benefit sharing and participation in decision making. The rights to save, use, exchange and sell farm-saved seeds and propagating material are addressed. Governments are free to choose the measures they deem appropriate, according to their needs and priorities. More information on the contents of these provisions and on the further decisions of the Governing Body of the ITPGRFA is available in Chapter 8 of this volume.

ITPGRFA's multilateral system of access and benefit sharing (MLS)

A central component of the ITPGRFA is its multilateral system of access and benefit sharing (MLS), as set out in Articles 10 to 13.[36] Whereas the ITPGRFA covers all PGRFA, its MLS covers 35 food crops and 29 forage plants that are in the public domain and under the management and control of the contracting parties. These genera are listed in Annex 1 to the ITPGRFA (the Annex 1 crops) and include major staple crops as well as a range of other plants widely used for food and agriculture.[37] The MLS has the following features:

1. It is a common pool of genetic resources, and into this pool all contracting parties (countries) place the genetic resources of Annex 1 crops that are in their public domain and under their control. According to this mandate, therefore, the contracting parties invite all holders of such material to deposit it into the MLS (Article 11.2). In practical terms, the material remains with the holders (normally gene banks), but it is accessible under the same terms and conditions as everywhere in the MLS. What makes this system multilateral is the fact that it involves a common pool of genetic resources overseen by the Governing Body of the ITPGRFA. Accessions of plant genetic resources that are outside the public domain, such as the resources held in private collections, are not included in the MLS, but countries are requested to take appropriate measures to encourage them to be included.
2. A SMTA is applied to all transfers of genetic material under the MLS. This mandate enables the expeditious transfer of PGRFA since no negotiations are required. The SMTA, which was adopted by the parties to the ITPGRFA in 2006, involves prior informed consent on mutually agreed terms among and between states in a standardized form. An increasing number of gene banks have established a Web-based 'click and wrap' system, which allows recipients of genetic material to enter into the SMTA simply by clicking in a box at the website of the gene bank to confirm their acceptance of the SMTA. This process makes the facilitation of access to the material

more efficient, which is in line with Article 12.3.b, which stipulates that access shall be provided expeditiously, without the need to track individual accessions. In the first eight months of operation, almost 100,000 transfers of genetic material took place within the MLS (ITPGRFA, 2007, 3). Since that time, the number has been steadily increasing.

3 Access is provided free of charge, or when a fee is charged, it is not to exceed the minimum cost involved (Article 12.3.b). All available passport data and related information are to be provided together with the material (Article 12.3.c).
4 IPRs are not allowed on material from the MLS or on its genetic parts and components, in the form it is received (Article 12.3.d). However, it is uncertain exactly how much the material must be modified before it is no longer regarded as being 'in the form received' under this article. The intention of this provision is to ensure that material in the MLS remains in the public domain, but how this will work in practice remains to be seen.
5 Monetary benefit sharing is fixed in terms of shares from the sales of products developed by the use of material from the MLS, as set out in the SMTA. If a product that has resulted from the use of material from the MLS is protected by patents, then a fixed share of the sales must be paid to the benefit-sharing mechanism.[38] If the product is not patent protected and is still available for use and further research and development, then benefit sharing is optional. This regulation is meant to be a further incentive to keep material from the MLS in the public domain – a point that is crucial for accessibility. Since the MLS is still new, and crop breeding takes time, it is still too early to expect to see much benefit from these provisions. It is also uncertain how much benefit the provisions will generate since patents are used only to a limited extent for PGRFA.[39] Thus, other forms of optional benefit sharing are taking place. For example, Norway is providing an annual contribution that is equivalent to 0.1 percent of the total sales of seeds in the country to the benefit-sharing mechanism. It urges other countries and multinational companies to do likewise, as it would substantially improve the capacity of the benefit-sharing mechanism. Discussions continue on how to further strengthen this mechanism, as set out in Articles 13.2.d and 13.6.
6 Nonmonetary benefit sharing is to be facilitated between the contracting parties independently of the transfer of material. This feature includes making available information on PGRFA, the transfer of technology for the conservation and sustainable use of PGRFA, and capacity building in terms of education and training, improvement of facilities, as well as research cooperation for the conservation and sustainable use of PGRFA (Article 13.2).
7 Benefits are to be shared with the custodians of PGRFA and not with the actual providers of specific material. This is an important difference between the CBD approach to ABS and the MLS. Under the MLS, benefits do not flow back to certain provider countries and certain providing

communities, as foreseen under the CBD, but rather are funnelled into a benefit-sharing mechanism. From there, they are distributed primarily to farmers, directly or indirectly, especially those in developing countries and countries with economies in transition who conserve and sustainably use PGRFA (Article 13.3). The first disbursement of benefits from the benefit-sharing mechanism was announced at the third session of the ITPGRFA Governing Body in June 2009. Eleven projects in developing countries were selected from a large number of applications to receive support for their contributions to the conservation and sustainable use of PGRFA. The total amount of money disbursed was approximately US$500,000.[40]

8 A third-party beneficiary of the agreement monitors compliance with the SMTA. The parties to the SMTA agree that the FAO, acting on behalf of the Governing Body of the treaty and its MLS, is the third-party beneficiary under the agreement. This third-party beneficiary monitors compliance, has the right to initiate dispute settlement and reports to the Governing Body of the ITPGRFA. Given the high numbers of transactions governed by the SMTA, the third-party beneficiary faces great challenges in terms of monitoring compliance with the SMTA. Procedures for the operation of the third-party beneficiary were adopted at the third session of the Governing Body under Resolution 5/2009 (ITPGRFA, 2009, 28–30). A list of mediation and arbitration experts has been made available, with representation from several regions.

9 A dispute settlement procedure has been established in the SMTA for cases of noncompliance. Any of the three parties – the provider, the recipient or the third-party beneficiary – may initiate dispute settlement procedures. The first step in these procedures is amicable negotiations, whereby the parties try to solve the dispute in good faith. If this does not work, mediation is the second step. For this purpose, the parties are required to select a neutral party as a mediator. The last step – if nothing else solves the conflict – is arbitration. In arbitration, the parties to the dispute must either agree on an appropriate international body to carry out the arbitration or the dispute will be settled under the Rules of Arbitration under the International Chamber of Commerce. The result of such arbitration is legally binding.

The MLS has been a success in terms of facilitating access to PGRFA. The benefit-sharing mechanism is, however, still weak and needs further development, as set out in Articles 13.2.d and 13.6. In the meanwhile, the Nagoya Protocol on Access to Genetic Resources and the Fair and Equitable Sharing of Benefits Arising from their Utilization to the Convention on Biological Diversity has been adopted by the Parties to the Convention on Biological Diversity (CBD). This is an international agreement which aims at sharing the benefits arising from the utilization of genetic resources in a fair and equitable way, including by appropriate access to genetic resources and by appropriate transfer of relevant technologies, taking into account all rights over those resources and to technologies, and by appropriate funding. It was adopted by the Conference of the Parties to the CBD at its 10th meeting on 29 October 2010 in Nagoya, Japan, and entered into force in 2014. The Nagoya Protocol entered into force on 12 October 2014, 90 days after the date of deposit of the fiftieth instrument of ratification.[41] Implications for the ITPGRFA and the MLS are discussed in the final section of this chapter.

Recent developments with regard to *ex situ* conservation

In 2004, the Global Crop Diversity Trust was established as a public-private partnership to raise funds from individual, corporate and government donors in order to establish an endowment fund to provide complete and continuous funding for key crop collections. The goal is to advance an efficient and sustainable global system of *ex situ* conservation by promoting the rescue, understanding, use and long-term conservation of valuable plant genetic resources. The trust is an independent organization closely interlinked with the ITPGRFA. During the first session of the Governing Body of the ITPGRFA in 2006, a relationship agreement between the Governing Body and the trust was formally approved. The agreement recognizes the trust as an essential element of the funding strategy of the treaty. It provides for the governing body to give policy guidance to the trust and to appoint four members of the Executive Board. It also recognizes the board's executive independence in managing the operations and activities of the trust. As of April 2010, the total amount of money pledged to the trust amounted to US$168,179,144, of which US$142,000,925 had been paid.[42] Also later developments show that the Global Crop Diversity Trust has become a great success in channelling funds for *ex situ* conservation.[43]

The Svalbard Global Seed Vault was opened in 2008 in the Arctic permafrost of Svalbard. It offers backup free of charge for the seed collections that are held in seed banks around the world. The seed vault has the capacity to store 4.5 million different seed samples. As each sample contains on average 500 seeds, approximately 2.25 billion seeds may be stored in the seed vault. It will therefore have the capacity to hold all of the unique seed samples currently held by the approximately 1,400 gene banks that are found in more than 100 countries all over the world. In addition, the seed vault will have the capacity to also store many new seed samples that may be collected in the future. If seeds are lost – for example, as a result of natural disasters, war, or simply a lack of resources – the seed collections may be reestablished using seeds from Svalbard. Each country or institution owns and controls access to the seeds it has deposited. The seed vault facility was built and is owned by Norway, and its operation is managed in partnership between the government of Norway, the Global Crop Diversity Trust and the Nordic Genetic Resource Centre (NordGen). By December 2009, more than 430,000 unique seed samples had arrived at the seed vault from seed banks all over the world.[44] The Svalbard Global Seed Vault provides insurance against the loss of PGRFA from the gene banks of the world and is, as such, an important component in the global effort to stop genetic erosion.

With these developments, much has been achieved with regard to *ex situ* conservation since the ITPGRFA entered into force. Although the challenges are still huge in many parts of the world with regard to *ex situ* conservation, the prospects for improvement are better than ever.

Conclusions and specific challenges with regard to benefit sharing

In this chapter, we have seen how the current international cooperation on the management of PGRFA took form in the FAO and how farmers' rights have evolved in this process. To understand farmers' rights, it is necessary to understand this context. How to reward farmers for their contributions to conserving, improving and making available PGRFA has been a central topic in the negotiations. An international fund for supporting and assisting farmers in their efforts has long been on the agenda. Discussions have also focused on how farmers' rights can balance breeders' rights, so as to ensure an equitable system that can facilitate farmers' continued access to – and free use of – PGRFA. It is also important to note that the rights holders were not seen to be single farmers or communities but entire peoples – that is, a form of collective right. The ITPGRFA represents the most important international instrument to change the trends of genetic erosion, in general, and to support and promote farmers' rights, in particular. We have seen that great progress has been achieved in the area of *ex situ* conservation and the facilitation of access to PGRFA through the MLS. As for *in situ* conservation and farmers' rights, much less has been achieved, which is a reflection of the relative infancy of the ITPGRFA as well as of the priorities that the Governing Body has set in its work plan during this period.

The treaty contains important provisions regarding *in situ* conservation and the sustainable use of PGRFA, which are of particular relevance for farmers' varieties and farmers' rights. For example, contracting parties are requested, as appropriate, to promote or support farmers and local communities in their efforts to manage and conserve PGRFA on farm (Article 5.1.c), to promote participatory plant breeding and to strengthen the capacity to develop varieties particularly adapted to the specific conditions on marginal areas (Article 6.2.c). Implementing these provisions would substantially improve the situation of those working with farmers' varieties and thus also improve farmers' rights. A more in-depth discussion of the provisions on farmers' rights is provided in Chapter 8 of this volume.

However, to conclude this chapter, I will now examine more closely the question of benefit sharing. As we have seen, the benefit-sharing mechanism of the MLS is still weak, even though some funds have been distributed to relevant projects in developing countries and countries with economies in transition, pursuant to Article 13. The important question now is how the benefit-sharing mechanism can be improved so as to create the balance between access and benefit sharing that was foreseen in the MLS and also realize the rights of farmers to benefit sharing as set out in Article 9 – which is, in turn, an important precondition for the conservation and sustainable use of crop genetic diversity.

According to the ITPGRFA, the Governing Body is to assess, within a period of 5 years from its entry into force, whether the mandatory payment requirement of the MLS shall also apply in cases where such commercialized products are available without restriction to others for further research and breeding (Article 13.2.d). These issues are being considered by the Ad Hoc Open-ended Working Group to Enhance the Functioning of the Multilateral System of Access and Benefit-sharing. The group convened in October 2015 and reported back to the Governing Body at its sixth session in 2015. The contracting parties are also to consider the modalities of a strategy of voluntary benefit-sharing contributions whereby food-processing industries that benefit from PGRFA would contribute to the MLS (Article 13.6). The Norwegian initiative to unilaterally pay an amount equivalent to 0.1 percent of all of its seed sales in the country to the benefit-sharing mechanism was meant to be a model in this regard and one way to seek to address this question. If other countries followed, the benefit-sharing mechanism would soon amount to US$40 million. It is absolutely necessary for the Governing Body to move towards strengthening the benefit-sharing mechanism of the treaty if the treaty objectives are to be successful.

In this context, it is also important to be aware of the developments under the Nagoya Protocol of the CBD. The interfaces between the two regimes have been analyzed in several publications during the past years, e.g. Halewood et al. (2013), Medaglia et al. (2013) and Oberthür and Rosendal (2014) (see also Andersen 2008, 2010; Bulmer, 2009; Young and Tvedt, 2009; Andersen et al., 2010). Thus, only the most important interface with regard to farmers' varieties and farmers' rights under the treaty will be taken up in this chapter.

The MLS applies only to PGRFA in the public domain and under the control of parties to the ITPGRFA. This means that the material that is covered in public gene banks and other public *ex situ* facilities shall be included in the MLS, whereas material in the possession of private companies, other nongovernmental institutions and material in farmers' fields and in their possession is not automatically included. The parties to the ITPGRFA are expected to take appropriate measures to encourage natural and legal persons within their jurisdiction who hold PGRFA listed in Annex 1 to include this material in the MLS (Article 11.3). However, while Article 11 is about the coverage of the MLS, Article 12 is about facilitated access within the MLS. Article 12.3.h states that the contracting parties, without prejudice to other provisions of Article 12, must agree that access to crop genetic resources found in *in situ* conditions will be provided 'according to national legislation or, in the absence of such legislation, in accordance with such standards as may be set by the Governing Body.' In other words, the PGRFA of Annex 1 plants may either be (1) invited into the MLS by the respective contracting parties; (2) placed in the MLS by way of legislation; or (3) accessed according to standards set by the Governing Body. There are so far few, if any, cases of the first two options. As for the third option, it seems unrealistic to arrive at any standards in this regard before the benefit-sharing mechanism is in full operation.

As long as the material in farmers' fields is not included in the MLS, the CBD and the Nagoya Protocol remain the international framework that applies with respect to access and benefit-sharing regulation. The Nagoya Protocol marks a substantial step forwards in the international regulation of access and benefit sharing to genetic resources. If implemented according to its intentions, it will provide greater legal certainty and transparency for both users and providers of genetic resources.

Depending on how it is implemented, the Nagoya Protocol may also create disincentives for the continued sharing of PGRFA and related knowledge among farmers and to the collection and conservation of PGRFA in gene banks. The Nagoya Protocol is based on a contractual mechanism for realizing ABS. It prescribes a system for providing access based on agreements between users and provider countries, and a system for enforcing those contracts in user countries. Such bilateral agreements between providers and recipients of genetic resources may lead provider countries authorities as well as farmers to expect potential future benefits if certain crops were to be 'discovered' by plant breeders. As a result, farmers may refrain from sharing their propagating material and related knowledge with other farmers, and the collection of propagating material and related knowledge for conservation purposes may become more difficult (see e.g. Andersen, 2005b, 2008). Therefore, for any access and benefit-sharing regime on PGRFA outside the MLS, and in particular for the crops listed in Annex 1 which are not in the public domain, it is important to consider how incentives are shaped. Especially, it is important to consider whether they are supportive of the continued sharing of these resources and related knowledge among farmers as well as for continued conservation efforts of local, national and international gene banks.

In this context, Article 10 of the Nagoya Protocol is of particular interest. It provides that the parties are to consider the needs and modalities for a multilateral benefit-sharing mechanism for resources and knowledge that occur in transboundary situations, or for which it is not possible to obtain prior informed consent. Most PGRFA have been developed through the exchange of seeds and propagating materials for millennia, and this applies in particular to the Annex 1 crops outside the public domain. Therefore, it could be argued that Article 10 would apply for such crops. If this is taken into account, the further development of Article 10 under the Nagoya Protocol could be promising for Annex 1 crops outside the MLS. The 11th meeting of the Conference of the Parties to the Convention on Biological Diversity decided on a process which includes a request to the Executive Secretary to conduct a broad consultation on Article 10 and an invitation to parties and stakeholders to submit targeted views on the issue (Decision XI/1) Furthermore, COP-11 requested the Executive Secretary to synthesize the views received for the consideration of an expert group to be convened subject to the availability of funds. The expert group submitted the outcomes of its work to the third meeting of the Intergovernmental

Committee for the Nagoya Protocol. The outcome of this process will show whether the Nagoya Protocol will offer a conducive solution for Annex 1 crops under the ITPGRFA that are not under the control of its parties.

Otherwise, gene banks seeking to collect material for conservation and plant breeding are obliged to follow regulations derived from the CBD/Nagoya Protocol for bilateral agreements – to the extent that such regulations have been implemented in the country in question – when *in situ* material has not been included in the MLS. As Annex 1 material that has been collected from farmers' fields and delivered to the gene banks is placed in the MLS, it means that any benefit-sharing arrangements beyond those of the MLS can only be asserted up front. Such arrangements could include guarantees that farmers will receive samples of their seeds from the gene bank in case of natural disasters; technology transfer through participatory selection breeding between collectors and farmers during collection work; and the possibility of access to other material held in the gene bank.

This interface between the Nagoya Protocol and the MLS of the ITPGRFA in terms of the Annex 1 crops is one important issue with regard to farmers' varieties and farmers' rights. It shapes the regulations and incentives for sharing seeds of widely used crops among farmers and for farmers' access to seed and propagating material from gene banks of such crops, as well as benefits related to the sharing of such resources for farmers.

However, there are other important questions related to all of the non–Annex 1 crops and PGRFA in countries that are not parties to the ITPGRFA (see Andersen et al., 2010). Many of these questions were addressed at the 12th regular session of the CGRFA in October 2009 where Resolution 1/2009 on Policies and Arrangements for Access and Benefit-Sharing for Genetic Resources for Food and Agriculture was adopted. Close cooperation between the relevant bodies of the FAO and the CBD was encouraged. Since then much work has been carried out to solve these questions and to develop further collaboration. Elaborating on this would however go beyond the scope of this chapter.

The ITPGRFA has great potential to promote the sustainable management of PGRFA and the development of farmers' rights related to crop genetic resources. The main challenges in the years to come include *in situ* conservation and the sustainable use of PGRFA, a functioning benefit-sharing mechanism and in particular the realization of farmers' rights. Whether the potential of the ITPGRFA will unfold is dependent on the political will of the contracting parties.

Notes

1 This figure is for rice.

2 It has been published continuously since then. Originally it was called *FAO Plant Introduction Newsletter.* Since 1978, it has been issued jointly by the FAO and the International Plant Genetic Resources Institute (IPGRI), now Bioversity International, as the *Plant Genetic Resources Newsletter.*

3 In 1974, the IBPGR was transformed into the IPGRI, which is now Bioversity International, a part of the CGIAR. The CGIAR was founded in 1971 on the initiative of the Ford and Rockefeller Foundations to unite privately funded international agricultural research centres (IARCs) into one network. As an informal association of public and private donors that support the IARCs, it is a donor-led group that has provided a forum for discussion of research priorities and coordination of funding (FAO, 1998, 248). As divisions of the network, the IARCs have their own governing bodies. The United Nations Environment Programme (UNEP), the FAO, the United Nations Development Programme (UNDP), and the World Bank co-sponsor the system, and the CGIAR is headquartered at the premises of the World Bank in Washington, DC (ibid., 249).

4 This picture has changed dramatically since then, and today the overwhelming majority of transfers of genetic material go to developing countries (see e.g. Fowler, Smale and Gaiji, 2001; System-wide Genetic Resources Programme, 2011).

5 Pat Roy Mooney and Cary Fowler founded the Rural Advancement Foundation International (RAFI) in 1984, together with Hope Shand. It was probably the most influential nongovernmental organization during the negotiations in the FAO on crop genetic resources in the 1980s and the early 1990s. RAFI has since changed its name to the Action Group on Erosion, Technology and Concentration (ETC Group). Mooney and Shand still front the organization.

6 Convention on Biological Diversity, 31 ILM 818 (1992) [CBD]; Agreement on Trade-Related Aspects of Intellectual Property Rights, in Annex 1C of the Marrakech Agreement Establishing the World Trade Organization, 15 April 1994, 33 ILM 15 (1994) [WTO Agreement].

7 International Undertaking on Plant Genetic Resources, 1983, online: <www.fao.org/waicent/faoinfo/agricult/cgrfa/IU.htm> (last accessed 15 June 2012).

8 It was established by FAO Conference Resolution 9/83. It was later renamed the Commission on Genetic Resources for Food and Agriculture (CGRFA), as its mandate was broadened (see discussion later in this chapter).

9 At that time, there were still only 74 signatories.

10 For a more comprehensive account of the history of farmers' rights, see Andersen (2005a).

11 The first use of farmers' rights as a political concept dates back to the early 1980s, when Pat Roy Mooney and Cary Fowler coined the term to highlight the valuable but unrewarded contributions of farmers to PGRFA. According to Fowler (1994, 192), the concept can be traced back to the work of, inter alia, Jack R. Harlan (1917–88), the renowned plant explorer, geneticist, and plant breeder, who spoke of farmers as the 'amateurs' who had in fact created the genetic diversity that had become subject to controversies.

12 General Agreement on Tariffs and Trade, 30 October 1947, 55 UNTS 194.

13 WTO Agreement, supra note 6.

14 This principle was first voiced at the 1972 United Nations Conference on the Human Environment in Stockholm in the form that states have sovereign rights to exploit their natural resources in accordance with their own environmental priorities (Stockholm Declaration on the Human Environment, 16 June 1972, 11 ILM 1416 [1972], Principle 21).

15 Among the best-known gene banks of the IARCs of the CGIAR are:

- Centro Internacional de Agricultura Tropical for beans, cassava, tropical forage crops, and rice in Latin America;
- Centro Internacional de Mejoramiento de Maiz y Trigo for maize and wheat in developing countries;
- Centro Internacional de la Papa for potato, sweet potato, and Andean root and tuber crops;

- International Crops Research Institute for the Semi-Arid Tropics for sorghum, pearl and finger millet, chickpea, pigeonpea, and groundnuts;
- International Institute of Tropical Agriculture for cassava, maize, cowpea, soybean, yam, banana, and plantain;
- International Rice Research Institute for rice in developing countries.

16 Such estimates depend on the data available as well as methods for calculation. In this case, the FAO argued that the extent of redundancy within the CGIAR base collections is low. Therefore, they included an estimated 20% duplicates, compared to an estimated total of 1–2 million unique accessions in the world. Three years later, Fowler, Smale and Gaiji (2001) calculated with a different method and concluded that the CGIAR centres maintain approximately 12% of the accessions held in *ex situ* conditions worldwide.
17 While landraces and wild relatives comprise only approximately 16% of national *ex situ* collections of PGRFA, they make up 73% of the CGIAR collections (FAO, 1998, 94).
18 International Treaty on Plant Genetic Resources for Food and Agriculture, 29 June 2004, online: <http://sedac.ciesin.org/pidb/texts-menu.html> (last accessed 15 June 2012) [ITPGRFA].
19 FAO, Global Plan of Action for the Conservation and Sustainable Utilization of Plant Genetic Resources for Food and Agriculture, adopted in 1996 by the International Technical Conference on Plant Genetic Resources, Leipzig, 17–23 June 1996 [Global Plan of Action].
20 According to the Treaty Reference Guide of the United Nations Office of Legal Affairs, the term agreement can be used for legally binding as well as nonbinding agreements (see <http://untreaty.un.org/ola-internet/Assistance/Guide.htm#agreements> (last accessed 15 June 2012)).
21 This section is based on Andersen et al. (2010, 3).
22 Bilateral in the sense that each material transfer agreement would have to be negotiated between two parties, based on the legislation in the relevant country (derived from the CBD in various ways from country to country).
23 Multilateral in the sense that the material transfer agreements for all transactions are standardized and involve a third-party beneficiary to monitor their implementation on behalf of the parties to the treaty.
24 FAO, Conference Resolution 3/95, 28th Sess., 1995.
25 FAO, Council Resolution 1/110, 1995.
26 Further components of the global system that have developed since 1983 leading up to the adoption of the ITPGRFA are analyzed in Andersen (2003). They are as follows:

- the International Fund for Plant Genetic Resources (which never materialized, although some minimal contributions were received);
- the International Network of *Ex Situ* Collections under the Auspices of the FAO (which no longer exists, as this material is covered by Article 15 of the ITPGRFA);
- the World Information and Early Warning System on Plant Genetic Resources for Food and Agriculture;
- the International Code of Conduct for Plant Germplasm Collecting and Transfer;
- the *State of the World's Plant Genetic Resources for Food and Agriculture* report;
- the Global Plan of Action for the Conservation and Sustainable Use of Crop Genetic Resources.

27 Leipzig Declaration, adopted by the International Technical Conference on Plant Genetic Resources in Leipzig, 17–23 June 1996, FAO, Rome.
28 Global Plan of Action, supra note 19.
29 The report was compiled and produced by a team at the FAO coordinated by Cary Fowler and David Cooper.
30 It has been followed up with a Second Report on the State of the World's Plant Genetic Resources for Food and Agriculture in 2010 (FAO, 2010).

31 Implementation of the Global Plan of Action is monitored under the CGRFA and regarded as a supporting component of the ITPGRFA, supra note 18, Article 14. A Second Plan of Action for Plant Genetic Resources for Food and Agriculture was adopted by the CGRFA in 2011, and its implementation is being monitored by the CGRFA. See <www.fao.org/docrep/015/i2624e/i2624e00.htm> (last accessed 17 January 2016).
32 A thorough analysis of the recognition of farmers' rights in the ITPGRFA is found in Batta Bjørnstad (2004). Further analyses of the ITPGRFA provisions on farmers' rights are provided by the Farmers' Rights Project, online: <www.farmersrights.org> (last accessed 15 June 2012). See also Moore and Tymowski (2005).
33 The Cartagena Protocol on Biosafety to the Convention on Biological Diversity, 29 January 2000, online: <https://bch.cbd.int/protocol> (last accessed 17 January 2016). It is not dealt with in this chapter, but, as a protocol to the CBD, it is a part of an already established regime.
34 The United States has also signed the CBD but has not ratified it.
35 An interesting analysis of the contents and prospects of the ITPGRFA is found in Fowler (2004). Explanations on the background and contents of the ITPGRFA are presented in Moore and Tymowski (2005).
36 This section is based on Andersen (2008) and Andersen et al. (2010).
37 For example, rice, wheat, maize, rye, potatoes, beans, cassava and bananas. Not included are other important crops, including soybeans, tomatoes, cotton, sugarcane, cocoa and groundnuts, as well as many vegetables and important tropical forage plants.
38 The ITPGRFA does not prohibit IPRs on products developed on the basis of material received from the MLS – only on material from the MLS in the form received.
39 Plant breeders' rights are most frequently used to ensure IPRs to plant varieties.
40 Information sheet from the Secretariat, online: <ftp://ftp.fao.org/ag/agp/planttreaty/news/news0009_en.pdf> (last accessed 15 June 2012).
41 For more information, see the official website of the Nagoya Protocol at <www.cbd.int/abs/> (last accessed 17 January 2016).
42 Global Crop Diversity Trust (2010): Global Crop Diversity Trust Pledges (as of 26 April 2010). An updated list of funds pledged and raised can be downloaded at <www.croptrust.org/main/funds.php> (last accessed 15 June 2012).
43 For more information on the Global Crop Diversity Trust, see <www.croptrust.org> (last accessed 15 June 2012).
44 Ministry of Agriculture and Food, online: <www.regjeringen.no/en/dep/lmd/campain/svalbard-global-seed-vault/news/svalbard-global-seed-vault-50-000-seed-s.html?id=588507> (last accessed 15 June 2012). For more information, see Global Seed Vault, online: <www.seedvault.no> (last accessed 15 June 2012).

References

Andersen, R. (2001). *Conceptualising the Convention on Biological Diversity: Why Is It Difficult to Determine the 'Country of Origin' for Agricultural Plant Varieties?* FNI Report no. 7/2001, Fridtjof Nansen Institute, Lysaker.

Andersen, R. (2003). 'FAO and the Management of Genetic Resources,' in Olav S. Stokke and Øystein B. Thommessen (eds.), *Yearbook of International Co-operation on Environment and Development 2003/2004,* Earthscan, London, pp. 43–53.

Andersen, R. (2005a). *The Farmers' Rights Project – Background Study 1: The History of Farmers' Rights: A Guide to Central Documents and Literature,* FNI Report no. 8/2005, Fridtjof Nansen Institute, Lysaker.

Andersen, R. (2005b). *The Farmers' Rights Project – Background Study 2: Results from an International Stakeholder Survey on Farmers' Rights,* FNI Report no. 9/2005, Fridtjof Nansen Institute, Lysaker.

Andersen, R. (2008). *Governing Agrobiodiversity: Plant Genetics and Developing Countries,* Ashgate, Aldershot, UK.

Andersen, R. (2012). 'The Plant Treaty – Crop Genetic Diversity and Food Security,' in S. Andresen, E. Lerum Boasson and G. Hønneland (eds.), *International Environmental Agreements: An Introduction,* London/New York, Routledge, pp. 134–50.

Andersen, R., M. Walløe Tvedt, O. Kristian Fauchald, T. Winge, G. Kristin Rosendal and P. Johan Schei (2010). *International Agreements and Processes Affecting an International Regime on Access and Benefit Sharing under the Convention on Biological Diversity Implications for Its Scope and Possibilities of a Sectoral Approach,* FNI Report no. 3/2010, Fridtjof Nansen Institute, Lysaker.

Batta Bjørnstad, S.-I. (2004). *Breakthrough for 'the South'? An Analysis of the Recognition of Farmers' Rights in the International Treaty on Plant Genetic Resources for Food and Agriculture,* FNI Report no. 13/2004, Fridtjof Nansen Institute, Lysaker.

Bulmer, J. (2009). *Study on the Relationship between an International Regime on Access and Benefit-Sharing and Other International Instruments and Forums That Govern Genetic Resources: The International Treaty on Plant Genetic Resources for Food and Agriculture and the Food and Agriculture Organisations' Commission on Genetic Resources,* Information Paper for the Seventh Meeting of the Ad Hoc Open-Ended Working Group on ABS, Doc. UNEP/CBD/WG-ABS/7/Inf/3/Part.1, Rome.

Evans, P., and J. Walsh (1994). *The EIU Guide to the New GATT,* The Economist Intelligence Unit, London.

Food and Agriculture Organization (FAO) (1985). *Report of the Commission on Plant Genetic Resources, First Session, Rome, 11–15 March 1985,* Doc. CPGR/85/REP, FAO, Rome.

Food and Agriculture Organization (FAO) (1986a). *Report of the Working Group of the FAO Commission on Plant Genetic Resources, 2–3 June 1986,* Doc. CPGR 87/3, FAO, Rome.

Food and Agriculture Organization (FAO) (1986b). *1986: Progress Report on the International Undertaking on Plant Genetic Resources,* Doc. CPGR/87/4, FAO, Rome.

Food and Agriculture Organization (FAO) (1987). *Report of the Second Session of the Commission on Plant Genetic Resources,* Doc. CL 91/14, FAO, Rome.

Food and Agriculture Organization (FAO) (1989). *Report of the Conference of FAO, Twenty-fifth Session, Rome, 11–29 November 1989,* Doc. C 1989/REP, FAO, Rome.

Food and Agriculture Organization (FAO) (1991). *Report of the Conference of FAO, Twenty-sixth Session, Rome, 9–27 November 1991,* Doc. C 1991/REP, FAO, Rome.

Food and Agriculture Organization (FAO) (1994). *Revisions of the International Undertaking: Mandate, Context, Background and Proposed Process,* Doc. CPGR-Ex1/94/3, FAO, Rome.

Food and Agriculture Organization (FAO) (1998). *State of the World's Plant Genetic Resources for Food and Agriculture,* FAO, Rome.

Food and Agriculture Organization (FAO) (2010). *Second Report on the State of the World's Plant Genetic Resources for Food and Agriculture,* FAO, Rome.

Fowler, C. (1994). *Unnatural Selection: Technology, Politics, and Plant Evolution,* Gordon and Breach, Yverdon, Switzerland.

Fowler, C. (2001). 'Protecting Farmer Innovation: The Convention on Biological Diversity and the Question of Origin,' *Jurimetrics* 41: 477–88.

Fowler, C. (2004). 'Assessing Genetic Resources: International Law Establishes Multilateral System,' *Genetic Resources and Crop Evolution* 51: 609–20.

Fowler, C., and T. Hodgkin (2004). 'Plant Genetic Resources for Food and Agriculture: Assessing Global Availability,' *Annual Review of Environment and Resources* 29: 143–79.

Fowler, C., and P. Mooney (1990). *Shattering: Food, Politics and the Loss of Genetic Diversity,* University of Arizona Press, Tucson, AZ.

Fowler, C., M. Smale and S. Gaiji (2001). 'Unequal Exchange? Recent Transfers of Agricultural Resources and their Implications for Developing Countries,' *Development Policy Review* 19(2): 181–204.

Halewood, M., E. Andrieux, L. Crisson, J. Rwihaniza Gapusi, J. Wasswa Mulumba, E. Kouablan Koffi, T. Yangzome Dorji, M. Raj Bhatta and D. Balma (2013). 'Implementing 'Mutually Supportive' Access and Benefit Sharing Mechanisms Under the Plant Treaty, Convention on Biological Diversity, and Nagoya Protocol,' *Law, Environment and Development Journal* 9(1): 68. Available at: www.lead-journal.org/content/13068.pdf (last accessed 19 February 2016).

Harlan, Jack R. (1975; second edition 1992). *Crops and Man,* American Society of Agronomy/Crop Science Society of America, Madison, WI.

International Treaty on Plant Genetic Resources for Food and Agriculture (ITPGRFA) (2007). *Report by the Secretary of the Plant Treaty to the Second Session of the Governing Body in 2007,* Doc. IT/GB-2/07/5, Rome.

International Treaty on Plant Genetic Resources for Food and Agriculture (ITPGRFA) (2009). *Report from the Third Session of the Governing Body of the Plant Treaty*, Doc. IT/GB-3/09/Report. Available at: ftp://ftp.fao.org/ag/agp/planttreaty/gb3/gb3repe.pdf (last accessed 19 February 2016).

Keystone Center (1991). *Oslo Plenary Session: Final Consensus Report of the Keystone International Dialogue Series on Plant Genetic Resources: Global Initiative for the Security and Sustainable Use of Plant Genetic Resources,* Third Plenary Session, 31 May–4 June 1991, Oslo, Norway.

Medaglia, J.C., M. Walløe Tvedt, F. Perron-Welch, A. Jørem and F.-K. Phillips (2013). *The Interface between the Nagoya Protocol on ABS and the ITPGRFA at the International Level: Potential Issues for Consideration in Supporting Mutually Supportive Implementation at the National Level.* FNI Report 1/2013. Lysaker, FNI, 2013. Available at: www.fni.no/doc&pdf/FNI-R0113.pdf (last accessed 19 February 2016).

Mooney, P.R. (1979). *Seeds of the Earth: A Private or Public Resource?* Canadian Council for International Co-operation and the International Coalition for Development Action, Ottawa.

Moore, G., and W. Tymowski (2005). *Explanatory Guide to the International Treaty on Plant Genetic Resources for Food and Agriculture,* IUCN Environmental Policy and Law Paper no. 57, International Union on the Conservation of Nature, Gland, Switzerland and Cambridge, UK.

Oberthür, S. and K. Rosendal (eds.) (2014). *Global Governance of Genetic Resources. Access and Benefit Sharing after the Nagoya Protocol,* Routledge, London/New York.

Palacios, X.F. (1998). *Contribution to the Estimation of Countries' Interdependence in the Area of Plant Genetic Resources,* Background Study Paper no. 7, Rev. 1, Commission on Genetic Resources for Food and Agriculture, Rome.

Pistorius, R. (1997). *Scientists, Plants and Politics: A History of the Plant Genetic Resources Movement,* International Plant Genetic Resources Institute, Rome.

Ten Kate, K., and S.A. Laird (1999). *The Commercial Use of Biodiversity: Access to Genetic Resources and Benefit-Sharing,* Earthscan, London.

Stokke, O.S., and Ø.B. Thommessen (2003). *Yearbook of International Co-operation on Environment and Development 2003/2004,* Earthscan, London.

System-wide Genetic Resources Programme (2011). Experience of the IARC of the CGIAR with the implementation of the agreements with the Governing Body, with particular reference to the use of the standard material transfer agreement for Annex 1 and non-Annex 1 crops. Available at www.planttreaty.org/sites/default/files/gb4i05e.pdf (last accessed 19 February 2016).

United Nations Environment Programme (UNEP) (1992). *Nairobi Final Act of the Conference for the Adoption of the Agreed Text of the Convention on Biological Diversity.* Available at: http://biodiv.org/doc/handbook/cbd-hb-09-en.pdf (last accessed 15 June 2012).

Wilson, Edward O. (1992). *The Diversity of Life,* Belknap Press of Harvard University Press, Cambridge, MA.

Young, T.R. and M. Walløe Tvedt (2009). *Balancing Building Blocks of a Functional ABS System.* FNI Report 7/2009. Lysaker, FNI. Available at: www.fni.no/doc&pdf/FNI-R0709.pdf.

Yusuf, A.A. (1998). 'TRIPS: Background, Principles and General Provisions,' in Carlos M. Correa and Abdulqawi A. Yusuf (eds.), *Intellectual Property and International Trade: The TRIPS Agreement,* Kluwer Law International, Dordrecht and London, pp. 3–22.

8 Farmers' rights

Evolution of the international policy debate and national implementation[1]

Regine Andersen

Realizing farmers' rights essentially means enabling farmers to maintain and develop crop genetic resources as they have done since the dawn of agriculture, and recognizing and rewarding them for this indispensable contribution to the global pool of genetic resources. The realization of farmers' rights is a precondition for maintaining crop genetic diversity, which is the basis of all food and agricultural production in the world. Plant genetic diversity is probably more important for farming than any other environmental factor, simply because it is *the* factor that enables farmers to adapt to changing environmental conditions, such as climate change (Esquinas-Alcázar, 2005; Andersen, 2008; Fujisaka, Williams and Halewood, 2009; United Nations, 2009). Since farmers are the custodians and developers of crop genetic resources in the field, their rights in this regard are crucial for enabling them to continue this role. For this reason, farmers' rights constitute a cornerstone in the International Treaty on Plant Genetic Resources for Food and Agriculture (ITPGRFA).[2] Achieving the first two objectives of the treaty – the conservation and sustainable use of crop genetic resources (Article 1) – depends to a large extent on farmers and their ability to maintain these resources *in situ* on their farms, which in turn depends on farmers' rights.

In this chapter, we look at how these rights are addressed in the ITPGRFA. We then proceed to examine the concept of farmers' rights and to identify the central content of these rights and achievements so far with regard to their realization. Examples of best practices will be presented, and an overview of the negotiations leading up to the treaty's ratification will be highlighted before conclusions are drawn.

Farmers' rights in the ITPGRFA

Farmers' rights in the ITPGRFA are strictly related to plant genetic resources for food and agriculture. Some nongovernmental organizations (NGOs) have criticized this mandate as being too limited since farmers' rights to land, water and other resources and services are also closely interlinked with their rights to seed and propagating material. In this respect, however, it should be borne in mind that farmers' rights under the treaty can only be related to the mandate of the treaty, which concerns specifically plant genetic resources for food and agriculture. This is not to say that other rights are not important, but rather that, in the context of the ITPGRFA, farmers' rights are necessarily related to crop genetic resources.

To understand farmers' rights in the ITPGRFA, the text of the provisions on farmers' rights forms a starting point. Article 9 is devoted to the realization of farmers' rights:

9.1 The Contracting Parties recognize the enormous contribution that the local and indigenous communities and farmers of all regions of the world, particularly those in the centres of origin and crop diversity, have made and will continue to make for the conservation and development of plant genetic resources which constitute the basis of food and agriculture production throughout the world.

9.2 The Contracting Parties agree that the responsibility for realizing Farmers' Rights, as they relate to plant genetic resources for food and agriculture, rests with national governments. In accordance with their needs and priorities, each Contracting Party should, as appropriate, and subject to its national legislation, take measures to protect and promote Farmers' Rights, including:

- a protection of traditional knowledge relevant to plant genetic resources for food and agriculture;
- b the right to equitably participate in sharing benefits arising from the utilization of plant genetic resources for food and agriculture; and
- c the right to participate in making decisions, at the national level, on matters related to the conservation and sustainable use of plant genetic resources for food and agriculture.

9.3 Nothing in this Article shall be interpreted to limit any rights that farmers have to save, use, exchange and sell farm-saved seed-propagating material, subject to national law and as appropriate.

Many other provisions are relevant for the realization of farmers' rights, and there are various angles from which implementation can be derived. For example, the ITPGRFA provides that countries shall promote or support, as appropriate, farmers' and local communities' efforts to manage and conserve on-farm their plant genetic resources for food and agriculture (Article 5.1(c)) and take steps to minimize or, if possible, eliminate threats to plant genetic resources for food and agriculture (Article 5.2). Article 6 states that the contracting parties shall develop and maintain appropriate policy and legal measures that promote the sustainable use of plant genetic resources for food and agriculture. A range of measures is listed for this purpose, among them 'reviewing, and, as appropriate, adjusting breeding strategies and regulations concerning variety release and seed distribution' (Article 6.2(g)). In addition, the ITPGRFA supports the implementation of the Global Plan of Action (Article 14), with its provisions on farmers' rights. Articles 7 and 8 provide for international cooperation and technical assistance, with a particular view to strengthening developing countries capabilities to implement the ITPGRFA.

Two other provisions (paras. 13.3 and 18.5) state that a funding priority is to be given to farmers who contribute to the maintenance of agrobiodiversity. The first paragraph states that farmers who contribute to maintaining plant genetic resources for food and agriculture are entitled to receive benefits arising from the multilateral system of access and benefit sharing that was established under the treaty. Paragraph 18.5 ensures that a funding priority will be given to those farmers in developing countries who implement agreed plans and programs to conserve and sustainably utilize plant genetic resources for food and agriculture. Finally, according to Article 21, the governing body is to ensure compliance with all of the provisions of the ITPGRFA (not only the obligations), and the preamble highlights the necessity of promoting farmers' rights at the national as well as the international levels.

Two approaches to understanding farmers' rights

One reason why the negotiators of the ITPGRFA were not able to agree on a definition of farmers' rights was that the situations of farmers differ so greatly from country to country, as do the perceptions of farmers' rights. With no official definition of farmers' rights, there was an uncertainty over what the concept involved and, in connection, how these rights could be realized. Thus, it was important to establish a common ground of understanding in order to develop a fruitful dialogue among the stakeholders on the measures that needed to be taken. A point of departure for developing such an understanding was the attempt to understand all of the different perspectives on the subject. These perspectives generally fall under one of two specific approaches, or somewhere in between:

- The ownership approach refers to the right of farmers to be rewarded on an individual or collective basis for genetic material that has been obtained from their fields and used in commercial varieties and/or protected with intellectual property rights. The idea is that such a reward system is necessary to enable the equitable sharing of benefits arising from the use of agrobiodiversity and to establish an incentive structure for the continued maintenance of this diversity. Access and benefit-sharing legislation and farmers' intellectual property rights are suggested as central instruments.[3]
- The stewardship approach refers to the rights that farmers must be granted collectively in order to enable them to continue as stewards and innovators of agrobiodiversity. The idea is that the 'legal space'[4] required for farmers to continue this role must be upheld and that farmers involved in the maintenance of agrobiodiversity – on behalf of their generation for the benefit of all mankind – should be rewarded and supported for their contributions (Andersen, 2005).[5]

Table 8.1 Goals for the realization of farmers' rights: two approaches

ITPGRFA measures	*Stewardship approach*	*Ownership approach*
Protection of farmers' traditional knowledge (para. 9.2(a))	The goals are to protect farmers' knowledge from extinction and thus to encourage its further use.	The goals are to protect farmers' knowledge from misappropriation and to enable its holders to make decisions over its use.
Equitable sharing of the benefits arising from the use of genetic resources (para. 9.2(b))	Benefits are to be shared between stewards of plant genetic resources and society at large – partly through the multilateral system and official development assistance.	Benefits are to be shared between purported 'owners' and 'buyers' of genetic resources upon prior informed consent on mutually agreed terms.
Participation in relevant decisions at the national level (para. 9.2(c))	Participation is important to ensure legal space and rewards for farmers' contributions to the genetic pool.	Participation is important to ensure adequate legislation on access and intellectual property rights.
Farmers' customary use of propagation material (saving, sharing and selling seeds) (para. 9.3)	The goal is to uphold the legal space to ensure that farmers continue to maintain plant genetic resources.	The goal is to balance intellectual property rights for farmers with breeders' rights.

In realizing farmers' rights according to the measures suggested under the ITPGRFA, the goals for each of these two approaches are very different (Andersen, 2006). Table 8.1 illustrates these two approaches in detail.

Protecting farmers' traditional knowledge

Protecting farmers' traditional knowledge can mean different things. Based on the ownership approach, it would mean offering ownership status to farmers with the right to act against misappropriation and to decide over the use of their knowledge and related plant genetic resources. In Norway, farmers stress that their traditional knowledge is about to disappear. Therefore, protection, as they understand it, must be about ensuring that such knowledge does not die out (Andersen, 2010). In order to ensure such a thing, knowledge must be shared in the broadest manner possible. An ownership approach to protection could provide disincentives to sharing knowledge between and among farmers – as has been seen among potato farmers in Peru (Andersen, 2005). In contrast, the stewardship approach mandates that agricultural plant varieties and related knowledge should be shared among farming communities, and its proponents insist that ownership in this context has been an alien idea among farmers and that it represents a profound break with traditional perceptions. Whether a stewardship approach, an ownership approach or a combination of the two is chosen, it is important to ensure that it does not provide any disincentives to the sharing of knowledge and genetic resources among farmers and that it does not contribute to genetic erosion or the loss of traditional knowledge. Such activities would be against the intentions of the ITPGRFA.

Ensuring equitable benefit sharing

Measures to ensure the equitable sharing of benefits arising from the use of genetic resources can be designed in many ways. Under an ownership approach, these measures would mandate the development of direct benefit sharing in which the benefits would be shared directly between the purported 'owners' and 'buyers' of genetic resources – based on a prior informed consent on mutually agreed terms (as set out in the Convention on Biological Diversity [CBD]).[6] In contrast, proponents of the stewardship approach would encourage an indirect means of benefit sharing – one in which the benefits are shared between 'entire peoples,' all stewards of plant genetic resources in agriculture and society at large. This line of thinking originates from the early days of negotiations in the Food and Agriculture Organization (FAO). The idea is that it is the legitimate right of farmers to be rewarded for their contribution to the global genetic pool, from which we all benefit, and that it is an obligation of the international community to ensure that they receive this reward. Benefit-sharing mechanisms would include the multilateral system of the ITPGRFA as well as official development assistance.

Proponents of the stewardship approach maintain that it would be difficult to identify exactly who should be rewarded if an ownership approach is used. In addition, they point out that the demand for farmers' varieties among commercial breeders is limited and so relatively few farmers would benefit, while most of the contributors to the global pool of genetic resources would remain unrewarded. They continue in saying that the ownership approach to sharing benefits could lead to disincentives to sharing seeds and propagating material among farmers because of benefit expectations.

Although several countries of the South – among them the Philippines, Peru, India and Ethiopia – have enacted legislation on direct benefit sharing, no countries so far have instituted direct monetary benefit sharing with regard to agrobiodiversity (Andersen, 2008, 2009, 6, addendum 3). In contrast, there are many examples of indirect benefit sharing, although these have usually been nonmonetary (see the next section in this chapter). It would seem that the transaction costs of establishing access and benefit-sharing legislation in many countries has been comprehensive. Thus, the ownership approach has not proven to be particularly promising with regard to benefit sharing, even though some stakeholders would say that it is the most fair and equitable approach. These concerns are serious and must be taken into account when designing measures to ensure that benefit sharing is in place that is in line with the intentions of the ITPGRFA.

Participation in decision making

Under the two approaches participation in relevant decision making is important, but for different reasons. With the stewardship approach, the most important objectives would be to ensure legal space for farmers to continue their practices as custodians and innovators of plant genetic resources and to establish reward mechanisms for farmers' contributions to the global genetic pool. Under the ownership approach, the goals would be to ensure appropriate legislation on access and benefit sharing as well as to safeguard farmers' intellectual property rights to the genetic resources in their fields and related knowledge. It is clear that these two sets of objectives could be conflicting. However, the overall objectives of the ITPGRFA to conserve, sustainably use and share benefits from crop genetic resources for sustainable agriculture and food security may serve as guiding principles. Measures that limit a farmer's ability to take part in these activities would go against the intentions of the treaty.

Farmers' rights to save, use, exchange and sell farm-saved seed

Farmers' customary use of propagating material – their right to save, use, exchange and sell farm-saved seed and propagating material – could likewise be handled in several different ways. Under the stewardship approach, it is vital to uphold the *legal space* for farmers to save, use, exchange and sell farm-saved seed and propagating material. Various forms of regulations, such as intellectual property rights legislation and plant variety release and seed marketing laws, are currently reducing this space, thereby threatening farmers' ability to maintain and breed plant genetic resources as well as to sustain their very livelihoods. Under an ownership approach, the most important goal is to provide farmers with intellectual property rights on the varieties in their fields that are on an equal footing with breeders' rights. Arguments related to this objective were discussed earlier in this chapter. India's 2001 Protection of Plant Varieties and Farmers' Rights Act represents a noteworthy example of an attempt at combining these two sets of objectives (Ramanna, 2006).[7] Undoubtedly, there are many other means of combining these two approaches in order to realize farmers' rights. What matters in this context is that the approach that is chosen must not conflict with the principles of the stewardship approach, which has been the primary goal of the FAO since the issue was first taken up as well as the rationale behind the ITPGRFA.

Contents of farmers' rights and experiences with their implementation

In the discussion on farmers' rights – as well as in the practice of realizing these rights – there is a growing understanding of the core issues and challenges, which often combine the stewardship and ownership approaches to a various extent. The next section will examine in detail the four issues addressed in Article 9, which are often referred to as the elements of farmers' rights – namely, protection of traditional knowledge; benefit sharing; participation in decision making; and the rights to save, use, exchange and sell farm-saved seed.

Farmers' rights related to the protection of traditional knowledge

Traditional knowledge is vital to understanding the properties of plants, their uses and how they are cultivated. Traditional knowledge includes knowledge of how to select seeds and propagating material, how to store them and how to use them for the next harvest. Thus, it comprises the basic necessities for farmers to be able to maintain crop genetic diversity in the fields. Article 9.2(a) is the only provision on traditional knowledge in the ITPGRFA; the treaty provides no further guidance on how this article can be interpreted and operationalized. However, since the objectives of the ITPGRFA are to be implemented in harmony with the CBD (Article 1), Article 8(j) of the CBD is also relevant in this context. According to this article, each contracting party shall – as far as possible and as appropriate and pursuant to national legislation – respect, maintain and preserve traditional knowledge, innovation and practices and promote their wider application. This activity should be done with the approval of the holders of such knowledge, innovations and practices. Moreover, the equitable sharing of benefits from its use should be encouraged.

Understanding the challenges that are related to the protection of traditional knowledge has significantly influenced the current views about how Article 9.2(a) of the ITPGRFA can be implemented. If we examine the contents of this right from a stewardship and an ownership approach, different possibilities appear:

1 *Protection against extinction* means ensuring that traditional knowledge is kept alive and can further develop among farmers. The best way to protect traditional knowledge from the threat of extinction is to share it – a widespread approach in the North – and, thus, the motto 'protection by sharing.' Measures for the sharing of traditional knowledge include:

   - seminars and gatherings among farmers to share knowledge;
   - seed fairs for the exchange of propagating material and associated knowledge;
   - documentation of knowledge in seed catalogues and registries;
   - documentation of knowledge in books, magazines and on websites;
   - documentation of knowledge in gene banks and making such knowledge accessible.

2 *Protection against misappropriation* is a different approach. It is based on the anticipation that farmers' varieties, together with associated knowledge, could be 'discovered' and developed by commercial actors as well as possibly by the use of intellectual property rights – without benefit-sharing mechanisms. Thus, under this approach, the sharing of knowledge should

not take place unless there are measures in place to avoid any misappropriation. This view is often accompanied by a widespread regret that the fear of misappropriation has made it necessary to be cautious concerning activities that are so vital to the availability of genetic resources and related knowledge. Measures for protection against misappropriation include:

- regulating access to genetic resources and associated traditional knowledge with measures on prior informed consent and mutually agreed terms;
- developing legal clauses in catalogues of genetic material and associated material in order to avoid misappropriation;
- introducing 'user country measures' such as conditions for intellectual property rights and certificates of origin for genetic resources and following the appropriate legal procedures for access to genetic resources in provider countries.

According to existing documentation, it would seem that, in developing new varieties, commercial plant breeders tend to use already improved varieties from their own stocks or from other plant breeders. Farmers' varieties are generally regarded as being difficult to work with due to their genetic heterogeneity. Only when particular traits are sought – traits that cannot be found in their own stocks or among other improved varieties – are farmers' varieties deemed to be necessary. When they are sought out, they are normally obtained from gene banks and not from the farmers' fields or markets. In the gene banks, there is normally not much traditional knowledge included in the passport data. Thus, at this point in time, traditional knowledge related to crop genetic resources is rarely used in commercial breeding. Whether this will remain the case in the future is difficult to predict. Generally, the genetic foundation for commercial plant breeding appears to be getting increasingly narrow (Esquinas-Alcázar, 2005, 948). This situation, together with the effects of climate change, may very well change the demand for landraces and farmers' varieties – together with their associated knowledge – and make them much more valuable in the future (ibid.).

Ultimately, the measures that are chosen should reflect the situation. What is most important today, with the rapid erosion of traditional knowledge, is to protect traditional knowledge related to crop genetic resources from becoming extinct. Relevant measures for avoiding misappropriation are second priority. What these secondary measures will be is another question, and to answer it we need to take a closer look at what misappropriation of traditional knowledge is actually about in the context of the ITPGRFA and the multilateral system on access and benefit sharing. There are essentially three forms of action that farmers tend to regard as misappropriation:

1 if farmers' varieties and related knowledge are used in commercial plant breeding without recognizing the farmers in question;

2 if plant breeders obtain intellectual property rights to farmers' varieties, thereby removing the varieties from the public domain and the traditional uses of farmers;
3 if plant breeders profit from the use of farmers' varieties and related knowledge without sharing the benefits with the farmers in question.

Measures to avoid such misappropriation could include:

- *Ensuring recognition:* Recognition is very important to many farmers, particularly in the South. Ways of showing recognition include naming varieties after the farmers or communities in question, providing information about the farmers on the wrapping of products and/or rewarding farmers for their contribution in terms of benefit sharing (see discussion later in this chapter) or with awards. With respect to the first measures, it may be difficult to identify the individual farmers in question since several farmers/communities/regions may have maintained a crop variety or contributed to its development. Awards are different in this regard since they can often be awarded for the maintenance of diversity and related knowledge as such, and not necessarily for specific varieties.
- *Countering breeders' claims to intellectual property rights on farmers' varieties:* Documenting plant varieties and their related knowledge is normally a useful way to establish prior art. It means that no one can claim intellectual property rights over those varieties in the form in which they are documented. This measure is to date the most promising means of ensuring protection against the misappropriation of genetic resources and associated traditional knowledge while, at the same time, promoting the sharing of knowledge. Plant variety registries have been established locally in many countries – for example, in the Philippines and in Nepal (Andersen and Winge, 2008).
- *Ensuring benefit sharing:* Under the ITPGRFA, benefit sharing is to take place according to the Standard Material Transfer Agreement (SMTA) in the multilateral system. The benefits should be shared with farmers in developing countries and in countries with economies in transition who conserve and sustainably use crop genetic diversity, not including any specific providers of genetic resources (and related knowledge). It should be noted, however, that there are many questions related to benefit sharing, which will be addressed in a later discussion.

There exist many useful and inspiring databases and catalogues on crop genetic resources and associated traditional knowledge around the world. These sources also establish prior art with regard to farmers' varieties and contribute to benefit sharing by making the knowledge accessible. Some of them also give explicit recognition to farmers. An impressive example is the potato catalogue from Huancavelica, Peru (Centro Internacional de la Papa and Federación Departemental de Comunidades Campesinas, 2006; see also Andersen and Winge, 2008, 23–25).

Other success stories include *in situ* conservation in Switzerland, which has combined on-farm conservation of a huge number of crop varieties with a range of measures for the dissemination of information regarding the varieties and the associated traditional knowledge; the community registry at Bohol, the Philippines, which is helping to keep traditional knowledge alive and accessible; and information and seminar activities in Norway that are helping to disseminate traditional knowledge. These models have succeeded in implementing farmers' rights with respect to traditional knowledge that is associated with crop genetic resources. However, they are only a beginning. Much more is needed to keep such knowledge alive among farmers and to promote its further development. In many countries, it would appear to be necessary to raise awareness about the importance of traditional knowledge related to crop genetic resources and to develop strategies on how to maintain and disseminate traditional knowledge in a systematic way before such knowledge is lost completely.

Farmers' rights to participate equitably in benefit sharing

Article 9.2(b) concerns a farmer's right to participate equitably in the sharing of benefits arising from the utilization of plant genetic resources for food and agriculture. To interpret this provision, some guidance can be found in Article 13 of the ITPGRFA on benefit sharing in the multilateral system. This article lists the most important benefits as: (1) facilitated access to plant genetic resources for food and agriculture; (2) the exchange of information; (3) access to and transfer of technology; (4) capacity building; and (5) the sharing of monetary and other benefits arising from commercialization. Moreover, it specifies that benefits arising from the use of plant genetic resources for food and agriculture that are shared under the multilateral system should flow primarily, directly and indirectly, to farmers in all countries – especially in developing countries and countries with economies in transition – who conserve and sustainably utilize plant genetic resources for food and agriculture.

Whereas these provisions all relate to the multilateral system and not directly to the provisions on farmers' rights in the ITPGRFA, they reflect a line of thought on benefit sharing that is relevant for interpreting Article 9.2(b) as a measure to protect and promote farmers' rights. First, it is clear that there are many forms of benefit sharing, of which monetary benefits are only a part. Second, the benefits are not only to be shared with those few farmers who happen to have plant varieties that are utilized by commercial breeding companies, but also with farmers in all countries that are engaged in the conservation and sustainable use of agrobiodiversity. This approach is consistent with the policy developed by the FAO after farmers' rights and benefit sharing were first officially recognized in 1989.[8] It differs from the bilateral and direct approach to benefit sharing that is mandated under the CBD, where benefits are to be shared between the purported 'owners' and buyers of the resources.

In the South, policies on benefit sharing – if there are any – are normally present in the laws and regulations on access to biological resources, which are sometimes found in the national legislation on the protection of biological diversity. Countries with legislation on indigenous peoples' rights often include provisions on benefit sharing in these laws, which then also cover indigenous farmers. Most of these regulations comprise forms of direct benefit sharing between the 'owners' and the 'buyers' of genetic resources, often based upon prior informed consent on mutually agreed terms, as set out in the CBD. However, despite all of these efforts, so far there have been no examples to date of direct monetary benefit sharing between the providers and recipients of plant genetic resources for food and agriculture as a result of such legislation.

There are, however, other ways of sharing benefits, which are often referred to as indirect methods of benefit sharing. These methods are in line with the FAO's mandate in the early days of negotiations on farmers' rights. As mentioned earlier, a basic principle was that benefits should be shared among 'entire peoples,' the stewards of plant genetic resources in agriculture and society at large (FAO, 1987, Appendix F, section 8). This principle is based on the idea that it is farmers' legitimate right to be rewarded for their contributions to the global genetic pool from which we all benefit, and it is an obligation of the international community to ensure that such recognition and reward is provided.

Where should the funds come from to enable such benefit sharing? First of all, as we have already noted, the benefit-sharing mechanism under the multilateral system specifies that the benefits from the system should flow primarily to farmers in all countries, especially in developing countries and countries with economies in transition, who conserve and sustainably use crop genetic resources (Article 13.3). However, it is uncertain how much funding can be generated by this mechanism and even whether this mechanism will be successful and make a substantial difference to the farmers it is supposed to be helping (see the discussion in Chapter 3 of this volume).

The funding strategy of the ITPGRFA (as set out in Article 18) is another important source in so far as it supports the implementation of conservation (Article 5), sustainable use (Article 6) and farmers' rights (Article 9), which would all greatly benefit diversity farmers. However, since there are to date no fixed mandatory contributions, it is uncertain how much money the fund can generate. Thus, for the time being, Article 7 on international cooperation and Article 8 on technical assistance are the primary documents on benefit sharing. In these articles, the contracting parties agree to promote the provision of technical assistance to developing countries and countries with economies in transition, with the objective of facilitating the implementation of the ITPGRFA.

The third source of benefit sharing, and the most successful at the present time, is official development assistance (Brush, 2005; Andersen, 2008). Official development assistance can be channelled through bilateral or multilateral cooperation or through NGOs. There are many examples of NGO-channelled support, which have greatly supported diversity farmers in the South and thus contributed to benefit sharing in many developing countries.

In an international stakeholder survey carried out in 2005, the most frequently mentioned nonmonetary benefits were:

- access to seeds and propagating material and related information;
- participation in the definition of breeding goals;
- participatory plant breeding with farmers and scientists collaborating;
- stronger and more effective farmers' seed systems;
- conservation activities, including local gene banks;
- enhanced utilization of farmers' varieties, including market access (Andersen, 2005).

This 2005 survey shows that, for many reasons, benefit sharing is more promising when the primary target for funding is the farming community that actually contributes to the maintenance of plant genetic diversity rather than the providers of genetic resources to commercial plant breeders. Still, the dominant view on benefit sharing in many countries, particularly in the South, is one of direct benefit sharing between the purported owners and buyers. While such an approach might seem to be fair and equitable as a point of departure, there are many difficulties with it, including that:

- it is difficult to identify exactly who should be rewarded;
- the demand for farmers' varieties among commercial breeders is limited, so relatively few farmers would benefit and most of the contributors to the global pool of genetic resources would remain unrewarded;
- the approach could lead to disincentives to share seeds and propagating material among farmers because of the expectations of personal benefit or the benefit to a community;
- although several countries in the South have enacted legislation on direct benefit sharing, no instances of such benefit sharing have been reported so far with regard to agrobiodiversity;
- in many countries, the transaction costs of establishing access and benefit-sharing legislation have been considerable.

Thus, the direct benefit-sharing approach has not proven to be especially promising so far, and these concerns must be taken into account when measures are designed to ensure benefit sharing that is in line with the intentions of the ITPGRFA.

According to the findings of the Farmers' Rights Project, three categories of measures appear to be particularly important when seeking to operationalize the concept of benefit sharing with regard to farmers' rights (Andersen, 2009). The first category ensures that *incentive structures* in agriculture favour farmers who conserve and sustainably use plant genetic resources for food and agriculture on an equal footing with, or more than, farmers who are engaged in the monoculture production of genetically homogeneous plant varieties. Such incentive structures might include extension services to support particularly the farmers of the first group, loans on favourable conditions for the purchase of farm animals and other necessary input factors, the facilitation of marketing products from diverse varieties and other infrastructure measures. A strategy that would cover the incentive structures for each of these areas in combination would substantially support farmers who conserve and sustainably use agrobiodiversity. This has not been done in any country so far. In fact, existing incentive structures have generally proven to be detrimental to farmers' customary practices. However, there are also many local-level initiatives that can provide good models of how incentive structures could be designed on a larger scale.

The second category would create *reward and support systems* that would enable farmers to benefit significantly from their contributions to the global genetic pool, through added value to the crops they grow and through improved livelihoods and increased income. There currently exist many small-scale programs and projects that demonstrate the enormous potential in this regard, such as community gene banks, seed fairs and registries (to ensure access), dynamic conservation programs coupled with participatory plant breeding, plant breeding and farmers' field schools, capacity building and various marketing activities. Today, however, the benefit of these programs reaches only a very limited number of farmers. A major challenge is to scale up these activities so that all farmers engaged in the maintenance of agrobiodiversity can share in these benefits.

The third category would ensure the *recognition of farmers' contributions* to the global genetic pool in order to show that their contributions are valued by society. One form of recognition that is often discussed is the procurement of intellectual property rights for farmers. There are strong views for and against such rights. Proponents claim that farmers should be granted intellectual property rights on an equal footing with breeders as a matter of fairness. Opponents stress that such a system would create disincentives for farmers to share their seeds because of the expectations that the seeds could prove to be economically valuable. Such a development could be harmful to traditional seed systems and could negatively affect farmers' rights to own, use and distribute their own seeds. However, since the idea of exclusive intellectual property rights for farmers is fairly new and largely unexplored (except for a few individual acts of legislation), we will not examine this topic in this chapter. Another means of recognizing farmers' contributions could be to provide some sort of remuneration to those farmers who register varieties in seed catalogues for free distribution among farmers (this idea was suggested by Maria Scurrah at the Lusaka Consultation; see Andersen and Berge, 2007, 26). However, once again, this method has not been attempted in the field. A more usual way of granting recognition to farmers and farming communities is through awards for innovative practices, as has been done in several countries. Yet, this is not to say that farmers are not be entitled to intellectual property rights. Rather, it indicates where the greatest potential for benefit sharing may lie and what dangers should be avoided if countries are seeking to establish intellectual property rights for farmers.

The Farmers' Rights Project, which revealed several success stories from the realization of farmers' rights, presents many good examples of indirect forms of benefit sharing, including incentive structures in the Philippines; community seed fairs in Zimbabwe; community gene banks and on-farm conservation in India; dynamic conservation and participatory plant breeding in France; participatory plant breeding in Nepal, which is adding value to farmers' varieties; capacity building for seed potato selection in Kenya; the development of a Peruvian Potato Park; and the reward for best farming practices in Norway (Andersen and Winge, 2008). These are all examples of programs and developments that provide models for the further implementation of farmers' rights. The major challenge today is to find ways and means to scale up such activities – for example, through the national agricultural extension service systems which are being planned in Nepal. However, such initiatives are heavily dependent on political will, which is often lacking. In order to increase the political will, it will be necessary to raise awareness in society in general on the vital importance of agrobiodiversity and farmers' rights (Andersen, 2005).

Farmers' rights to participate in decision making

Article 9.2(c) deals with the right of farmers to participate in decision making at the national level on matters related to the conservation and sustainable use of plant genetic resources for food and agriculture. However, no further guidance is provided in the ITPGRFA as to how such decision making can be implemented in practice. To operationalize this measure, it will be necessary to specify the 'relevant matters' in which farmers can have the right to participate as well as the way in which they can participate.

The development of laws and regulations related to the management of plant genetic diversity in agriculture is clearly relevant for farmers' participation. At the current time, there are numerous examples of such laws and regulations, including seed acts, seed certification regulations, other regulations regarding seed distribution and trade, plant variety protection laws, patent laws, bioprospecting laws or regulations, laws on the conservation and sustainable use of biodiversity in general or crop genetic resources in particular (as well as on several specific crops) and legislation on the rights of indigenous peoples and traditional knowledge. In addition, it is also important to consider any legislation that regulates mainstream agriculture since such legislation tends to produce incentive structures that are often detrimental to farmers' rights without providing any compensation. The extensive use of hearings at various stages in the process is an important measure to ensure participation. It is particularly important to ensure that farmers who are engaged in the management of plant genetic diversity are aware of the processes and are explicitly invited to participate through their organizations.

The implementation of laws and regulations is also relevant to farmers' participation. The way in which these regulations are interpreted and implemented often has an enormous influence on a farmer's management of these resources and also on his or her livelihood. Normally, such acts and regulations establish boards and institutions to oversee and/or administer implementation. Farmers' representation and participation in these bodies is therefore integral, and the means by which farmers are selected for membership is of crucial importance. If they are appointed by a government official, for example, they can hardly be said to represent the farmers of the country. If, however, they are appointed by farmers through their own organizations, it is more likely that they will be regarded as true representatives of the farming community – depending on the number of farmers that they represent and the process by which they were appointed. Again, it is essential to ensure that farmers are actually represented and engaged in agrobiodiversity conservation – there are too few success stories in this regard. In addition, the development of policies and programs in agriculture, particularly in relation to the management of plant genetic resources for food and agriculture, also requires farmer participation. In order to create policies and programs that are valuable for farmers, they have to be targeted specifically at the situations that farmers are in, taking farmers' perspectives as points of departure.

Ultimately, then, the implementation of farmers' rights requires farmers' participation. This is not only because of their unquestioned right in this regard, according to the ITPGRFA, but also because they are the ones who can best define the needs and priorities of farmers in the context of farmers' rights and they are also the central actors in the implementation process. Comprehensive consultative processes of various kinds are relevant: the better represented farmers are, the greater legitimacy the results will have and the more likely it is that they will constitute effective measures for the realization of farmers' rights. In particular, it is important for farmers to actually be involved in the management of plant genetic diversity in order to participate in such processes since they constitute the main target group of the ITPGRFA. The most comprehensive consultative process on the implementation of farmers' rights to date was carried out in Peru in 2008, and it involved 180 farmers from many different regions as well as numerous central decision makers. It resulted in a report that currently forms the basis for the implementation of farmers' rights in Peru (Scurrah, Andersen and Winge, 2008).[9]

There are two major preconditions for the increased participation of farmers in decision making. First, decision makers need to be aware of the role that is played by farmers in conserving and developing plant genetic resources for food and agriculture, and thus in contributing to national food security, in order to understand why their participation is so important. Second, without prior capacity building, many of the world's farmers would not be in a position to participate effectively in complicated decision-making processes. Hence, it is essential to raise awareness among decision makers on the role of farmers in agrobiodiversity management and to build the capacity of farmers' organizations. While there is not much evidence of the former to date, there has been much more activity in the latter goal.

In general, we find few examples of legislation on farmers' participation, although some countries in the South have extensive legislation on farmers' participation in decision making (Andersen, 2005). All the same, the actual participation of farmers in decision-making processes seems marginal and is often limited to large-scale farmers who are normally not engaged in the maintenance of plant genetic diversity. In the North, the participation of farmers in decision-making processes is more common, even if diversity farmers are rarely represented, but such participation does not usually involve specific laws or policies. It should be noted that farmers in the North claim that their influence is now decreasing, due to their countries' commitments to regional and international organizations and agreements such as the World Trade Organization (WTO) and the European Union (EU) (ibid.). While the process of implementing participation has been slow, there have been a few success stories, including the implementation of farmers' rights in Peru (see the earlier discussion); various capacity-building measures to prepare farmers for participating in decision making in Malawi, Zimbabwe, the Philippines and Peru; and several successful advocacy campaigns regarding the implementation of elements of farmers' rights, where farmers have been directly involved (e.g. in India, Norway and Nepal). More successes are on the way.

Farmers' rights to save, use, exchange and sell farm-saved seed

The IPGRFA is vague on farmers' rights to save, use, exchange and sell farm-saved seed. Section 9.3 of the treaty states that nothing in the relevant article (Article 9 on farmers' rights) 'shall be interpreted to limit any rights that farmers have to save, use, exchange and sell farm-saved seed, subject to national law and as appropriate,' but this article does not really offer much direction, except for labelling these practices as 'rights.' The preamble notes that 'the rights recognized in this Treaty to save, use, exchange and sell farm-saved seed and other propagating material . . . are fundamental to the realization of Farmers' Rights.' This statement indicates the importance of these rights, but it does not give much guidance since the rights to which it refers are only vaguely addressed. Despite this lack of precision, the general line of thought would seem clear. It is important to grant their rights in this subject area, but individual countries are free to define the legal space that they deem to be sufficient for farmers regarding their rights to save, use, exchange and sell farm-saved seed.

The freedom to define such legal space for farmers is also restricted by other international commitments. Most countries in the world are members of the WTO and are thus obliged to implement the WTO Agreement on Trade-Related Aspects of Intellectual Property Rights (TRIPS Agreement).[10] According to the TRIPS Agreement, all WTO member countries must protect plant varieties either by patents, by an effective *sui generis* system (a system of its own kind), or a combination of both (Article 27.3.b). The limits to a *sui generis* system and the meaning of an 'effective' *sui generis* system are not explicitly defined in the text. In other words, countries have to introduce some sort of plant breeders' rights.

The Union for the Protection of New Varieties of Plants (UPOV) explains that the most effective way to comply with the provision concerning an effective *sui generis* system is to follow the model of the International Convention for the Protection of New Varieties of Plants (UPOV Convention).[11] There are several versions of the UPOV model. The most recent one (the 1991 Act of the UPOV Convention) provides that plant breeders are to be granted comprehensive rights – to the detriment of farmers' customary rights to save, reuse, exchange and sell seeds. It is still possible to make exceptions for small-scale farmers in order to enable them to save and reuse seeds, but only within strict limits. The exchange and sale of seeds among farmers is prohibited. It should be noted, however, that all of these regulations apply only to seeds protected by plant breeders' rights and not to traditional varieties.

The UPOV model has met with resistance from some countries and many organizations that fear that their ratification of the treaty would be detrimental to the rights of farmers to save and share propagating material. The TRIPS Agreement provides only minimum standards, leaving enough scope for the development of other solutions that are more compatible with the demand for farmers' rights. The challenge in the context of the ITPGRFA is thus for WTO member countries to meet their TRIPS obligations regarding plant breeders' rights, while at the same time maintaining the necessary legal space to realize farmers' rights to propagating material. It will be questionable how much room will be left for countries to manoeuvre within the framework of their international obligations in order to grant farmers the right to save, use, exchange and sell seeds.

A further constraint to farmers' rights in many countries is the introduction of seed laws that affect all propagating material, whether it is protected with intellectual property rights or not. The most important factor is that these laws also affect traditional varieties and farmers' varieties. They require that all varieties be officially approved for release and that seed and propagating material be certified before they are offered on the market. The original reason for these regulations was to ensure plant health and seed quality. However, in many countries, the regulations have gone so far that they now hinder the maintenance of crop genetic resources in the fields in two ways. First, since traditional varieties are normally not genetically homogeneous enough to meet the requirements for approval and certification, these varieties are excluded from the market and gradually disappear from active use when those farmers who currently use them begin to give them up. Second, many seed laws also stipulate that only authorized seed shops are allowed to sell seeds, and they prohibit all other seed exchange (sometimes with exceptions for horticultural plants or certain other species). This is the case in most of Europe. Such regulation means the end of a 10,000-year-old tradition of seed exchange that made possible the development of today's rich agrobiodiversity.

When combined, these two processes – restrictions on plant variety release and seed marketing laws – may constitute serious obstacles to the implementation of the ITPGRFA in terms of *in situ* on-farm conservation and sustainable use as well as to farmers' rights. It is a paradox that rules originally intended to protect plant health have, in fact, contributed to removing the very basis for ensuring plant health in the future – namely, the diversity of genetic resources. Seed laws, together with strict plant breeders' rights, represent a major obstacle to farmers' rights to save, use, exchange and sell seeds. What possibilities are there to make such laws more compatible with these customary rights of farmers, which are so crucial to the maintenance of agrobiodiversity for food security, today and in the future? The EU has tried to solve the problem with a specific directive on conservation varieties. However, EC Directive 62/2008 Providing Certain Derogations for Acceptance of Agricultural Landraces and Varieties Which Are Naturally Adapted to the Local and Regional Conditions and Threatened by Genetic Erosion and for Marketing of Seed and Seed Potatoes of Those Landraces and Varieties (EU Conservation Varieties Directive) is not adequate to solve these new hurdles for the implementation of the ITPGRFA. This directive is inadequate for the following reasons:

1. seed exchange and sale is still prohibited among farmers under the new directive;
2. only varieties deemed interesting for conservation and sustainable use by certain authorities can be covered by the system, which limits diversity;
3. the variety release and certification criteria are still too strict to allow for the release of many traditional and farmers' varieties;
4. the marketing and use of the varieties are limited to the regions of origin;
5. only limited quantities may be used;
6. the conservation varieties may not be further developed by farmers.

These provisions do not encourage the conservation and sustainable use of crop genetic diversity, and they pose serious barriers to the implementation of Articles 5, 6 and 9 of the ITPGRFA.

An ultimate objective from the perspective of farmers' rights would be to grant the rights to save, use, exchange and sell farm-saved seed, whether it is from varieties protected with intellectual property rights or not. Other solutions would be needed in order to compensate plant breeders for their efforts and to solve the issue relating to plant health concerns. Due to the different forms of existing legislation, however, the challenge should rather be to uphold or reestablish sufficient legal space for farmers to continue their crucial role as custodians and innovators of crop genetic diversity within the existing legal framework on plant breeders' rights, variety release and seed distribution.

Generally, legislation on intellectual property rights, variety release and seed certification are most restrictive in the North and least restrictive in Africa, while countries in Asia and Latin America fall somewhere in the middle. In the EU, for example, farmers are not allowed to use farm-saved seed from protected varieties on their own holdings or they must pay a license fee to do so. With respect to nonprotected varieties, they are not allowed to exchange seed or even to give it away. These are major hurdles for the proponents of farmers' rights and the implementation of the ITPGRFA in terms of on-farm conservation and sustainable use. The Farmers' Rights Project of the Fridtjof Nansen

Institute has gathered several pertinent stories on how legal space for farmers' rights can be established and maintained in order to allow farmers to maintain their traditional practices and innovation in agriculture (see e.g. Andersen and Winge, 2008). The report provides several examples, including India's 2001 Protection of Plant Varieties and Farmers' Rights Act,[12] and Norway's 'No' to stricter plant breeders' rights in order to maintain the balance with farmers' rights and the ways in which farmers are circumventing the law in the Basque Country in Spain. Nevertheless, establishing and maintaining legal space for farmers' rights to save, use, exchange and sell farm-saved seed constitutes the main barrier to implementing the ITPGRFA in terms of the conservation and sustainable use of crop genetic diversity and of the realization of farmers' rights. Solutions are urgently needed.

Achievements at the international level

A resolution on farmers' rights was adopted at the third session of the Governing Body to the ITPGRFA in Tunis on 4 June 2009. The resolution marks a substantial step forwards in the implementation of Article 9 of the treaty. This achievement was not only due to the contents of the resolution (see discussion later in this chapter) but also because of the broad consensus that was reached among the contracting parties at an early stage in discussions on the proposed text. The issue of farmers' rights has proven to be a difficult topic at earlier Governing Body sessions. In fact, at the second session of the Governing Body in 2007, there was substantial resistance among several industrialized countries to the resolution text proposed by the developing countries (the G-77 and China). Only through intense negotiations in a contact group was it possible to arrive at a consensus text in that session. By the time of the third session, however, the situation had changed. The only substantial resistance against the resolution on farmers' rights came from Canada, whereas all other contracting parties were largely united. This marks the beginning of an emerging understanding across the regions of the crucial importance of farmers' rights for the implementation of the ITPGRFA as well for other various challenges in the field.

The 2009 resolution was proposed by Brazil on behalf of Africa, Latin America and the Caribbean countries. It contained the following operational provisions:

- The contracting parties are invited to consider reviewing and, if necessary, adjusting their national measures affecting the realization of farmers' rights as set out in Article 9 of the ITPGRFA in order to protect and promote farmers' rights.
- The contracting parties and other relevant organizations are encouraged to continue to submit views and experiences on the implementation of farmers' rights as set out in Article 9 of the ITPGRFA, involving, as appropriate, farmers' organizations and other stakeholders.

- The Secretariat is requested to convene regional workshops on farmers' rights, subject to the agreed priorities of the program of work, the budget and the availability of financial resources, that will be aimed at discussing national experiences on the implementation of farmers' rights as set out Article 9 of the ITPGRFA, involving, as appropriate, farmers' organizations and other stakeholders.
- The Secretariat is requested to collect the views and experiences submitted by contracting parties and other relevant organizations and the reports of the regional workshops, as a basis for an agenda item that is to be considered by the Governing Body at its fourth session and to disseminate relevant information through the website of the ITPGRFA, where appropriate.
- The Governing Body appreciates the involvement of farmers' organizations in its further work, as appropriate, according to the rules of procedure established by the Governing Body.

The 2009 resolution and later resolutions following in the same lines have great potential for the realization of farmers' rights. If the contracting parties review and adjust their seed regulations and other national measures to promote farmers' rights, and thereby enable farmers to continue to conserve and sustainably use crop genetic diversity, it would represent a major step forward. If they present these views to the Governing Body, it may provide a solid basis for identifying further steps that could be made at the fourth session. In addition, regional workshops will be instrumental in promoting the realization of farmers' rights at the national level as well in providing input to the Governing Body. If the Secretariat makes all of the submissions and reports from the regional workshops available on the Internet, it will facilitate an exchange of experiences across countries and regions and enable an external analysis of potential steps that can be taken in the future – steps that could feed into the Secretariat's own preparations for the agenda item on farmers' rights at the fourth session. The participation of farmers' organizations at the Governing Body sessions will be important to demonstrate that farmers are participating in the decision making on farmers' rights. Whether these potentials will materialize depends, as always, on the political will of the contracting parties and on the engagement of the involved organizations.

Conclusions

Over the last few years, discussions in the Governing Body, as well as various informal consultations in various forums, have contributed to shaping the elements of a common understanding of farmers' rights. There is a general recognition that farmers need legal space to continue to perform their role as custodians of crop genetic diversity, although there are many different opinions on what this legal space should cover. There is also a common understanding that farmers need to be recognized and rewarded for their contribution to the global genetic pool and that they have a right to participate in decision making. As we have seen in this chapter, there are many efforts around the world in this direction, in line with Article 9 of the ITPGRFA. These are often small-scale local initiatives, but many of them provide models that have the potential to develop into something bigger. A key challenge is to find ways and means to scale them up to the national level.

Nevertheless, there are also substantial obstacles in the way of these goals. Overcoming these hurdles will require the development of various common solutions:

- Variety release and seed certification regulations pose serious hurdles to farmers' rights to exchange and sell farm-saved seed as well as to the marketing of landraces and many farmers' varieties. This constitutes a serious obstacle to on-farm conservation and the sustainable use of crop genetic diversity. In order to overcome this hurdle, shared norms need to be developed on how seed laws can be designed so as to ensure adequate legal space for farmers. The 2009 resolution on farmers' rights by the Governing Body of the ITPGRFA marks a promising start in this regard.
- Intellectual property rights constitute hurdles to the realization of farmers' rights to various degrees. In some countries, the balance between farmers' and breeders' rights is seen as being acceptable, such as in India and Norway. In other countries, however, plant breeders' rights and patents are more problematic since they prohibit customary uses of seed. It is necessary to discuss what kind of legal space farmers should be ensured with regard to plant breeders' rights and patents, with a view to developing shared norms. Norway's decision in regard to plant breeders' rights may provide some inspiration.
- Fear of misappropriation of farmers' varieties and associated traditional knowledge has led to protectionism with regard to seeds and knowledge among farmers in several countries. Such a tendency is detrimental to the sharing of seed and knowledge among farmers as well as to *ex situ* conservation measures. Ways and means must be found to ensure that farmers do not need to fear misappropriation. One challenge is to identify efficient measures to establish *prior art* for landraces and farmers' varieties in order to ensure that these cannot be made subject to intellectual property rights. Another challenge is to include provisions in intellectual property rights legislation on the disclosure of origin of resources and legal provenance in order to ensure that no misappropriation takes place. Norms and rules in this regard need consideration.
- There are many good examples of farmers' rights being realized, and many of these have the potential to be scaled up to the national level, such as through extension service systems. To date, however, there have been no examples of such efforts being made. More consideration is required in order understand how to facilitate such an effort.
- Participation in decision making is an issue with many facets. The general picture is that in countries where farmers are granted some sort of participation, farmers engaged in the conservation and sustainable use of crop genetic diversity are often not represented. Ways of identifying such farmers and involving them in decision-making processes are needed.

Much has already been achieved with regard to developing a joint understanding of farmers' rights, their importance and the steps required for their realization, and there are many success stories. These are all important achievements. However, much still remains to be done to ensure that these rights are realized on the same scale that is required to enable farmers to continue to maintain and further develop the crop genetic diversity that is the basis of local and global food security and to recognize and reward them for their contributions to the global genetic pool. Awareness of the challenges, the political priority, and international cooperation are required to make farmers' rights a reality.

Notes

1 This chapter is based on the results of the Farmers' Rights Project of the Fridtjof Nansen Institute, an international project designed to support the implementation of farmers' rights as they are addressed in the ITPGRFA. Started in 2005, it has been a long-term project with many different components, comprising research and surveys as well as policy guidance, facilitation of consultations, information, and capacity building. For an overview of the research reports and activities, see <www.farmersrights.org>. The author of this chapter led the Farmers' Rights Project until she went on leave from the Fridtjof Nansen Institute. She maintains the responsibility of the website from her new affiliation.

2 International Treaty on Plant Genetic Resources for Food and Agriculture, 29 June 2004, online: <www.planttreaty.org/texts_en.htm> (last accessed 15 June 2012) [ITPGRFA].

3 Ownership is used as a term here because it is regarded as the basis for a reward system (benefit sharing).

4 This concept was first used in connection with this requirement in Andersen (2006). It will be further explained later in this chapter.

5 Stewardship is used as a term here, although it does not sufficiently cover the innovative work that farmers are doing as breeders of plant genetic resources. Since no other term was found that could sufficiently cover farmers' maintenance work and innovations, the term was kept, with this footnote as an explanation of its contents.

6 Convention on Biological Diversity, 31 ILM 818 (1992).

7 Protection of Plant Varieties and Farmers' Rights Act, online: <http://agricoop.nic.in/seeds/farmersact2001.htm> (last accessed 15 June 2012).

8 FAO Conference Resolution 5/89.

9 Progress is slow, however, due to a lack of resources and political attention.

10 Agreement on Trade-Related Aspects of Intellectual Property Rights, Annex 1C of the Marrakech Agreement Establishing the World Trade Organization, 15 April 1994, 33 ILM 15 (1994).

11 International Convention for the Protection of New Varieties of Plants, 2 December 1961, online: <www.upov.int/en/publications/conventions/index.html> (last accessed 15 June 2012).

12 Protection of Plant Varieties and Farmers' Rights Act, 2001, online: <http://agricoop.nic.in/seeds/farmersact2001.htm> (last accessed 15 June 2012).

References

Andersen, R. (2005). *The Farmers' Rights Project – Background Study 2: Results from an International Stakeholder Survey on Farmers' Rights,* FNI Report no. 9/2005, Fridtjof Nansen Institute, Lysaker, Norway.

Andersen, R. (2006). *Realising Farmers' Rights under the International Treaty on Plant Genetic Resources for Food and Agriculture, Summary of Findings from the Farmers' Rights Project (Phase 1),* FNI Report no. 11/2006, Fridtjof Nansen Institute, Lysaker, Norway.

Andersen, R. (2008). *Governing Agrobiodiversity: Plant Genetics and Developing Countries,* Ashgate, Aldershot, UK.

Andersen, R. (2009). *Information Paper on Farmers' Rights Submitted by the Fridtjof Nansen Institute, Norway, Based on the Farmers' Rights Project,* input paper submitted to the Secretariat of the ITPGRFA, 19 May 2009, Doc. IT/GB-3/09/Inf, Rome.

Andersen, R. (2010). *Farmers' Rights in Norway: A Case Study,* Fridtjof Nansen Institute, FNI Report no. 11/2011 (Norwegian edition) and FNI Report no. 17/2012 (English edition), Fridtjof Nansen Institute, Lysaker, Norway.

Andersen, R., and B. Gunnvor (2007). *Informal International Consultation on Farmers' Rights,* 18–20 September 2007, Lusaka, Zambia, Report no. M-0737 E, Norwegian Ministry of Agriculture and Food, Oslo.

Andersen, R., and W. Tone (2008). *Success Stories from the Realization of Farmers' Rights Related to Plant Genetic Resources for Food and Agriculture,* FNI Report no. 4/2008, Fridtjof Nansen Institute, Lysaker, Norway.

Brush, Stephen B. (2005). 'Protecting Traditional Agricultural Knowledge,' *Washington University Journal of Law and Policy* 17: 59–109.

Centro Internacional de la Papa and Federación Departemental de Comunidades Campesinas (2006). *Catálogo de Variedades de Papa Nativa de Huancavelica, Peru,* Centro Internacional de la Papa and Federación Departemental de Comunidades Campesinas, Lima.

Esquinas-Alcázar, J. (2005). 'Protecting Crop Genetic Diversity for Food Security: Political, Ethical and Technical Challenges,' *National Review of Genetics* 6: 946–53.

Food and Agriculture Organization (FAO) (1987). *Report of the Second Session of the Commission on Plant Genetic Resources,* Doc. CL 91/14, FAO, Rome.

Fujisaka, S., D. Williams and M. Halewood (eds.) (2009). *The Impact of Climate Change on Countries' Interdependence on Genetic Resources for Food and Agriculture,* Preliminary version presented as Background Study Paper no. 48, Commission on Genetic Resources for Food and Agriculture, FAO, Rome.

Ramanna, A. (2006). *The Farmers' Rights Project – Background Study 4: Farmers' Rights in India – A Case Study,* FNI Report no. 6/2006, Fridtjof Nansen Institute, Lysaker, Norway.

Scurrah, M., R. Andersen and T. Winge (2008). *Farmers' Rights in Peru: Farmers' Perspectives,* FNI Report no. 16/2008, Fridtjof Nansen Institute, Lysaker, Norway.

United Nations (2009). *Seed Policies and the Right to Food: Enhancing Agrobiodiversity and Encouraging Innovation,* Interim report of the UN Special Rapporteur on the Right to Food, Olivier De Schutter, and transmitted by the Secretary General to the members of the General Assembly of the UN for its sixty-fourth session, United Nations, New York.

Part IV

Identifying discrete farmers' varieties in law

9 *Sui generis* protection for farmers' varieties

Carlos M. Correa

Introduction

Farmers' varieties are a vital source of diversity in plant breeding. Much has been written about the development of *sui generis* regimes for the protection of plant varieties as an alternative to the dominant model of protection enshrined in the International Convention for the Protection of New Varieties of Plants (UPOV Convention).[1] This chapter briefly discusses, first, the evolution of intellectual property protection in the area of plant varieties and, second, some of the fears that have been voiced over the years concerning the implications of plant variety protection (PVP) and plant patents, as well as some of the expectations about the benefits that could accrue to farmers through the development of *sui generis* forms of protection that cover farmers' varieties. Against this backdrop, the chapter considers the main elements that may be present in *sui generis* regimes that depart from the model of the UPOV Convention, as illustrated by the *sui generis* systems adopted in India, Thailand and Malaysia as well as by the model legislation approved by the Organization of African Unity (OAU) in 2000. The requirements and other conditions of protection under these systems are examined, particularly as they apply to farmers' varieties that do not comply with the uniformity or stability standards.

The main argument presented in this chapter is that although the design of *sui generis* regimes for the protection of plant varieties that do not apply the UPOV model has been on the agenda of many developing countries, nongovernmental organizations (NGOs) and academics for at least 20 years, little progress has been made in finding solutions to the complex conceptual and technical problems that are involved. Despite the experiences in a few developing countries, there is little evidence about what such regimes have achieved. Indeed, reliable models that can be followed do not seem to exist yet, and considerable work is still necessary to design a national regime that effectively addresses the needs of farming communities in a particular national context.

Evolution of PVP systems

Until the emergence of professional breeding at the beginning of the twentieth century, the improvement, production and exchange of seeds was entirely dependent on farmers' practices. When breeding became a business activity of its own, breeders organized themselves to obtain some protection on the new plant varieties that they were creating.[2] Farmers' traditional practices of saving, replanting, exchanging or selling seed from their own harvest made it difficult to recoup investments in breeding. Both in the United States and Europe, early attempts were made to extend patent protection to plant varieties, but this possibility raised doubts – namely because of the incremental type of innovation that characterizes plant breeding – and fears began to develop regarding possible distortions of the patent system (Dutfield, 2003, 186). The Lisbon Diplomatic Conference on the Revision of the Paris Convention, which was held in 1958, considered the possible allowance of patents in this field, but no action was taken since the general view was that a 'special law' was needed to protect new plant varieties (Dhar, 2002, 4).

The legislative movement towards a special form of protection for plants was pioneered by the United States. In 1930, the Plant Patents Act was passed, which allowed for the protection of asexually reproduced varieties (except tubers).[3] In Europe, efforts were made to develop a legal system that was adapted to the characteristics of innovation in plant varieties. The first legislation on PVP was introduced in the Netherlands in 1942, followed by Germany in 1953 (Van Overwalle, 1999, 161). The Association Internationale pour la Protection de la Propriété Intellectuelle and the Association Internationale des Sélectionneurs pour la Protection de Obentions Végétales (ASSINSEL) took the lead in the search for a specific legal means of protection. ASSINSEL requested the French government to organize what became the International Conference for the Protection of New Varieties of Plants, which was eventually convened in May 1957 in Paris.[4] This conference laid down the basic principles of plant breeders' rights that were later reflected in the 1961 UPOV Convention (Dutfield, 2003, 186–87).

Although the model of protection for breeders' rights that is enshrined in the UPOV Convention and in the UPOV-based PVP laws has been influenced by patent law, it has also received a significant amount of influence from seed certification legislation.[5] The incorporation of concepts derived from such legislation (notably the uniformity and stability standards) has led to important differences between PVP and patent law.

By the 1960s, three European countries had introduced breeders' rights laws. Eight more nations followed suit in the 1970s. Also in 1970, the United States passed the Plant Variety Protection Act.[6] Thereafter, plants could be protected in that country both by the 1930 Plant Patents Act and the 1970 Plant Varieties Protection Act – a possibility that was, as a matter of principle, excluded under the UPOV Convention. This situation prevented the United States from acceding to the UPOV Convention until the ban on the accumulation of protections was lifted by the revision of the convention in 1978 for countries that were already practicing it (Article 37). The ban was finally eliminated entirely by another revision to the convention in 1991. As a result, under 1991 UPOV Convention, it is possible to accumulate patent and plant variety protection.[7]

While special plant patents, based on the Plant Patent Act, have been available in the United States since 1930 and breeders' rights have been in existence since 1970, the landmark decision by the US Supreme Court in *In re Chakrabarty* opened the way for the issuance of utility patents for plants.[8] The first patent to cover plants or segments thereof was issued by the US Patent and Trademark Office in *Ex parte Hibberd.*[9] This patent covered genetically engineered maize with high levels of tryptophan. Thereafter, a large number of patents were granted covering any of the following subject matter:

- DNA sequences that code for a certain protein
- isolated or purified proteins
- plasmids and transformation vectors containing a gene sequence
- seeds
- plant cells and plants
- plant varieties, including parent lines[10]
- hybrids
- processes to genetically modify plants
- processes to obtain hybrids.[11]

European countries followed a narrower approach than the United States in regard to the patentability of plants. The 1973 Convention on the Grant of European Patents (European Patent Convention) excluded plant varieties from patent protection as well as the essentially biological processes for their production (Article 53(b)).[12] These differences became apparent during the negotiation of the Agreement on Trade-Related Aspects of Intellectual Property Rights (TRIPS Agreement).[13] While the United States argued for patents on plants, the European Economic Community proposed that the agreement maintain the restrictive approach of the European Patent Convention. The lack of consensus on the matter led to a compromise that left open considerable options for the members of the World Trade Organization (WTO). Article 27.3(b) of the TRIPS Agreement provides that members may exclude from patentability

> plants and animals other than microorganisms, and essentially biological processes for the production of plants or animals other than non-biological and microbiological processes. However, Members shall provide for the protection of plant varieties either by patents or by an effective sui generis system or by any combination thereof. This provision shall be reviewed four years after the entry into force of the WTO Agreement.

Consistent with this provision, many developing countries excluded plant varieties from patentability. Some also excluded DNA sequences and amino acid sequences that corresponded to the peptides or proteins produced by naturally occurring organisms (Boettiger et al., 2004, 1093).

In accordance with Article 27.3(b), all WTO members are bound to protect plant varieties, but there is flexibility in regard to the form of protection.[14] This flexibility was, as mentioned earlier, a reflection of the lack of consensus among the industrialized countries rather than a North-South divide. The precarious nature of the agreement reached on Article 27.3(b) is indicated by the fact that it was the only provision in the entire TRIPS Agreement that was subject to an early revision – 4 years after the agreement's entry into force. This period was even shorter than the transitional period contemplated for developing countries and economies in transition (Article 65). While the review of Article 27.3(b) started in 1999, so far no outcome has been achieved and little interest has been shown by developed country members to make any progress on the matter.

Despite the fact that the UPOV Convention is not mentioned in the TRIPS Agreement, a UPOV-based breeders' rights regime may constitute 'an effective *sui-generis* system.'[15] More than 70 countries that are already members of the International Union for the Protection of New Varieties of Plants (UPOV), as well as others that generally follow the UPOV Convention model without being members of the union,[16] intend to comply in this way with the TRIPS Agreement. However, the ability to grant patents,[17] combine patents with breeders' rights or develop other types of *sui generis* regimes for the protection of plant varieties, provided only that such regimes are 'effective,' has created considerable space for national legislations to design the modalities of protection in this area. In fact, the introduction of the concept of *sui generis* regimes in the TRIPS Agreement has triggered the interest of many developing countries, NGOs and academia in finding new modalities of protection for plant varieties specifically adapted to the needs of developing countries.

The latitude of Article 27.3(b) is such that the scope, requirements and rights conferred under a *sui generis* regime, as discussed later in this chapter, do not need to conform to those prescribed under patent law or under the UPOV Convention's model of protection. Moreover, in some of the most recent legislation on PVP, it is apparent that other concepts have begun to have an influence, notably the notion of farmers' rights as well as the principles of benefit sharing contained in the Convention on Biological Diversity (CBD)[18] and in the International Treaty on Plant Genetic Resources for Food and Agriculture (ITPGRFA) (Correa, 2000).[19] The introduction of these concepts has been decisive in modelling new *sui generis* systems on the matter in some countries, as elaborated later in this chapter.

As a result, there is a range of *sui generis* modalities of PVP, depending on the type of requirements imposed and the type of rights conferred. Such modalities include one or more of the following elements:

- new or relaxed requirements of protection that extend protection to varieties that do not currently conform with the conventional requirements of PVP as stated by the UPOV Convention;

- the expansion of rights conferred to farmers with regard to the use of saved seeds;
- the addition of benefit-sharing provisions, with or without the registration of plant varieties.

However, it is difficult to define a typology of the existing *sui generis* regimes since many of them present a combination of these various elements. For instance, the Indian law, which is discussed in greater detail later in this chapter, extends PVP to some varieties that do not conform with the conventional requirements of PVP, expands the rights conferred to farmers under PVP and also includes benefit-sharing provisions.

Concerns about PVP

PVP that is based on the model of the UPOV Convention has raised two types of fears. On the one hand, some consider that PVP only benefits commercial breeders by creating private ownership rights to biodiversity, to the detriment of farmers/breeders and traditional communities that have ensured the conservation of plant biodiversity and varietal improvement for centuries. The recognition of PVP would only reward those at the very end of a more complex system of innovation and seed production and eventually limit farmers' and communities' rights to biodiversity and even reduce their space to innovate. A number of disadvantages for developing countries that choose to use UPOV-based PVP models have been identified, including the following:

- PVP encourages monopolies in genetic materials for specific traits;
- the plant variety holder may produce less seed than the demand to increase prices and profits;
- PVP inhibits the free exchange of materials;
- PVP increases the prices of seeds, which the poor farmer may not be able to afford;
- PVP will essentially benefit commercial breeders and not farmers or traditional communities;
- national breeders and local seed companies will be bought out by foreign companies;
- companies in the North will get full commercial control over the communities' germplasm and knowledge;
- the criteria for protection will exacerbate the erosion of biodiversity, leading to harvest loss and further food insecurity;
- PVP will reduce information and germplasm flows and act as a disincentive to research;
- PVP under the UPOV Convention conflicts with the CBD benefit-sharing principles (GAIA/GRAIN, 1998; Chawla, 2003).[20]

One of the most often mentioned fears is that the high level of uniformity required by PVP does not provide an agronomic advantage, but rather may erode the diversity in plant germplasm and negatively affect agricultural development and food security in the long run. It should be noted, however, that while greater uniformity may indeed be induced by the commercial breeders' need to comply with such a requirement, it is often the result of a demand by farmers for yield and quality maximization, the desire by urban consumers and processing industries for quality and a need to comply with seed certification legislation in order to obtain an authorization to commercialize seeds.

From a very different perspective, doubts have been expressed about the effectiveness of PVP as a method of promoting investment in plant improvement. For instance, a study on the application of PVP in the United States found that

> whereas plant variety protection was initially designed as the primary (or even exclusive) form of intellectual property protection for seed-grown plants, the coming of plant biotechnology, and the dawning acceptance of utility patents for plants, has relegated plant variety protection to a secondary role. Modest statutory amendments to the PVPA have shown no real promise of lifting the PVPA up from this secondary status. Second, our empirical assessment of licensing and enforcement activities concerning U.S. plant variety protection certificates confirms that the PVPA regime as presently constituted plays only a marginal role in stimulating plant breeding research in the United States. Our assessment strongly suggests that the PVPA does not provide patent-like ex ante innovation and investment incentives and that the PVPA has not generated substantial ex post licensing and enforcement activity. Instead, its role in the United States appears to be very modest: it may serve as a marketing tool; it may provide some non-propagation licensing rights akin to contractual shrink-wrap rights, enforceable against those who deal in 'saved' seeds; and it may provide a superior alternative to trade secret protection – for example, for seeds whose secret parent lines might otherwise be revealed through reverse engineering.
>
> (Kesan and Janis, 2002, pp. 776–777)

Other studies have also indicated a modest impact of PVP on private investments in research and development and on the number of varieties released as well as a tendency to focus on high value/low volume crops. In general, the literature assessing the impact of PVPs is largely inconclusive, particularly about the effects of such protection in developing countries.[21] However, other studies have reported positive outcomes from the implementation of the UPOV Convention in several countries, such as a series of studies conducted by the UPOV Secretariat (see Box 9.1).

Box 9.1 Impact of the UPOV Convention in selected developing countries

Argentina

Argentina introduced a PVP system in 1973 and acceded to the UPOV Convention in 1994. The following effects have been noted:

- The average annual number of titles granted to domestic breeders was 26 in 1982–91, which more than doubled to 70 in 1992–2001 (267 percent).
- The average annual number of titles granted to foreign breeders was 17 in 1984–93, which more than trebled to 62 in 1994–2003 (355 percent).
- The improved performance of new, protected varieties is indicated, for example, in crops such as wheat and soybean, where the demand for new, protected varieties is shown by their increased proportion of the certified seed area, which rose from 18 percent to 82 percent and from 25 percent to 94 percent, respectively, since the introduction of the UPOV-based PVP law and accession to the UPOV Convention.
- An increase in the number of domestic breeding entities was seen, for example, in soybean and wheat, most of which occurred in the private sector.
- An increase of horizontal cooperation in the seed industry was identified, involving foreign seed companies and agreements for technology transfer between national research institutes and breeding entities with other national companies (technological relationship agreements), and this cooperation has resulted in the more rapid movement of germplasm.

China

The PVP system became operational in 1999, and China also became a member of the UPOV Convention in 1999. China's PVP systems have only been in operation for 5 years and for only a limited number of genera and species, and it is not yet possible to evaluate their full impact. Nevertheless, the following effects have been observed:

- A rapid uptake by farmers of new, protected varieties such as maize and wheat has been seen, for example, in the province of Henan.
- New, protected varieties have been introduced for major staple crops (such as rice, maize and wheat), horticultural crops (such as rose, Chinese cabbage and pear), including traditional flowers (such as peony, magnolia and camellia), and forest trees (such as poplar).

- New, foreign varieties, particularly ornamental varieties, have been recently introduced.
- Commercial breeding activities have been stimulated in domestic public research institutes and domestic seed companies, with an increase in the number of breeders (e.g. of maize and wheat in Henan Province) linked to an increased number of PVP applications.
- An increase in the income of breeders has been seen, including public research institutions and agricultural universities, and further investment in plant breeding has been encouraged.

Kenya

In Kenya, the PVP scheme started to operate in 1997, and the country acceded to the 1978 Act of the UPOV Convention in 1999. Kenya grants plant breeders' rights for all plant genera and species other than algae and bacteria. The following impacts have been observed:

- Significantly higher number of varieties of various agricultural crops were developed and released in the six-year period following the introduction of PVP (1997–2003), compared to the previous six-year period (1990–96), particularly for maize.
- There has been an increased introduction of foreign varieties, especially in the horticultural sector, which contributes to the diversification of the horticultural sector (e.g. the emergence of the flower industry) and supports the competitiveness of Kenyan products in global markets (cut flowers, vegetables and industrial crops).
- There has been an increased introduction of foreign germplasm in the form of new, protected varieties (especially of horticultural crops), which have been used by Kenyan breeders for further breeding.
- There has been an increase in the number of Kenyan-bred varieties of agricultural crops with improved performance (e.g. in yield, pest and disease tolerance, nutritional qualities, early maturity, and tolerance to abiotic stresses) for local farmers, including subsistence farmers. PVP titles for many Kenyan-bred varieties are in the hands of public institutions, and local farmers can use the propagating material of the new, protected varieties under privileged conditions (e.g. subsistence farmers have been permitted to exchange seed among themselves).
- Public/private partnerships for plant breeding have been facilitated, including partnerships between international research institutes and Kenyan seed companies, and there has also been an emergence of new types of breeders (university researchers, private farmer-breeders, and so on).

Source: Based on International Union for the Protection of New Varieties of Plants (2005).

In sum, the adoption of PVP generates fairly divergent views ranging from the perception of a strong influence on seed production and use to opinions that minimize the possible impact on the generation of new varieties. Those aligning with the first approach generally advocate either non-intellectual property right (IPR) protection or the development of *sui generis* regimes. For those sharing the second view, the solution would be provided by the application of utility patents. The following sections focus on some of the expectations raised by the adoption and the characteristics of *sui generis* regimes.

Expectations about *sui generis* regimes

A significant number of proposals, with a varying degree of detail, have been made for the protection of plant varieties under *sui generis* regimes that are not based on the UPOV model. The general aim of these proposals is to reward or otherwise protect the interests of commercial breeders as well as of farmers/communities that are contributing to the improvement of plant varieties and, at the same time, to promote the conservation and sustainable use of plant biodiversity. The proposals are generally based on one or more of the following considerations.

Equity

It is expected that a *sui generis* regime would allow for the recognition of the innovations made by farmers/communities, including compensation for a third party's use of farmers' varieties for commercial purposes. Such recognition would arguably bring equity into the relations between farmers and commercial breeders, which have been marked so far by a significant asymmetry.

Conservation

A common rationale for a *sui generis* regime of protection is the assumption that it may contribute to the conservation of farmers' varieties and plant biodiversity in the fields. The legal recognition of and compensation for farmers' contributions would encourage them to preserve their knowledge and practices, which are essential for a sustainable agriculture. Thus, the argument has been made that

> vesting legally recognized ownership of knowledge in communities through sui generis IPRs will raise the profile of that knowledge and encourage respect for it both inside and outside the knowledge holding communities. This will make the learning and development of such knowledge a more attractive prospect for the younger members of such communities, thus perpetuating its existence.
>
> The possibility of economic returns for the use of that knowledge by third parties acts as a further incentive for community members to respect their knowledge and continue to engage in practices in which that knowledge is used and generated.
>
> (Crucible Group II, 2001, 68–69)

Preventing misappropriation

Another important justification for a *sui generis* regime is that it may erect a barrier against the use or appropriation by third parties of innovations made by farmers/communities without their consent and without benefit sharing. This rationale is explicitly or implicitly present in most proposals for *sui generis* protection and is often based on the assumption that such regimes may counterbalance breeders' rights (Genetic Resources Policy Initiative, 2006, 19).

Dissemination of knowledge

The existence of a *sui generis* regime of protection would encourage the dissemination of innovations by farmers/communities. In line with this argument, it has been observed that

> indigenous and local knowledge holders will be more willing to disclose otherwise secret knowledge once they know sui generis laws can give then control over how their knowledge gets used. In this way, IP laws encourage the disclosure, use and proliferation of knowledge that might otherwise be lost.
>
> (Crucible Group II, 2001, 69)

Farmers' freedom to save and sell seeds

Occasionally, the development of a *sui generis* regime has been seen as being instrumental in allowing farmers to continue to use their traditional varieties and to sell the seeds of these varieties (Genetic Resources Policy Initiative, 2006, 10).[22] It has also been regarded as being necessary to preserve farmers' freedom to utilize, save or even sell varieties protected by third parties that have been derived from their own varieties.

Incentive

Finally, a *sui generis* regime is often deemed to be an incentive for farmers/communities to produce new varieties and thereby widen the sources for further variation as well as to encourage them to share them with other farmers/communities or conventional breeders. In many cases, the proposals for the creation of *sui generis* regimes have not made their objectives explicit, and it is difficult to ascertain what the rationale for the protection is. In addition, it is not always clear whether the design of the proposed regimes is such that they could be successful in reaching what seems to be their intended objective. As noted by the Crucible Group II, merely using a law to make something into property that was previously part of the public domain 'does not suddenly save it, conserve it, make people respect it or want to use it . . . Fencing off their knowledge does nothing to protect it from being even more eroded, undermined, or ignored or at risk of being lost' (Crucible Group II, 2001). Achieving some of the possible objectives of the *sui generis* regimes may not require any new or additional legislation, whereas other objectives could require amendments to seed legislation rather than the establishment of a new IPR (Genetic Resources Policy Initiative, 2006, 10).[23] Some of the possible disadvantages of a *sui generis* regime that assigns property rights are indicated in Box 9.2.

One of the basic issues to be addressed under a *sui generis* regime is whether it is necessary or convenient and whether the potential benefits brought about by property rights over farmers' varieties would offset the potential costs derived from the establishment of private rights. There is a great deal of controversy over this issue. Many proposals have been made to provide a *sui generis* protection that would cover new commercial varieties as well as farmers' varieties, and some national laws have already implemented this dual approach. There is also a great deal of opposition to the belief that conferring private rights on farmers' varieties would be beneficial to the farmers/communities. Thus, it has been argued that

> where communities are concerned and where the innovation is the result of a collective process and collective action, assigning property rights to an individual or corporate holder may well lead to reduced availability of germplasm. It may also have the effect of further marginalizing those plant genetic resources and knowledge processes that are not covered by

Box 9.2 Possible disadvantages of a sui generis regime

- IPRs are meant to promote the invention of new things, such as plants. Protecting existing plants is not in line with this basic idea.
- IPRs could be a disincentive to the exchange of genetic resources between farmers and could thus lead to genetic erosion.
- With totally exclusive IPRs, farmers would probably tend to isolate commercially successful varieties, which could lead to the erosion of diversity.
- Since only the commercially attractive varieties are protected, it would not be not sufficient to support the maintenance of all diversity.
- The allocation of the rights could lead to intercommunity distress and social conflict because often the ownership of these rights is not clear.
- Transaction costs could be high.

Source: Genetic Resources Policy Initiative (2006, 19).

> the property rights regime. Indeed, the mere existence of such rights may still not offer sufficient incentives to develop markets that adequately capture the value of biodiversity, again, because of the public goods nature of many of the benefits. In particular, assigning exclusive property rights to germplasm might reduce the ability of poorer farmers to access seed germplasm, given that often less informed, less educated, and marginalized rural populations are at a disadvantage in claiming ownership . . . The granting of exclusive intellectual property rights over germplasm might reduce access to plant genetic material to everyone, including poorer farmers. Thus, even the seemingly positive benefits of granting intellectual property rights to local communities may lead to unintended negative consequences.
>
> (Eyzaguirre and Dennis, 2007, 1495)

An aspect that is often disregarded in the proposals for *sui generis* regimes that encompass farmers' varieties is that, as noted in the preceding quotation, individual farmers or communities may face a significant burden in complying with the formalities for acquiring rights such as registration because of both the complexity and possible cost of the procedures.[24] Most importantly, enforcing any conferred rights would require considerable effort and resources. Most farmer-bred varieties are unlikely to capture a significant share in the commercial seed market, and when a variety occasionally does achieve such a share, it may be very difficult for the rights holder to monitor whether an infringement has occurred (Salazar, Louwaars and Visser, 2007, 1523). Moreover, if an infringement has been identified, bringing a legal action in court would generally be very costly and beyond the reach of farmers/communities. The outcome of litigation may also be uncertain, especially when more than one farmer or community claims ownership of a particular plant variety.

Most importantly, all of these considerations probably underestimate the fact that many communities might not accept the concept of property rights over germplasm and might not wish to exercise community rights against their neighbouring communities. Assigning ownership for financial or other economic returns may run against farmers' spirit of free exchange. These cultural motives may prevent the widespread application of new *sui generis* regimes more than the legal problems emerging from any attempt to bring farmers' varieties under current IPR systems (Salazar et al., 2007, 1523).

Even if rights holders were willing and able to exercise the conferred rights, a further uncertainty is the state's capacity to implement *sui generis* legislation, especially to assess whether a particular variety meets the requirements of eligibility for protection. As mentioned later in this chapter, administrative authorities must establish whether a variety for which protection is sought is novel, distinct, uniform and/or stable, or at least identifiable, depending on the national law. This assessment requires technical competence, which is missing or insufficient in many developing countries, particularly if the legislation applies to a broad range of crops.[25] Countries adhering to the UPOV Convention may rely on the technical assistance provided by or through its Secretariat. There is no organization playing a similar role for those countries opting for a *sui generis* regime.

Whatever the expectations are about the goals that *sui generis* regimes may achieve, they cannot be empirically confirmed or dismissed so far, since only a few countries have introduced such regimes and they have only done so in the recent past. As noted by Pablo Eyzaguirre and Evan Dennis (2007, 1495), the current debates on the feasibility and benefits of establishing intellectual property protection in this area 'are often stymied or sterile due to the lack of empirical evidence and experiences of local communities and indigenous groups with established rights to local biological and biocultural resources.'

Scope and conditions of protection under *sui generis* regimes

A *sui generis* intellectual property regime is a set of rules tailored to the particular characteristics of the subject matter that it is intended to protect. The specificity of a *sui generis* regime is determined by the application of different requirements to obtain protection, including the subject matter that is being protected (e.g. commercial varieties and/or farmers' varieties); the scope of the rights that are conferred; the conditions that are imposed on the applicants or rights holders (e.g. disclosure of the source of the material); and the recognition of farmers' rights as defined in the ITPGRFA.[26]

In the literature, *sui generis* regimes for plant varieties are often deemed to be those that differ from the model established by the UPOV Convention, although, as noted earlier, the latter may also be regarded as a *sui generis* system, given its significant differences with the patent regime. In the next sections, some of the special features of *sui generis* systems for plant varieties that diverge from the UPOV model are examined.

Coverage

While PVP focuses on new plant varieties, *sui generis* regimes may cover other categories of varieties that are not necessarily novel or that do not comply with one or more of the requirements under the UPOV Convention. For instance, the Indian Protection of Plant Varieties and Farmers' Rights Act (PPVFR Act), which was adopted in 2001, applies to: (1) new plant varieties; (2) extant (domestic and existing) varieties; and (3) farmers' varieties.[27] Under the PPVFR Act, farmers' varieties are a subset of extant varieties.[28] They include varieties that have 'been traditionally cultivated and evolved by the farmers in their fields' and those that are a 'wild relative or landrace of a variety about which the farmers possess the common knowledge' (Article 2(1)). Thailand also adopted a *sui generis* regime in 1999. The Plant Varieties Protection

Act (PVP Act) applies to new and local varieties as well as to local domestic and wild varieties with a differentiated regime, as discussed later in this chapter.[29] In accordance with this Act:

- A 'local domestic plant variety' means 'a plant variety which exists only in a particular locality within the Kingdom and has never been registered as a new plant variety and which is registered as a local domestic plant variety under this Act.'
- A 'wild plant variety' means 'a plant variety which currently exists or used to exist in the natural habitat and has not been commonly cultivated.'
- A 'general domestic plant variety' means 'a plant variety originating or existing in the country and commonly exploited and shall include a plant variety which is not a new plant variety, a local domestic plant variety or a wild plant variety' (section 3).

In Malaysia, new commercial and 'traditional' varieties are covered by the legislation adopted in 2004 (Protection of New Plant Varieties Act 2004),[30] but the law incorporates the concept of 'plant variety' found under legislation modelled in accordance with the UPOV Convention.

Protection requirements

A new plant variety may be protected under the PVP laws that follow the UPOV Convention model if it meets the requirements of novelty, distinctness, uniformity and stability (NDUS). These requirements were essentially transposed to PVP regimes from the early seed certification laws enacted in Europe. They allowed for the differentiation of the plant protection regime from patent protection, which was generally deemed to be inappropriate for plant varieties at the time that PVP was first introduced.[31] *Sui generis* systems may not apply one or more of the NDUS requirements. For instance, the Indian PPVFR Act allows for the registration of extant and farmers' varieties that are not novel, but it requires that they conform to the distinctness, uniformity and stability (DUS) requirements (Article 15(2)). Dropping the novelty requirement (which is essential under PVP legislation) may significantly expand the range of varieties eligible for protection since those varieties that have been offered for sale or commercialized at any time before an application for protection is filed are eligible for protection.[32] The Thai PVP Act does apply the NDUS requirements (except for local domestic plant varieties that need not comply with the novelty requirement), but with a modified distinctness criterion.[33]

Section 14.2 of the Malaysian Protection of New Plant Varieties Act 2004 provides that plant varieties bred or discovered and developed by a farmer, local community or indigenous peoples are protectable if they are new, distinct and *identifiable*. While only 'new' varieties may be protected, this provision introduces a significant departure from the NDUS standards since uniformity and stability are not required.[34] This criterion is easier to apply to asexually reproducing crops that have built-in uniformity and stability than to other crops.[35]

The African Model Law for the Protection of the Rights of Local Communities, Farmers and Breeders and for the Regulation of Access to Biological Resources (OAU Model Law), which was approved by the OAU in 2000, also proposes to grant protection to varieties that may be identified without relying on the NDUS requirements.[36] Article 25(2) of the OAU Model Law states that

> a variety with specific attributes identified by a community shall be granted intellectual protection through a variety certificate which does not have to meet the criteria of distinction, uniformity and stability. This variety certificate entitles the community to have the exclusive rights to multiply, cultivate, use or sell the variety, or to license its use without prejudice to the Farmers' Rights set out in this law.

Hence, the OAU Model Law seems to replace the NDUS requirements by the following concept: 'specific attributes identified by a community.' It is unclear, however, which attributes would be considered and how they would be determined. The absence of general criteria to establish eligibility for protection might lead to significant uncertainty and competing claims about ownership.

Farmers' varieties generally are composed of a number of different genotypes and are not subjected to a process of selection to increase uniformity since uniformity would pose a high risk to small farmers (Salazar et al., 2007, 1523).[37] Different types within one farmer's variety may develop in order to respond to particular growing conditions or may tend to predominate as a result of biological or abiotic stresses (ibid.). The lower degree of uniformity, in turn, means that new farmers' varieties are less stable over generations than other varieties that meet the NDUS requirements.[38]

Views differ about the extent to which the uniformity and stability standards may be left out. One view is that their absence may become very problematic since different communities may make multiple claims over the same variety, especially for crops that outbreed. Another view is that the uniformity standard may be compromised in protecting farmers' varieties but that the stability requirement should definitely be preserved (Genetic Resources Policy Initiative, 2006, 11). According to still another opinion, an even more flexible approach is possible since the role played by the DUS requirements under intellectual property protection is not the same as it is under seed certification laws. These issues were discussed extensively at a meeting entitled 'Exploring Legal Definitions of Farmers' Varieties in Strategies to Promote Farmers' Rights,' sponsored by the Genetic Resources Policy Initiative (GRPI) in Hanoi and held 26–28 October 2006. Participants at the GRPI meeting in Hanoi concluded that the distinctness standard (or the ability to identify) was the most important condition for a *sui generis* protection of farmers' varieties[39] and that the two other requirements could be lowered:

> The group felt that the standards of uniformity set out in the UPOV guidelines could be loosened to take into account the special nature of farmers' varieties (particularly with respect to out-breeders). Alternative standards from those set out in the guidelines created by UPOV could be developed . . . Stability is not so important as a condition for intellectual property protection . . . [t]he overseeing authority can always retract protections when a variety shifts as a consequence of instability . . . It could be argued that stability, for example, is an important criterion for the purposes of advancing the public protection policy goal of a seed law. However, it is not so relevant when one is discussing the conditions under which the state may grant a form of monopoly over the use of that same material.
>
> (Genetic Resources Policy Initiative, 2006, 20–21)

In fact, there seems to be a good argument for allowing the relaxation of the uniformity and stability requirements, to the extent that it would not affect the identification of the subject matter of protection. The strict application of such requirements may lead to the exclusion of farmers' varieties from the possible coverage of a *sui generis* regime since very few varieties (except in the case of asexually reproducing plants) would qualify for protection. A standard of identifiability (as already adopted by the Malaysian *sui generis* regime) may overcome the problem posed by the heterogeneity of farmers' varieties as well as of extant varieties.[40] It should be borne in mind, however, that allowing for a relaxation of the DUS standards may significantly complicate the enforcement of conferred rights in the case of disputes about the 'title-hood' of a particular variety or the infringement of rights.

The possible adoption of a standard of identifiability (associated with novelty and distinctness) was considered by the Crucible Group as one of the options for the protection of farmers' varieties under a *sui generis* regime. In explaining this option, the Crucible Group stated that

> this element replaces the relatively strict requirements of uniformity and stability with the looser condition of 'distinctness and identifiability' (DI) . . . A DI protection requirement would not comply with the UPOV Conventions. This would not be a problem, of course, for countries that are not signatories to the UPOV Conventions. Despite not satisfying the UPOV standards, the criterion of identifiability may well satisfy TRIPs Article 27.3(b), which includes no obligation on WTO member countries to follow the UPOV model or to become members of UPOV. Being the widest in scope, Element 3 could be used as a national baseline criterion. Varieties that satisfy the stricter criteria could qualify for stronger and/or longer protection.
>
> (Crucible Group II, 2001, 148)

At the GRPI workshop in Hanoi, the participants recommended that it would be better to relax, rather than abandon, the uniformity requirement. It concluded that a *sui generis* regime could be based on 'nDu' standards, where

> n = *sui generis* novelty involving non-commercialization outside of the local setting of use of the farmers' varieties in question (i.e. a variety would be deemed 'new' despite its use in a particular area, if not commercialized outside it); D = distinct as understood and identified in the UPOV Convention; and u = a relaxed standard of uniformity, taking into account the less uniform nature of many varieties bred by farmers.
>
> (Genetic Resources Policy Initiative, 2006, 29)

Under this proposal, the novelty and uniformity requirements are less strict than under legislation based on the UPOV Convention, while the stability requirement is left out altogether. Only distinctness is preserved as it is provided for under such legislation, in recognition of the key importance of determining it for the operation of any plant variety protection system. Of course, the coverage of protection under these requirements would be much broader than under the UPOV Convention. Its implementation would require some safeguards to ensure that only varieties developed by a certain category of farmers (e.g. traditional farmers) are able to be protected since, otherwise, much of the output of commercial breeding may also be subject to protection. Further, although it would generally be in the interest of commercial breeders to have a strict set of conditions for PVP, such as they exist under the UPOV Convention, they may be in a better position than farmers/communities to use a broadly defined system of protection based on more relaxed requirements.

Rights conferred

Sui generis regimes may differentiate the rights granted in accordance with the type of plant varieties covered (NDUS-compliant varieties, farmers' varieties and so on). They may also provide exceptions that are different from those admitted under UPOV-based legislation. Thus, in Thailand, the PVP Act differentiates the protection that is accorded to the specific categories of new or local plant varieties from the protection that is granted to the general categories of domestic and wild varieties. The new plant varieties may be protected under exclusive rights in the same way that they are under UPOV-type regimes. A similar protection scheme applies to registered local community varieties.[41] General domestic and wild varieties are not eligible for exclusive rights, but rather are covered under a benefit-sharing mechanism. Permission by government officials is required in order to access these varieties for commercial purposes.[42]

In Malaysia, all protected varieties are subject to the exercise of exclusive rights generally available under PVP. In the case of India, extant varieties (including farmers' varieties) may be granted exclusive rights in a way that is similar to that of new varieties.[43] It is important to note, however, that Article 39.1(iv) of the Indian PPVFR Act provides an exception for the reuse of seeds that is broader than what is mandated in the UPOV Convention. It allows farmers to save, use, sow, resow, exchange, share and even sell farm produce, including the seed of a variety protected under the Act, provided that the seed is not branded.

There are reasons to think that the provision of exclusive rights that are equivalent to those available under conventional PVP but subject to less strict requirements of protection may lead to a great deal of legal uncertainty and litigation. Hence, if the standards of uniformity and/or stability are relaxed, the corresponding rights should generally be narrower (Genetic Resources Policy Initiative, 2006, 10). In addition, the exercise of exclusive rights might defeat the very purpose of some of the *sui generis* regimes, to the extent that farmers and communities may be prevented from continuing with the practices of exchange that are so important to conserve plant diversity and ensure a sustainable agriculture.

Prior consent and benefit sharing

Some of the *sui generis* systems proposed or adopted so far combine, in varying ways, exclusive rights with elements of benefit-sharing regimes aimed at the recognition and eventual compensation of farmers' innovations in accordance with the principles of the CBD. For instance, the Indian PPVFR Act provides for a detailed procedure for claiming compensation for benefit sharing (see Box 9.3).[44]

Box 9.3 Benefit sharing under Indian law

Rights of communities

41. (1) Any person, group of persons (whether actively engaged in farming or not) or any governmental or nongovernmental organisation may on behalf of any village or local community in India, file in any centre notified, with the previous approval of the Central Government by the Authority in the Official Gazette any claim attributable to the contribution of the people of that village or local community as the case may be in the evolution of any variety for the purpose of staking a claim on behalf of such village or local community.

(2) Where any claim is made under subsection (1), the centre notified under that subsection may verify the claim made by such person or group of persons or such governmental or nongovernmental organisation in such manner as it deems fit and if it is satisfied that such village or local community has contributed significantly to the evolution of the variety which has been registered under this Act, it shall report its findings to the Authority.

(3) When the Authority, on a report under subsection (2) is satisfied, after such enquiry as it may deem fit, that the variety with which the report is related has been registered under the provision of this Act, it may issue notice in the prescribed manner to the breeder of that variety and after providing opportunity to such breeder to file objection in the prescribed manner and of being heard, it may subject to any limit notified by the Central Government, by order, grant such sum of compensation to be paid to a person or group of persons or governmental or nongovernmental organisation which has made claim under subsection (1) to the Authority, as it may deem fit.
(4) Any compensation granted under subsection (3) shall be deposited by the breeder of the variety in the Gene Fund.
(5) The compensation granted under subsection (3) shall be deemed to be an arrear of land revenue and shall be recoverable by the Authority accordingly.

Source: PPVFR Act, available at: http://agricoop.nic.in/PPV&FR%20Act,%202001.pdf.

While the Indian law contains benefit-sharing elements, this law is perhaps closer to the PVP regimes than it is to access legislation. Other systems rely more heavily on the mechanisms introduced by the latter legislation. Thailand, for instance, 'has sought to provide other forms of incentives to breeders of domestic and farmers' varieties (i.e. it is closer to a liability regime than a property rights regime' (Robinson, 2007, 19).[45] Moreover, under the OAU Model Law, farmers would be given the right to 'obtain an equitable share of benefits arising from the use of plant and animal genetic resources' (Article 26).

The benefit-sharing mechanisms to which the Indian law refers may be orchestrated through bilateral agreements between the providers and users of genetic resources. Yet, in many cases, the transaction costs may be too high and the system too complex for farmers and communities. An alternative solution is to ask that payments be made into a fund that would subsequently distribute compensation to farmers or otherwise support them and their communities. For instance, under the Indian PPVFR Act,

> the farmer who is engaged in the conservation of genetic resources of landraces and wild relatives of economic plants and their improvement through selection and preservation shall be entitled in the prescribed manner for recognition and reward from the National Gene Fund.
>
> (Article 39.1(iii))

The Thai PVP Act also establishes a PVP fund. Its income is generated from the collection, use, research or commercialization of general domestic or wild varieties, registration fees, and so on. Similarly, the OAU Model Law envisions a 'community gene fund' (Article 66).

The application of this kind of benefit-sharing mechanism for farmers' varieties rather than the application of exclusive rights that are generally conferred under PVP has several advantages, including the fact that the varieties remain available for use and exchange by any farmer or breeder and that the procedures to obtain benefits would presumably be simpler than negotiating and enforcing a voluntary license case by case. While contributions to the existing pool of plant varieties will be rewarded, their diffusion would not be blocked or retarded on the grounds of infringement of such exclusive rights.

Conditions for the application of protection

Sui generis regimes may include obligations on applicants or rights holders that are not required under UPOV-based laws. Outstanding examples are the obligation to disclose information about the source of a plant variety for which protection is sought as well as the obligation to obtain the prior informed consent of the traditional farmers/communities who have developed/conserved the materials of origin. Thus, the Thai PVP Act makes the registration of a variety conditional upon the disclosure of the origin of either the new plant variety or the genetic materials used in the breeding of that variety (section 19(3)). In India, a breeder or other person making an application for registration of any variety must disclose information regarding the use of genetic material conserved by any tribal or rural families in the breeding or development of such a variety (Article 40(1)).[46] In Malaysia, applications for the registration of a new variety are subject to the 'prior written consent of the authority representing the local community of the indigenous peoples in cases where the plant variety is developed from traditional varieties' (section 12(1)(f)). In Egypt, the PVP law requires disclosure of the source of breeding material and traditional knowledge as well as the prior informed consent of the traditional knowledge holders.[47]

These types of conditions seem to be instrumental to the objectives of *sui generis* regimes that are aimed at preventing the misappropriation of genetic resources by commercial entities and at respecting the rights of local communities with respect to the plant varieties that they have developed and conserved.

Rights holders

A common criticism of PVP legislation has been the fact that rights can only be claimed by legally recognized natural or juridical persons and not by communities. This problem has been specifically addressed in the referred to *sui generis* regimes. For instance, under the Indian law, registration of a variety may be applied for individually or by a 'community of farmers claiming to be the breeder of the variety' (section 16(d)). In Thailand, section 44 of the PVP Act provides that a '*sui juris* person, residing and commonly inheriting and passing over culture continually, who takes part in the conservation or development of the plant variety' should appoint a representative for registration of the variety. The application should, inter alia, include the names of members of the community and 'the landscape together with a concise map showing the boundary of the community and adjacent areas.' Further, in accordance with section 45,

> when a plant variety only exists in any particular locality and has been conserved or developed exclusively by a particular community, that community shall have the right to submit, to the local government organization in whose jurisdiction such community falls, a request for initiating an application for registration of the local domestic plant variety in the name of such community.

Problems with the attribution of rights to a community need also to be addressed in a *sui generis* regime that covers a broad range of categories of plant varieties, including farmers' varieties.

Conclusions: developing a *sui generis* system

Establishing a *sui generis* regime for plant varieties poses significant technical, administrative and political challenges since many complex issues need to be addressed, and the adopted legislation will diversely affect different interest groups. Any attempt to develop a *sui generis* regime should be based on a careful assessment of the characteristics of the seed supply system, the role of private and public breeders, the structure of farm production, the farmers' capacity to potentially use and enforce a new legal system and other relevant evidence. The process followed to develop such a regime (including consultations with potential rights holders) may be crucial to understanding the different issues at stake and to drafting a set of rules that is efficient and instrumental for achieving its intended objectives. In undertaking this task, it would be crucial to clarify from the outset the rationale for granting such protection (beyond compliance with the obligation under Article 27.3(b) of the TRIPS Agreement) as well as the way in which the provided set of rights may maximize possible benefits and minimize the costs of introducing a new modality of IPRs.

The effectiveness of a legal regime based on intellectual property concepts (such as granting exclusive rights) to ensure the conservation, sustainable use and improvement of farmers' varieties should neither be presumed nor overstated. Such a regime would be irrelevant if other conditions are not met, particularly if farming communities are not able to keep their land and traditional practices. Indeed, too much emphasis on a solution based on IPRs may divert attention away from the factors that actually matter more to the preservation of plant diversity in the fields. In addition, granting exclusive rights may, under certain circumstances, be detrimental to the traditional practices of exchange and use of plant varieties and reduce, rather than promote, plant diversity and food security.

A key question to be addressed in designing a *sui generis* regime of plant variety protection that is intended to support farmers in the conservation, use and improvement of farmers' varieties is the extent to which different components of the overall national policy framework actually encourage such activities or, rather, stimulate the incorporation and use of commercial varieties. This would be the case, for instance, if the national policy actively promoted production for the supply of local or foreign markets demanding uniform agricultural products. In such cases, a *sui generis* regime of protection is very unlikely to provide by itself sufficient incentives to keep farmers' varieties in the fields. In particular, the interaction between different legal regimes may lead to unintended effects. Thus, while a *sui generis* regime for farmers' varieties may rely on a relaxed uniformity standard, farmers may be induced to develop more uniform varieties if this is required by the applicable seed certification law as a condition for the sale of seeds in the open market.

Even if the national policy framework were supportive of, or neutral to, the conservation, use and improvement of farmers' varieties, an outstanding question is whether a *sui generis* regime would be compatible with the culture and perceptions of its potential beneficiaries and, if such were the case, whether the possible financial benefits derived therefrom would be greater than the costs of acquiring, maintaining and, in particular, enforcing any rights conferred. Another important question is whether the required capacity exists within the country to set up and administer a new and complex system of rights.

Defining the taxonomy of varieties to be protected, the requirements of protection, the scope of rights and who may claim the conferred rights are some of the complex technical issues that need to be addressed in a coherent way in designing a *sui generis* regime. As noted, only a few national *sui generis* regimes have been adopted so far. Although some of them were enacted as early as 15 years ago, still little information has been made available from which to assess if they have been effective in attaining their objectives. It is also noticeable that the *sui generis* OAU Model Law did not make its way into the national laws of the various African countries. Yet this should not discourage governments from designing new *sui generis* regimes at the national or regional level; it only indicates that some caution is needed to embark on such an exercise. It should be borne in mind that the very nature of a *sui generis* regime requires consideration of the set of issues referred to earlier – namely, to have in view the particular context in which each regime is bound to apply. The issues at stake are too important to make decisions based on simple emulation or on unsubstantiated discourses of fear or hope about the impact of such regimes.

It has not been the purpose of this chapter to recommend a particular modality of *sui generis* regime nor the type of requirements that could be applied. As a general rule, however, it may be suggested that the more the NDUS standards are relaxed – which is necessary, in one way or another, in order to develop a *sui generis* regime – the narrower the rights conferred should be. It should also be borne in mind that a well-defined system of benefit sharing based on remuneration rights, without granting exclusionary rights, may suffice to compensate farming communities for their contributions to the conservation and improvement of plant varieties.

Notes

1 International Convention for the Protection of New Varieties of Plants, 2 December 1961, online: <www.upov.int/en/publications/conventions/index.html> [UPOV Convention].
2 In the United States, nearly 600 seed companies were operating by 1890. The American Seed Trade Association was established in 1883.
3 Plant Patent Act, 35 U.S.C. §§ 161–164.
4 With the participation of West Germany, Austria, Italy, Belgium, Spain and the Netherlands, with Denmark, Norway, and Switzerland as observers.
5 A basic difference between patent protection and plant variety protection (PVP) is that the latter allows for the use of a protected variety to develop and commercialize a new variety (breeder's exception). The 1978 UPOV Convention also allowed what is known as the 'farmers' privilege – that is, the right to save and use seeds obtained from the cultivation of protected varieties.
6 Plant Variety Protection Act, 7 U.S.C. §§ 2321–2382.
7 The United States has promoted this approach in the free trade agreements that have been signed with a number of developed and developing countries since 2000 (see, for example, Correa, 2009).
8 *In re Chakrabarty,* 571 F.2d 40, 197 USPQ 72 (CCPA), *cert. dism'd sub nom.*, 439 U.S. 801 (1978), *vacated*, 444 U.S. 1028, *aff'd*, 447 U.S. 303 (1980).
9 *Ex parte Hibberd*, 227 USPQ 443 (Bd. Pat. App. 1985).
10 Patents have been granted on the basis of claims relating to phenotypic characteristics or to a combination of phenotypic and genotypic characteristics. A trait identified or bred into plant lines may be claimed either phenotypically or genotypically.
11 Patent grants in the United States including plant cell and tissue culture technologies, enabling plant biotechnologies, genetic traits and germplasm showed strong growth trends after this decision (Boettiger et al., 2004, 1093).
12 Convention on the Grant of European Patents, online: <www.epo.org/patents/law/legal-texts/html/epc/1973/e/ma1.html>. The scope of this exclusion has been circumscribed by EC Directive 98/44 on the Legal Protection of Biotechnological Inventions, which states that 'inventions which concern plants or animals may be patented if the practicability of the invention is not technically confined to a particular plant or animal variety' (Article 4.2). In particular, the European Patent Office has allowed the patentability of transgenic methods and plants.
13 Agreement on Trade-Related Aspects of Intellectual Property Rights, Annex 1C of the Marrakech Agreement Establishing the World Trade Organization, 15 April 1994, 33 ILM 15 (1994).
14 It is to be noted that Article 27.3(b) obliges members to grant patents on microorganisms (provided they meet the corresponding patentability standards) but not on cells or subcellular parts, such as genes.
15 The International Seed Federation 'considers that the UPOV Convention, and particularly its 1991 Act, is an effective *sui generis* system for the protection of plant varieties' (International Seed Federation, 2003).
16 The admission of new members is subject to prior verification of compliance with the obligations under the UPOV Convention.
17 Since new plant varieties incorporate incremental improvements on existing varieties, they will rarely be patentable, unless the novelty and inventive steps are relaxed and the disclosure requirements are adapted. For instance, in Canada, a patent claim for a soybean variety applied by Pioneer Hi-Bred was rejected by the Supreme Court on the ground that no description of the method was available, although the seeds of the variety were deposited (see Judgments of the Supreme Court of Canada, online: <http://csc.lexum.umontreal.ca/en/1989/1989scr1–1623/1989scr1–1623>). The option of patenting plant varieties, thus, may in practice lead to the protection of a very small number of such

varieties. However, patent holders would enjoy stronger exclusive rights than under PVP. In addition, the peculiarities of patent laws (such as the mixed relative/absolute novelty requirement that was applied under US law) may allow the acquisition of patents over plant varieties developed by traditional farmers, thereby leading to one form of 'biopiracy' (Correa, 2002). In 2001, the US Supreme Court decided in *J.E.M. AG Supply, Inc. v Pioneer Hi-Bred Int'l, Inc.*, 534 U.S. 124 (2001), that sexually reproduced plants are statutorily proper subject matter for full utility patents.

18 Convention on Biological Diversity, 31 ILM 818 (1992).

19 International Treaty on Plant Genetic Resources for Food and Agriculture, 29 June 2004, online: <www.planttreaty.org/texts_en.htm>.

20 Many of these arguments have been contested in the literature on the subject (see e.g. Dutfield, 2000, 50–53). The concerns presented in this section are intended to provide the reader with a broad picture about the debates regarding this issue. The discussion on the merits of the various arguments is beyond the purpose of this chapter.

21 See an illustrative list of relevant literature in Annex I of this chapter.

22 Based on the intervention by Dan Leskien at the workshop held by the Genetic Resources Policy Initiative (GRPI) in Hanoi on 26–28 October 2006. Currently, however, there is no restriction to do so with farmers' varieties.

23 Based on a presentation by Dan Leskien at the workshop held by the GRPI in Hanoi on 26–28 October 2006.

24 On the cost of acquiring PVP on cultivars in some jurisdictions, see Tripp, Louwaars and Eaton, 2007, 363.

25 In recognizing the problems associated with the implementation of PVP for a broad number of crops, Article 4 of the 1978 UPOV Convention only required the gradual coverage of different crops. When the convention entered into force for a country, only a minimum of five crops had to be covered. This changed with the 1991 revision, which required new members to protect 15 genera or species upon accession and all genera and species within 10 years.

26 The treaty stipulates the following:

> Article 9.1 The Contracting Parties recognize the enormous contribution that the local and indigenous communities and farmers of all regions of the world, particularly those in the centres of origin and crop diversity, have made and will continue to make for the conservation and development of plant genetic resources which constitute the basis of food and agriculture production throughout the world.
>
> Article 9.2 The Contracting Parties agree that the responsibility for realizing Farmers' Rights, as they relate to plant genetic resources for food and agriculture, rests with national governments. In accordance with their needs and priorities, each Contracting Party should, as appropriate, and subject to its national legislation, take measures to protect and promote Farmers' Rights, including:
>
> a protection of traditional knowledge relevant to plant genetic resources for food and agriculture;
> b the right to equitably participate in sharing benefits arising from the utilization of plant genetic resources for food and agriculture; and
> c the right to participate in making decisions, at the national level, on matters related to the conservation and sustainable use of plant genetic resources for food and agriculture.
>
> Article 9.3 Nothing in this Article shall be interpreted to limit any rights that farmers have to save, use, exchange and sell farm-saved seed/propagating material, subject to national law and as appropriate.

27 Protection of Plant Varieties and Farmers' Rights Act, online: <http://agricoop.nic.in/seeds/farmersact2001.htm> [PPVFR Act]. On the various approaches and proposals

that influenced this legislation (including, in particular, the draft Plant Variety Recognition and Rights Act of the M.S. Swaminathan Foundation), see Dhar and Chaturvedi, 1998, 248–49.

28 Article 2(j): 'extant variety' means a variety available in India which is:

(i) notified under section 5 of the Seeds Act, 1966 (54 of 1966);
(ii) farmers' variety;
(iii) a variety about which there is common knowledge; or
(iv) any other variety which is in the public domain.

29 Plant Variety Protection Act, online: <www.grain.org/brl_files/thailand-pvp-1999-en.pdf> [PVP Act].
30 Protection of New Plant Varieties Act 2004, online: <www.grain.org/brl/?docid=657&lawid=1404>.
31 In the 1950s, the 'AIPPI [International Association for the Protection of Intellectual Property] opposed the patenting of plant varieties on the grounds that doing so would stretch basic patent law concepts like inventiveness to the point of undermining the credibility of the patent system' (Dutfield, 2003, 186).
32 It is worth noting, however, that unlike the concept of novelty under patent law, a variety continues to be 'novel' in the context of PVP even if it has been known for a long time as long as it was not sold or commercialized with the consent of the breeder for some specified periods before the filing date of the application for protection (see e.g. Article 6 of the 1978 UPOV Convention).
33 Section 12(2) of the act requires that distinctness be 'related to the feature beneficial to the cultivation, consumption, pharmacy, production or transformation, including the distinctness from the following plant varieties: (a) plant varieties already registered and protected, whether in or outside the Kingdom, prior to the date of filing the application; (b) plant varieties in respect of which application for registration has been made in the Kingdom and which will subsequently have been registered.'
34 The concept of novelty differs, however, from that contained in the 1991 UPOV Convention as disposal of the variety only affects its possible protection if made 'on a commercial basis' (section 14(3)(a)). This change may permit the protection of existing local varieties insofar as they have not been commercialized.
35 See the presentation by Lim Eng Siang at the GRPI meeting in Hanoi (Genetics Resources Policy Initiative, 2006, 17–18).
36 African Model Law for the Protection of the Rights of Local Communities, Farmers and Breeders and for the Regulation of Access to Biological Resources, online: <www.cbd.int/doc/measures/abs/msr-abs-oau-en.pdf>.
37 Many commercial varieties are bred with the deliberate aim of obtaining a high degree of uniformity in order to respond to market demands and/or to comply with the seed certification or PVP requirements.
38 Landraces, which constitute a major component of the pool of plant genetic resources available to farmers, have been defined by the Food and Agricultural Organization 'as an early, cultivated form of a crop species, evolved from a wild population, and generally composed of a heterogeneous mixture of genotypes.' See Biotechnology in Food and Agriculture, online: <www.fao.org/biotech/find-formalpha-n.asp>.
39 This is not problematic for farmers' varieties, which may be differentiated following conventional methods.
40 This standard means that each generation of a plant variety must be identifiable as the same distinct plant variety, without necessarily being uniform in all of its characteristics (Leskien and Flitner, 1997).
41 A community may be granted exclusive rights to conserve, use, research, sell and commercialize a registered plant variety.

42 In accordance with section 52 of the law, 'a person who collects, procures or gathers general domestic plant varieties, wild plant varieties or any part of such plant varieties for the purposes of variety development, education, experiment or research for commercial interest shall obtain permission from the competent official and make a profit-sharing agreement under which the income accruing therefrom shall be remitted to the Plant Varieties Protection Fund.'

43 This solution has raised concerns since some farmers may hypothetically exclude other farmers from using widely diffused varieties that are not novel and thereby jeopardize the traditional exchange of seeds (Robinson, 2007, 24).

44 The PPVFR Act, supra note 27, issued by the Ministry of Agriculture in December 2006, implements benefit sharing for farmers and communities, in cases where their genetic resources have contributed to third parties' new variety development (Form 1, Part 10(c)).

45 Under a liability regime no exclusive rights are granted; hence, any party may use the protected subject matter against payment of a remuneration to the title holder.

46 The PPVFR Act, supra note 27, requires information about the origin of the variety including geographical source and farmer/village/community/institution/organization (Form 1, Part 10(b)).

47 Owing to this obligation, the International Union for the Protection of New Varieties of Plants has refused to recognize Egypt as being in compliance with the UPOV Convention. However, the obligation is a condition for the application and not an additional condition for protection (Genetic Resources Policy Initiative, 2006, 18).

References and bibliography

Boettiger S., G. Graff, P.G. Pardey, E. van Dusen and B.D. Wright (2004). 'Intellectual Property Rights for Plant Biotechnology: International Aspects,' in P. Christou and H. Klee (eds.), *Intellectual Property Rights for Plant Biotechnology: International Aspects. Handbook of Plant Biotechnology,* John Wiley and Sons, Chichester, Chapter 56.

Chawla, H.S. (2003). *Introduction to Plant Biotechnology*, Springer Publishers, The Netherlands.

Correa, C. (2000). *Options for the Implementation of Farmers' Rights at the National Level,* working paper, South Centre, Geneva.

Correa, C. (2002). *Protection and Promotion of Traditional Medicine: Implications for Public Health in Developing Countries,* South Centre, Geneva.

Correa, C. (2009). *Trends in Intellectual Property Rights Relating to Genetic Resources,* Food and Agriculture Organization, Rome. Available at: ftp://ftp.fao.org/docrep/fao/meeting/017/k533e.pdf (last accessed 15 June 2012).

Crucible Group II (2001). *Seeding Solutions, Volume 2: Options for National Laws Governing Control over Genetic Resources and Biological Innovations,* International Development Research Centre, Ottawa.

Dhar, B. (2002). *Sui Generis Systems for Plant Variety Protection: Options under TRIPS.* Available at: www.qiap.ca/documents/SGcol1.pdf (last accessed 15 June 2012).

Dhar, B., and S. Chaturvedi (1998). 'Introducing Plan Breeders' Rights in India. A Critical Evaluation of Proposed Legislation,' *Journal of World Intellectual Property* 1: 245–62.

Dutfield, G. (2000). *Intellectual Property Rights, Trade and Biodiversity,* Earthscan, London.

Dutfield, G. (2003). *Intellectual Property Rights and the Life Science Industries,* Ashgate, London.

Eyzaguirre, P., and E. Dennis (2007). 'The Impacts of Collective Action and Property Rights on Plant Genetic Resources,' *World Development* 35(9): 1489–98.

GAIA/GRAIN (1998). 'Ten Reasons Not to Join UPOV,' *Global Trade and Biodiversity in Conflict*. Available at: www.grain.org/briefings/?id=1 (last accessed 15 June 2012).

Genetic Resources Policy Initiative (2006). *Exploring Legal Definitions of Farmers' Varieties in Strategies to Promote Farmers' Rights,* 26–28 October, Hanoi, Vietnam.

International Seed Federation (June, 2003). *View on Intellectual Property,* Bangalore. Available at: www.worldseed.org/isf/on_intellectual_property.html (last accessed 15 June 2012).

International Union for the Protection of New Varieties of Plants (2005). *Report on the Impact of Plant Variety Protection,* UPOV Publication no. 353(E), Geneva.

Kesan, Jay P., and Mark D. Janis (2002). *U.S. Plant Variety Protection: Sound and Fury?* University of Illinois Working Paper no. LE03–002. Available at: http://papers.ssrn.com/s013/papers.cfm?abstract_id=384140 (last accessed 15 June 2012).

Leskien, D., and M. Flitner (1997). *Intellectual Property Rights and Plant Genetic Resources: Options for a Sui Generis System,* Issues in Genetic Resources no. 6, International Plant Genetic Resources Institute, Rome.

Perrin, R.K., K.A. Hunnings and L.A. Ihnen (1983). *Some Effects of the U.S. Plant Variety Protection Act of 1970,* Economics Research Report no. 6, Department of Economics and Business, North Carolina State University, Raleigh, NC.

Robinson, D. (2007). *Exploring Components and Elements of Sui Generis Systems for Plant Variety Protection and Traditional Knowledge in Asia,* International Centre for Trade and Sustainable Development, Geneva.

Salazar, R., N. Louwaars and B. Visser (2007). 'Protecting Farmers' New Varieties: New Approaches to Rights on Collective Innovations in Plant Genetic Resources,' *World Development* 35(9): 1515–28.

Singh, H. (2006). *Emerging Plant Variety Legislations and Their Implications for Developing Countries: Experiences from India and Africa.* Available at: www.iprsonline.org/ictsd/docs/ResourcesTRIPSharbir_singh.doc (last accessed 15 June 2012).

Tripp, R., N. Louwaars and D. Eaton (2007). 'Plant Variety Protection in Developing Countries: A Report from the Field, *Food Policy* 32: 354–71.

Van Overwalle, G. (1999). 'Patent Protection for Plants: A Comparison of American and European Approaches,' *IDEA – Journal of Law and Technology* 39(2): 143–94.

Annex I: selected bibliography on the impact of PVP

Alston, J., and R. Venner (2000). *The Effects of the U.S. Plant Variety Protection Act on Wheat Genetic Improvement.* Environment and Production Technology Discussion Paper no. 62, International Food Policy Research Institute. Available at: www.ifpri.cgiar.org/divs/eptd/dp/eptdp62.htm (last accessed 15 June 2012).

Butler, L.J., and B.W. Marion (1985). *The Impacts of Patent Protection on the US Seed Industry and Public Plant Breeding,* Food Systems Research Group Monograph no. 16, University of Wisconsin, Madison.

Carew, R., and S. Devadoss (2003). 'Quantifying the Contribution of Plant Breeders' Rights and Transgenic Varieties to Canola Yields: Evidence from Manitoba,' *Canadian Journal of Agricultural Economics* 51: 371–95.

Commission on Intellectual Property Rights (2002). *Integrating Intellectual Property Rights and Development Policy,* September, London. Available at: www.iprcommission.org (last accessed 15 June 2012).

Foster, W.E., and R. Perrin (1991). *Economic Incentives and Plant Breeding Research,* Faculty Working Papers, Department of Agricultural Economics, North Carolina State University, Raleigh.

Frey, K.J. (1996). *National Plant Breeding Study. Volume 1: Human and Financial Resources Devoted to Plant Breeding Research and Development in the United States in 1994,* Special Report no. 98, Iowa Agriculture and Home Economics Experiment Station, Iowa State University, Ames.

Jaffe, Adam B., and J. Lerner (2004). *Innovation and Its Discontents: How Our Broken Patent System Is Endangering Innovation and Progress, and What to Do About It,* Princeton University Press, Princeton, NJ.

Knudson, M., and C. Pray (1991). 'Plant Variety Protection, Private Funding, and Public Sector Research Priorities,' *American Journal of Agricultural Economics* 73(3): 882–6.

Kolady, D.E., and W. Lesser (2009). 'But Are They Meritorious? Genetic Productivity Gains under Plant Intellectual Property Rights,' *Journal of Agricultural Economics* 60(1): 62–79.

Kuyek, D. (2001). *Intellectual Property Rights: Ultimate Control of Agricultural R&D in Asia,* Grain. Available at: www.grain.org/briefings/?id=35 (last accessed 15 June 2012).

Marion, B.W. (1996). 'Plant Breeders' Rights in the United States: Update of a 1983 Study,' in J. Van Wijk and W. Jaffé (eds.), *Intellectual Property Rights and Agriculture in Developing Countries,* University of Amsterdam, Amsterdam, pp. 17–33.

Moschini, G., and O. Yerokhin (2007). *The Economics Incentive to Innovate in Plants: Patents and Plant Breeders' Rights,* Staff General Research Papers no. 12895, Department of Economics, Iowa State University, Ames.

Naseem, A., J.F. Oehmke and D.E. Schimmelpfenning (2005). 'Does Plant Variety Intellectual Property Protection Improve Farm Productivity? Evidence from Cotton Varieties,' *AgBioForum* 8: 100–7.

Perrin, R.K., K.A. Hunnings and L.A. Ihnen (1983). *Some Effects of the U.S. Plant Variety Protection Act of 1970,* Economics Research Report no. 6, Department of Economics and Business, North Carolina State University, Raleigh, NC.

Rangnekar, D. (2000). *Intellectual Property Rights and Agriculture: An Analysis of the Economic Impact of Plant Breeders' Rights,* Action Aid UK. Available at: www.actionaid.org.uk/_content/documents/ipr.pdf (last accessed 15 June 2012).

Rangnekar, D. (2000). *Access to Genetic Resources, Gene-based Inventions and Agriculture, Access to Genetic Resources, Gene-based Inventions and Agriculture, Commission on Intellectual Property Rights,* Study Paper 3a. Available at: www.iprcommission.org/graphic/documents/study_papers.htm (last accessed 15 June 2012).

Singh, H. (2006). *Emerging Plant Variety Legislations and Their Implications for Developing Countries: Experiences from India and Africa.* Available at: www.iprsonline.org/ictsd/docs/ResourcesTRIPSharbir_singh.doc (last accessed 15 June 2012).

Smale, M., J. Hartell, P.W. Heisey and B. Senauer (1998). 'The Contribution of Genetic Resources to Wheat Production in the Punjab of Pakistan,' *American Journal of Agricultural Economics* 80: 482–93.

Srinivasan, C.S, B. Shankar and G. Holloway (2003). *An Empirical Analysis of the Effects of Plant Variety Protection Legislation on Innovation and Transferability.* Available at: http://ecsocman.edu.ru/images/pubs/2003/11/29/0000135447/039–049-srinivasanx2cx20shankarx2cx20holloway.pdf (last accessed 15 June 2012).

Tripp, L., and D. Eaton (2007). 'Plant Variety Protection in Developing Countries: A Report from the Field,' *Food Policy* 32: 354–71.

van Wijk, J. (1996). 'How Does Stronger Protection of Intellectual Property Rights Affect Seed Supply? Early Evidence of Impact,' *ODI Natural Resources Perspectives,* no 13, Overseas Development Institute, London.

van Wijk, J., and W. Jaffé (1995). *The Impact of Plant Breeders' Rights in Developing Countries: Debate and Experience in Argentina, Chile, Colombia, Mexico, and Uruguay*, University of Amsterdam, Amsterdam.

van Wijk, J., and W. Jaffé (1996). *Intellectual Property Rights and Agriculture in Developing Countries*, University of Amsterdam, Amsterdam.

10 Variety registration

The evolution of registration systems with a special emphasis on agrobiodiversity conservation[1]

Niels Louwaars and François Burgaud

Introduction

Variety registration procedures serve various purposes, and their origin and current application in different countries have a differential impact on the number of varieties available to farmers and the total genetic diversity that they represent. Most of the models for variety registration that have been developed to date do not recognize specifically farmers' varieties such as traditional landraces or the new varieties developed by farmers in participatory plant breeding. Seed regulation and variety registration, in particular, were developed with the objective of increasing crop production by safeguarding farmers from purchasing bad quality seed of poorly performing varieties. When the archetype seed laws were originally developed – the same ones that many developing countries have copied – policy makers were not concerned with diversity conservation or how the use of farmers' traditional varieties could play a role in the economic development of rural areas and specific local markets. The development of variety registration systems shows that they have evolved together with farmers' changing needs, including current concerns about diversity.

This chapter deals with the origins of variety registration regulations, their implementation and their impacts on the use and improvement of farmers' varieties. In this chapter, we document how variety registration systems have developed in Europe and the United States and how developing countries have adopted these models. Then we discuss how variety registration and *sui generis* models for the protection of plant varieties are linked. In addition, we discuss the operation of the current systems and how strict compulsory registration systems may outlaw the use, or at least the marketing, of the seeds from farmers' varieties. We then examine attempts in Europe to correct this restriction and to merge the objectives of the seed and biodiversity policies. While these initiatives appear to have the potential for a positive impact on the use of farmers' varieties, it is too early to assess their actual impact.

The origins of variety registration

The development of plant breeding

Agriculture has been a central element of human history for some 10,000 years. Through millennia of selecting, sowing and harvesting, plants have been domesticated and bred to have traits that respond to the consumption needs (e.g. larger seeds) and the requirements of farming (e.g. nonshattering, erect plant architecture) – traits that distinguish the crops from their wild ancestors (Zeven and de Wet, 1982). By the early 1800s, agricultural techniques and cultivars had been improved, and seed production had started to become a business in several countries. In 1900, Gregor Mendel's work on trait inheritance in peas was rediscovered, which created the foundation for the science of genetics and modern plant breeding; this sped up the creation of new varieties and formed the basis of hybrid technology during the early twentieth century.

Variety registration procedures developed along with advances in plant breeding. Prior to the rise of commercial seed production, which developed as farmers and farmer groups specialized in seed production, there had been no seed laws and no variety registration laws in particular. Of course, farmers have always been equally concerned about obtaining good seed and/or varieties with desirable traits. However, since farmers have traditionally obtained seed from saved stocks, through exchange with neighbours and relatives, or through local markets, issues related to 'consumer protection' and free-riding copycat seed supply did not merit the regulatory apparatus that later developed. Not surprisingly, therefore, the organization of plant breeding and seed sectors, as well as the policy makers that oversee these sectors, have determined to a large extent how these regulations have been framed.

Specialized plant breeding developed in Europe in the mid-nineteenth century, along with the emergence of seed production as a specialized business. Those selling seed started to promote good quality seed (i.e. germination capacity and purity). Since farmers cannot distinguish most of the qualities of good seed simply by looking at it, countries started to establish seed-testing stations from the 1860s onward in order to provide an independent quality assessment. In addition, the emerging seed companies sought to create added value by selling distinguishable selections (varieties) with useful agronomic qualities. These efforts by the seed producers had built upon the methods that farmers had used for years to select particular traits in their crops – that is to say, mass selection. In the latter part of this century, however, they engaged in what was then a new approach to plant breeding – pedigree selection and cross-breeding – which allowed for quicker advances and more uniform crops (see Box 10.1).

Box 10.1 Plant breeding and registration in France

The history of wheat breeding in France

The history of wheat breeding in France retraces the way in which the agricultural world has been collectively responding to multiple demands that have evolved over the years.

In 1944, Jean Bustarret, chief inspector and then director of the Institut National de la Recherche Agronomique (INRA), noted that at the beginning of the nineteenth century about 40 varieties of wheat existed in France that were 'more or less heterogeneous, often of an unknown origin, well adapted as well as ill-adapted to the regional environment' (Bustarret, 1944). Called 'domestic wheat,' these varieties were the double fruit from natural selection and the conscious mass selection carried out by farmers either in the field or when sorting the seeds. At this time, the main concern was to obtain varieties that were sufficiently productive and, above all, had yield stability. The best varieties of wheat yielded 12–15 metric quintals per hectare (i.e. 1.2–1.5 tonnes).

Starting from the first half of the nineteenth century, one variety (Noé) had established itself as the principal genitor of a wide variety of subsequently developed varieties. Noé had been discovered in 1826 in a field sowed in Nérac with seed from Odessa (in present-day Ukraine). It was cultivated in the winter and spring. The farmers realized that it was sturdier and less sensitive to the rain than the domestic varieties of wheat. Through cross-breeding, Noé became one of the most important progenitors of the varieties known as the 'wheat of Aquitaine,' including the Rouge de Bordeaux (red of Bordeaux), the Japhet, and the Gros Bleu (big blue).

From 1860 to 1880, France was the setting for one of the first agricultural revolutions, initiated by Henry de Vilmorin. He was responsible for an entire series of new varieties (Dattel, Bordier, Massy and Trésor), whose progenitors included the wheat of Aquitaine and varieties of wheat of British origin (Victoria white, Chiddam, red of Schotland [Rouge d'Ecosse] and Prince Albert), which were cultivated essentially in the north of France. As Robert Mayer, the former leader of the genetics and plant improvement station at Inra-Versailles stated: 'What was truly a determining factor for the progress of plant improvement was the method utilized, which, at that time, was revolutionary: the creation of populations resulting from artificial hybridizations after a judicious selection of genitors.'[2]

At the beginning of the twentieth century, these cultivated varieties were thus no longer the simple outcome of the observation of the farmers, but rather were developed by seed selectors who utilized the most advanced techniques of the times. These new varieties were readily adopted, and in 1926 a survey by Emile Schribaux indicated that the Vilmorin varieties of wheat represented 39 percent of the national wheat cultivation, compared to domestic varieties, which represented 35 percent. However, the breeders, captivated by these new French varieties and ignoring the foreign genitors, were making very little progress. Only Emile Schribaux provided any innovative ideas. He proved to be very concerned with this genetic impoverishment and wrote in 1928: 'We do not adequately realize the paramount importance that the material drawn from abroad could have for us, which is a real gold mine in which it would often be possible to discover, and almost without any effort, varieties that are likely to be the most useful and most precious for us' (Schribaux, 1928). Along with Charles Crepin, he introduced cross-breeding material from central and southern Europe as well as from North America, such as Oro, Thatcher and Martin. This diversification was also pursued after the Second World War, in particular by the Maison Desprez. In the 1980s, Nickerson, which was at that time a subsidiary of Royal Dutch Shell, put in place a pattern of selection of European varieties of wheat by setting up stations in Great Britain, France, Germany and Spain. During the same period, the INRA introduced resistances to the diseases originating from species other than wheat (*Aegilops ventricosa* and *Triticum cathlicum*) into its new varieties of wheat. In this manner, each important stage of progress in the evolution of varieties has been initiated by the contribution of genetic characteristics coming from lines that were very different from the ones present in the local territory.

The history of the variety catalogue in France

During the same period that wheat breeding was evolving in France, there was a parallel, incremental development in recordkeeping and the registration and marketing of plant varieties. Initially, these efforts were carried out by private experts in an attempt to systematize what they saw going around them 'in the field' as a means of recording a baseline and taking note of 'genetic progress.' In the mid-nineteenth century, M. de Vilmorin (and later his son) published a descriptive and comparative catalogue of wheat. The catalogue was based on existing European wheat types and proposed a first classification. These descriptions were useful for plant breeders in their breeding programs and for farmers who needed to choose wheat that was adapted to local environmental conditions.

On 8 December 1922, the public sector got involved in order to create transparency in a market where the customer was not able to identify the qualities of the product (seed) by simply looking at it. The first 'register of selected plants' on cereals was created by a ministerial decree, fixing the registration conditions of new varieties. Only new species or varieties that were identifiable and that showed 'unquestionable progress' could be registered. Independent field tests were carried out by the INRA to evaluate the nature and the value of the innovation through an agronomic evaluation. The committee that managed the register also established a 'synonym' catalogue.[3] This evaluation involved 2 years of experimentation, and even though the concept of distinctness, uniformity and stability was not officially used, the concept of distinguishing a new variety from the known ones was being applied.

The INRA was also mandated to develop an administrative application examination. It was specified that the applicant should pay for the expenses incurred by the varietal examination. The decree ruled that the registration of a variety would last for 6 years and be renewable. A committee was installed, composed of eight permanent members from the administration and from the public, and 12 'private' members who were nominated for 5 years, including four representatives from commercial seed companies. Any person who had bred, invented or discovered a species or a variety could apply for registration. The decree thus provided important information for farmers on the innovative characteristics of the varieties and their agronomic values, which was based on official tests. At the same time, it created a 'brand' protection for the breeder at a time when intellectual property rights were not being applied to varieties. It gave two rights to the applicant: (1) exclusive use of the variety denomination and (2) the exclusive ability to put 'seeds registered within the register of selected plants' on the seeds. Registration was not compulsory.

A ministerial decree of 1925 was developed to prevent fraud in the wheat seed trade. It required seed to be transported, marketed and sold with packaging that included (1) the variety name and average seed number, (2) the source of the seeds (French area name or country and area of import), (3) the guarantee that the seed would not contain more than 1 percent of another variety. The variety name had to be similar to the 'usual denomination in conformity with local, honest and constant uses' or in conformity with the synonym catalogue or the register of selected plants. The germination rate could not be less than 85 percent. The methods of sampling and analysis were fixed and could not be changed. Any other seed trade was forbidden. The report to the president of France, which introduced the decree, stated clearly:

> The agriculture administration does its best efforts to improve the yield of wheat crop production . . . The farm production is essentially linked with the quality of seeds . . . In itself the regular use of good seeds would be sufficient to increase our production with 10 to 15% and this enormous increase would be 100% beneficiary because the cultivation of a good variety is not more costly than a bad one![4]

In 1932, it became mandatory to register varieties in order to market seeds.[5] Requests for registration in the catalogue had to provide a detailed description of the variety, enumerating its agricultural characteristics and indicating the origin and the breeding process for the species or the variety and providing a sample (seed, cutting, tuber, bulb or graft). When necessary, the applicant had to give the synonym name according to fair and traditional practices. In 1933, the Register for Wheat Varieties was published. In the years that followed, registers for other species were published – in 1935–37 for major field crops and from 1944 onwards for a growing number of vegetable species and later for fruits and grapevines. A ministerial decree in 22 January 1960 instituted a catalogue of species and varieties of crop plants, which identified the varieties that were available in the market and characterized the innovations that had resulted from the various breeding programs.

Meanwhile, the certification of seed had also become commonplace and finally compulsory. Certification involved field inspection, seed sampling, and testing and labelling requirements. In order to perform seed certification, a variety had to be identifiable and thus stable (meaning that it was identifiable over time). In order for a variety to be stable, it had to be sufficiently uniform since genetically diverse landraces are bound to develop further based on natural and farmers' selection. The variety registration requirements went hand in hand with seed certification leading towards identifiability (distinctness), stability and thus uniformity.

Meanwhile, there were many efforts to harmonize variety registration rules at the European regional level. France implemented the relevant European Union directives related to seed and seedling trade by virtue of a national decree, issued by the minister of agriculture in May 1981.[6] This decree established, among other things, that in order to be registered in the national catalogue, varieties had to be distinct, sufficiently uniform and stable, based on standardized criteria that had been developed and standardized, in the meantime, for breeders' rights by the International Convention for the Protection of New Varieties of Plants (UPOV Convention), established in 1961.

One underlying objective of all of these developments – as the catalogues went from private contributions to mandatory registration schemes with legally established criteria for registration – is that they offered transparency in the seed market by linking variety names to well-described varieties that had particular agronomic and use values.

These methods were applied with the aim of maximizing the positive alleles in the population – that is, creating more (cross-fertilizing crops) or very uniform (self-pollinators) varieties. The results of breeding could only be verified when the varieties were able to be distinguished from one another. An account of the agricultural fair in Malmö, Sweden, in 1898 states that the breeders had created 'hundreds of distinct varieties' (Nilsson, 1898). Although plant breeding originated in the private sector, primarily through cooperatives (e.g. Svalöf in Sweden), it was in the public sector that it became a thriving undertaking in most European countries in the early twentieth century. This development, however, did not occur quickly. For example, the Agricultural School in Wageningen, the Netherlands (later Wageningen University), established its Institute for Plant Breeding in 1912 – well over a decade after the rediscovery of Mendel's laws of inheritance by Hugo de Vries and the recognition of these laws by the scientific community.

This trend towards uniform varieties became even more pronounced after Mendel's laws of inheritance were applied to self-fertilizing cereals and legumes in the early twentieth century. It took some time before knowledge about quantitative genetics could explain the behaviour of these populations of cross-fertilizing crops. The discovery of hybrid vigour in maize in the 1920s – that is, the advantages of first generation offspring of different parents based on the value of genetic diversity (alleles) within plants – spurred commercial seed production in the United States, beginning with H. Wallace's establishment of the Pioneer Hi-Bred Company in 1929. The uniformity of varieties and hybrids became more important with the gradual mechanization of agriculture throughout North America and Europe from the 1930s onward.

Unlike the development of plant breeding in Europe, specialized plant breeding originated in the United States in the public sector, following the creation of the land grant colleges in 1887 – in particular, the experimental stations (see Box 10.2).

Box 10.2 Plant breeding and registration in the United States

Specialized plant breeding did not start in the United States in the private sector as it did in Europe, but was established in the public sector following the Land Grant College Act of 1862, which established institutions for agricultural education on a total of 11 million acres of public land. This legislation was followed in 1887 by the Hatch Agricultural Experiment Stations Act, which included a system of federal budget allocations for research activities in each state. The first federal seed regulations were issued in 1905, but a full federal seed act was published in 1939. This act, which was opposite in many ways to its European counterparts, dealt primarily with labelling requirements for seed. These requirements included information about the supplier, the seed quality (germination percentage and so on) and even the variety's name. With regard to 'variety,' the only limitation that the law specified was that a group of plants had to be distinct from other such groups.

At the state level, however, different rules were developed in the different states, and seed certification and testing systems were gradually initiated. In most states, these systems are voluntary compared to the compulsory systems that are predominant in Europe. This means, in practice, that seed producers who see a commercial benefit in attaching a certification label to their seed will need to go through the cost and trouble of having their crop fields inspected and their seeds sampled and tested by a registered certification agency. An association of seed certification agencies from several states (and also the one in Canada) was formed in 1919 in order to share experiences; this led to the harmonization of their methodologies. In addition, varieties had to be distinct and sufficiently uniform in order to certify their identity at the end of the generation system. This standard, however, is a result of implementation rules from the certification agencies rather than of the seed law. Criteria of distinctness, uniformity and stability, in line with the UPOV Convention, are applied, however, if a breeder applies for breeders' rights.

The origins of variety registration

Variety registration started in Europe and the United States in order to create transparency in the marketplace. The different seed suppliers used the same names for different varieties: when a variety became popular, it was tempting to sell any seed under that popular name. The opposite also happened: using different names for the same variety was popular to create a brand name for the seed company based on a variety that was bred by a competitor. This confusion led to a call from the farmers' associations to the governments to develop a system that would protect them from misrepresentation. In order to create a variety list, it was necessary to be able to identify the different varieties and to validate the claims of the seed producers with respect to their value for farmers. This need led to coordinated variety trials for agronomic value organized by farmers' associations (the Netherlands) or by the government (France), combined with some form of morphological description. Thus, variety registration was able to serve an additional objective – namely, to provide an independent source of information for farmers on the value of the plant for cultivation and use. Variety testing became a specialized procedure separate from plant-breeding activities. The first official catalogues of recommended varieties appeared in the 1920s in different European markets. To our knowledge, however, the first seed law was the federal seed law of the United States in 1905, which concentrated entirely on consumer (farmer) protection by introducing compulsory labelling.

In order to implement this transparency in the market, farmers had to be sure that when they found a registered name on the label of the seed bag, the seed would indeed be of that variety. Seed-testing services for germination and physical purity began to develop in the 1860s, yet guarantees on the varietal identity and purity through what became known as seed certification emerged only in the 1950s. Grow-out tests to validate the variety identity (post control) and field inspections preceded the concept of certification through a regulated generation system that continues today. The generation system (from breeder's seed to certified seed) with a strong emphasis on variety maintenance by the breeder aims at securing varietal identity and uniformity throughout the seed multiplication phases. In order to perform a reliable seed certification, it is extremely important to have a detailed morphological description for each variety. The identification of off-types (plants that do not belong to the variety) can only be done well in uniform crops and, thus, the process of seed certification provides an additional institutional incentive for varietal uniformity.

The compulsory systems for variety registration and seed certification that were developed in Europe for political goals (mainly food security) and for consumer (farmer) protection were maintained even after the farmers started to become more educated and business oriented. One of the main reasons why these systems were maintained was that they created a level playing field for the seed companies. Since the seed companies were able to obtain an independent and publicly available review of the value of their varieties, they were less dependent on the marketing powers of the larger seed companies.

These trends were different in most parts of the United States where most varieties were bred and named by the public breeding stations at the universities, which did their variety trials as part of the breeding process. Following the hybrid revolution in maize, however, the system began to change, and there was an increased development of commercial plant-breeding companies that began to develop their own hybrid maize varieties. The public sector in the United States still has a significant role in breeding cereals and legume crops. Developments in other industrialized countries (notably Australia, New Zealand and Japan) are in most aspects somewhere in the middle of the United States and Europe.

These differences in origin between the two regions may be regarded as significant in explaining the different approaches taken on opposite sides of the Atlantic. In Europe, the governments were called to protect farmers through obligatory variety lists (and seed certification) from commercial seed producers misrepresenting the value of their varieties. In the United States these quality control systems remained voluntary initially because it was felt that the public breeding programs did not require the same level of scrutiny to protect farmers' interests. In addition, it is important to realize that such regulation was felt to be unnecessary in the free market paradigm that was much stronger in the United States than it was in Europe in the twentieth century.

We can conclude that the history of seed regulation in Europe is a history of a joint concern by governments, farmers and the emerging seed producers for food security and sustainable rural development. Food security was a vital element in US policies in the nineteenth century when experimental stations were first established. In Europe, the security of food remained important due to the famines that occurred during the Second World War. Above all, the variety regulations were developed to advance agricultural production by (1) creating transparency in the emerging seed market; (2) making available to farmers information on crop varieties; and (3) creating a level playing field for the emerging seed industry.

Variety registration trends in developing countries

Copying variety registration regulations

As was the case in Europe and the United States, seed regulatory frameworks in developing countries have commonly been developed after the emergence of a formal seed sector. From the 1970s onwards, there was an especially vigorous development in order to make it possible for as many farmers as possible to take advantage of the benefits of the varieties emerging from the Green Revolution. The Seed Industry Development Programme, led by the Food and Agriculture Organization (FAO), assisted many countries in setting up seed farms, contract grower schemes and seed conditioning plants for their major food crop seeds (Feistritzer, 1984). These formal seed systems subsequently developed specialized in-house or independent seed quality control institutions – similar to the official seed certification agencies in the North – to create quality awareness among both seed producers and customers and to safeguard the interests of farmers. In an era when public institutions were being readily privatized, such as at the end of the 1980s, these seed quality control institutions became a driving force behind the development of seed legislation (Louwaars and van Marrewijk, 1996). Such legislation was intended to provide these institutions with the legal mandate and backing that was considered necessary for them to perform their task of policing seed production and marketing, especially in relation to the emerging private seed sector. Variety release systems, which had initially been developed as a kind of final stage of the breeder's selection process, became regulated as well, partly because varietal identity was an important basis for seed certification.

Since the seed programs in developing countries were built using the effective formal seed systems in the North as their example, many national seed laws were also developed on the basis of these examples. In his discussion of this development, L. Bombin (1980) shows that the first Latin American seed laws were based on the Spanish example; that francophone African countries derived legislation from the French seed law; and that Commonwealth country seed laws resembled British or US regulations depending on which country was supporting their seed sector (the first Indian law adopted the US system). Subsequent changes, in some cases, have included issues that are specifically important for regulating seed systems in development, but the majority of seed laws in developing countries have focused on the formal sector only, making (support to) informal seed systems illegal (Louwaars, 2005).

Specific needs

In most developing countries, the formal seed systems provide only a small percentage of the seeds that farmers use, with the exception of some cash crops (e.g. vegetables) and hybrids (maize and pearl millet). This limited use is a result of several different factors that are relevant to this chapter. First, the greater use of informally produced seed is due to the limited scope of the breeding programs, which often do not take into account traits that farmers consider important (e.g. the value of straw and the specific taste) or which are unable to select good varieties for every ecological niche. (Note that the use of fertilizers, pesticides and irrigation may reduce ecological diversity and thus facilitates breeding for a larger 'recommendation domain.') As a result, specifically adapted varieties are very important for smallhold farmers. Second, poverty may reduce the market potential for seeds from the formal sector; farmers may simply not be able to afford to purchase such seeds. Third, there may be instances when the quality factors of formally produced seed may not exceed those of farm-saved seed; a good farmer can produce good seed. It is very common that modern variety seeds are saved on the farm and shared between family, friends and neighbours.

Variety registration in developing countries could serve the same purposes that it did earlier in the industrialized world – that is, it could create transparency in the seed market and provide information on the value of the varieties for cultivation and use. However, such benefits can only be achieved when the system is effective and efficient. Robert Tripp and Niels Louwaars (1998) identify efficiency, standards, participation and transparency as key elements in effective variety controls. These are often limiting factors when official variety testing suffers from under investment, when the acceptance of new varieties is based on statistical analyses of yield data only and when an effective voice from stakeholders is lacking. When seed regulations are tacked on to bureaucratic structures and imposed upon both the seed producers and users, chances are that they will obstruct, rather than support, development. For instance, varieties may be released that do well under high-input regimes that most farmers cannot implement or varieties that have a specific adaptation to a particular region or use may not be identified at all (Ceccarelli, 1996). In some countries, varieties were not released at all for over a decade simply because the variety release committee did not meet in order to approve the release (e.g. in Yemen in the 1990s). As a result of all of these different factors, there is only a very narrow choice of varieties available in the market and there is a significant delay in the time it takes good varieties to reach farmers (see Louwaars, 2002, 2005, for more elaborate analyses of the seed laws of developing countries).

Compulsory or voluntary systems

Most developing countries have adopted compulsory seed certification and variety registration systems that are similar to the European models. However, some countries have a voluntary seed quality control system (such as is found in the United States, Australia, and India) or they use a variety of systems, depending on the species. The main reason for using a compulsory system for both variety registration and seed certification is that farmers can simply look at the seed tag and know that it is a quality-controlled seed (either the seed will have a tag or it will not). In a voluntary system – one in which there is an obligation to print the quality parameters on the label – illiterate farmers are unable to read the labels and thereby distinguish the quality parameters. This is one of the reasons why the new Indian seed bill of 2004 switched to a compulsory system. However, due to pressure from the emerging private sector, other countries have been relaxing the role that government plays in variety registration, particularly with respect to the inclusion of company data in the value for cultivation and use (VCU) system.

The advantage of voluntary registration is that a much wider array of varieties may be allowed in the seed market, which provides an enormous opportunity to develop commercial markets for both farmers' varieties and imported seeds. A disadvantage of voluntary registration is that it provides a reduced transparency in the naming of varieties (and therefore less protection for the farmer).

Variety registration regulations and plant breeders' rights

The emergence of scientific breeding in the first decades of the twentieth century and the later emergence of specialized breeders led to a call for protection for the breeders, similar to the protection of industrialized inventors in the nineteenth century. Living organisms were excluded from patentability because of ethical, legal and political reasons (Le Buanec, 2006). The variety registration procedure did succeed in fixing a variety to its name, but it did not give any exclusive rights to the breeder. This shortcoming was identified as early as 1919 at the Horticultural Congress in Paris, when the need to protect the commercial investments of breeders was first discussed (Bos, 1920). As a result, the Plant Patent Act was legislated in the United States in 1930, offering protection to breeders of vegetatively propagated horticultural crops.[7] It only applied to these crops because they were considered to be absolutely uniform and stable, thus providing a clear description of the protected subject matter. This development is relevant to this discussion because it has had a marked impact on variety registration procedures and standards (notably uniformity). Edible roots and tubers were excluded from this protection in order to avoid the privatization of food security crops (Kloppenburg, 1988).

In the years that followed, more *sui generis* protection systems for plant varieties developed in Europe. For example, Germany protected plant breeders with a kind of trademark – a 'breeder's seal' (Leskien and Flitner, 1997). Finally, in 1941, the first breeders' rights law was established in the Netherlands and described in detail the various requirements of distinctness, uniformity and novelty. Other European countries followed suit, and five countries of the then

European Community for Coal and Steel harmonized their systems in the UPOV Convention and became the first members of the International Union for the Protection of New Varieties of Plants (UPOV).[8] In this harmonization process, the five requirements for protection were formalized: distinctness, uniformity, stability, novelty and denomination (DUS-NN), and the standardized methodologies for establishing these requirements were developed. The Act of this convention was revised in 1972, 1978 and 1991, not only gradually strengthening the rights of the breeder but also maintaining the two exemptions: the breeders' and the farmers' exemptions. These criteria were derived from the distinctness and uniformity criteria in existing registration systems. These criteria were much more practical than the inventive step and use requirements of the patent system.

Membership in UPOV gradually expanded until, in the 1990s, many more countries joined following the entry into force of the World Trade Organization's (WTO) Agreement on Trade-Related Aspects of Intellectual Property Rights (TRIPS Agreement).[9] The TRIPS Agreement provides for minimum levels of protection to be included in national intellectual property rights laws in all WTO member countries. It has a specific clause in Article 27(3)(b) on the protection of plant varieties that provides an option to protect them through 'a patent, an effective *sui generis* system or a combination thereof.' The UPOV system was mentioned in early drafts of the TRIPS Agreement (Dhar, 2000), but it was not included in the final text. It is generally considered that the UPOV Convention provides an effective and internationally harmonized *sui generis* system and that it is better that it is rooted in the agricultural tradition than in the industrial patents that are available on plant varieties in a few countries.

The African Organisation for Intellectual Property (OAPI), with 17 member countries, joined UPOV on June 10, 2014. Prior to that, 26 developing or emerging countries were already UPOV members.

The legal protection of varieties thus led to the creation of yet another national variety list in addition to the lists of registered and recommended varieties (see Box 10.3). Both protection and registration for market regulation required that a clear description of the variety be provided so that the variety could be easily identified (and the protected subject matter be clearly demarcated). Both also required that the variety did not change after repeated reproduction (it had to be stable), and uniformity is generally considered to be the best measure for stability. As a result of these requirements, the DUS standards eventually became the same as the registration standards. This consistency makes application for both purposes easy – one set of variety trials may serve both for registration and protection procedures. These standards became more stringent over time, mainly as a result of the numbers of varieties that were entering the market – the more varieties that have to be identified, the stricter the distinctiveness standards have to be – and as a result of various technological developments, which made it possible to breed more uniform varieties, particularly for cross-fertilizing species.

Box 10.3 Summary of different variety registration systems

- Variety registration is used to create transparency in the market. A national variety list fixes the variety to a single name.
- Variety registration may also be used to provide information to the users of varieties. A national list of recommended varieties lists those varieties that have proven to have specific agronomic or use values.
- Variety registration may also be used to provide the breeder with legal protection (breeders' rights). A national list of protected varieties is maintained for this reason.

Any registration system is valuable only if implemented effectively and efficiently.

Harmonization of variety registration

The existence of different registration procedures in different countries could potentially harm the international seed trade, particularly if registration procedures were time-consuming and bureaucratic and delayed the availability of foreign-bred varieties in the market. To combat these issues, the Organisation for Economic Co-operation and Development (OECD) developed its 'seed schemes,'[10] and different regions in the world simultaneously developed their own harmonized systems. The most advanced system is the 'common catalogue' of the European Union (EU). The common catalogue includes all of the varieties that have been released in any of the member countries. In this way, national registration becomes EU-wide, which means that seed may then be traded throughout the union unless a country explicitly refuses to use such a variety. This means that a European farmer potentially has access to 50,000 varieties of species produced in Europe. These are of course not all suitable for all farmers, which means that farmers mainly rely on their own national lists. These lists are more common for field crops, but there are also recommended lists for vegetables, fruits and forest trees.

In 1970, the EEC Directive 70/457 on the Common Catalogue of Varieties of Agricultural Plant Species and the EC Directive 70/458 on the Marketing of Vegetable Seed set up the European Community catalogues for field crops and vegetables, respectively. Details were provided in 1972 with regard to the examination methods and standards for varieties to be included in the catalogue. For all of the crops listed in these catalogues, new varieties have to be distinct, uniform and stable; for field crops, their value for cultivation and use also has to be assessed. The EU also harmonized its plant breeders' rights systems. At the regulatory level, the UPOV Convention had already provided the basis for a harmonized system, so the EU formed a harmonized system for implementation by establishing the Community Plant Variety Office, which administers applications and organizes centralized testing.

Regional harmonization is also being increasingly pursued in developing countries, notably in Africa, with the same aim of creating larger markets for the seed industry and a potentially broader choice for farmers (Muhhuku, 2002). This is likely to benefit the local seed companies in the various regions, including the ones with their bases in the North, and it may stimulate the development of a wider range of varieties with specific adaptation to local needs, on the basis of which smaller local companies may build their businesses. Harmonization was only completed in 2009 for the area encompassing the Economic Commission for West African States (ECOWAS), and it is too soon to judge whether there have been concrete results. Harmonization is also close to completed for the Southern African Development Community (SADC).

Current variety registration procedures

Variety registration developed in history as part of an attempt to create transparency in the naming of varieties in the market (Box 10.1). For this reason, varieties had to be identifiable and thus distinct from one another, and they had to be stable because the name had to represent the same variety over time. Increased uniformity was the result of early methods of plant breeding (pedigree selection), and it serves as a good proxy for stability. A genetically diverse variety is likely to change with time due to segregation and genetic drift. When official variety lists based on field trials (VCU) were introduced in the 1920s, the variety descriptions became even more formal. These descriptions included both morphological and agronomic characteristics. The former became more and more important for the purpose of variety listing when more (and more similar) varieties were developed. When rights were being granted on these varieties, the morphological description based on characters that are least influenced by the growing conditions of the plant became the basis of the DUS testing methods and standards that were harmonized by the UPOV Convention.

Name

A variety has to be named in such a way that it does not create confusion in the market. It should not be too similar to existing names, should not consist of numbers only (except where this is established practice), should not identify specific qualities and should not be disrespectful to the morals of the local community. When the varieties are actually selections from a known population ('umbrella variety'), the variety name commonly consists of the name of the umbrella group plus a name to identify the specific variety (e.g. the name 'Chantenay Red Cored' carrot was chosen for a variety of the Chantenay type). Some countries have detailed rules for this naming process, but most use the rules established by the UPOV Convention. One challenge for a region with different languages such as the EU is to know whether a variety name may have a specific meaning in another language. In this case, synonyms are accepted.

Distinct, uniform and stable (DUS)

Through the DUS criteria, the catalogue supports the traceability of the variety during the seed production phases. In order to be registered, a new variety must be distinct from all varieties of common knowledge (internationally). It also has to be sufficiently uniform in its essential characteristics and highly stable after repeated multiplication. 'Common knowledge' is interpreted narrowly in the United States.

The UPOV Convention outlines the DUS testing procedures for many crops. Most countries that apply these tests require the breeder of a new variety to complete a form from the official registration office. The office then requests that an independent institute test for DUS in field trials. In some countries, the breeder can provide the DUS test results obtained from his own trials. In this case, the office simply performs an administrative check.

The standards for DUS are fairly complex and flexible. Distinctness standards depend on what can be observed on the basis of a standard list of descriptors. If, on the basis of this list of descriptors, two varieties appear to be the same but are clearly different with respect to another trait, the registrar may accept an additional trait in the description. Distinctness is thus defined as a relative standard.

According to the UPOV Convention, uniformity is also a relative criterion. New varieties have to be uniform in their main characteristics, taking into account the reproduction system of the variety, and the standard is measured relative to the average uniformity of the existing set of varieties. This means that – together with the development of more and more varieties that have to be distinguished – uniformity standards technically tend to become stricter over time.

Novelty

There is no novelty criteria for the registration except the distinctness. It means that a variety may be considered as 'new' and registered even if this variety is already known in another country. It is very different for the list of protected varieties: novelty is an essential criterion. Novelty checks have to assess whether the variety has been commercialized anywhere in the world (including those countries that do not offer such protection), and if so, for how long. This check is done on the basis of seed company catalogues and using the expert knowledge of people who know the market very well.

Value for cultivation and use (VCU)

The VCU testing system is meant to support the use of improved varieties and, if performed well, is a key factor in a farmer's decision to buy a particular seed.

The compulsory system of variety testing for agronomic and use values was introduced in Europe in order for farmers to have an independent comparison of the yield, quality and value of the grain that they were producing (e.g. tests of baking quality of wheat). The official tests make the farmer less dependent on the promotion of the varieties by the breeding companies. At the same time, it is an agricultural development tool for the government. Since the tests are commonly performed under 'good farmer' conditions, the trials demonstrate the yield potential under 'good management' conditions and, thus, create incentives for farmers to use such practices. Another important aspect of general interest is the testing of disease resistance in varieties, which should serve to reduce epidemics. In addition, the tests are of particular interest for breeders since they can also evaluate the competitors' varieties. Finally, small companies may benefit from the system since they would normally be unable to bear the cost of conducting their own trials throughout their country.

VCU experiments are organized through multilocal networks, which recognize the diversity of soil and climatic conditions in order to measure the productivity, agronomic (cold, disease or lodging resistance), and technological characteristics (oil, sugar and protein content). Some bonus points or penalties can be allotted to varieties that present favourable or unfavourable characteristics. These methods are regularly reexamined to take into account (1) new varietal innovations, (2) modification of crop cultivation practices and (3) new user needs.

Obviously, such tests only create benefits for both farmers and the seed sector when trials are performed effectively and efficiently and when decisions are made wisely. However, when trials are not performed well or decisions are made unwisely, the VCU regulations tend to create the opposite effect: there are fewer varieties available, there is less agrobiodiversity and there is less agricultural development. Hence, an important prerequisite for developing official VCU trials is that they represent the actual farming conditions. It is therefore essential that there is agreement between those planning and performing the trials and the farmers for whom these trials are to be laid out. It is important that they are consulted on what conditions should be (e.g. fertilizer levels) and which characteristics are to be observed. Countries differ widely in terms of farmer involvement, both with respect to the voice of their representatives and the representation structure itself. Too often, 'professional' farmer representatives, especially in developing countries, have insufficient links with the actual reality of crop production. The effectiveness of the system also depends on sufficient funds being available in the particular regions to do the work properly. A chronic lack of funds in certain areas may be one of the reasons why UK farmers now tend to select their varieties based on voluntary trial schemes rather than on the regulated trials.

In countries where VCU testing is not legally compulsory for variety registration (such as the United States), such variety trials are highly valued by farmers and often demanded by farmers' organizations or by the extension service of a particular state. The big difference is that the outcome of these trials is not binding on the decision of whether to market the seeds (to release them). Theoretically, this means that many more varieties are available to farmers, but, in practice, most farmers will only choose the varieties that perform well in the trials. Having said this, particular groups of farmers (e.g. organic farmers who do not want to use chemical fertilizers and pesticides) are free to conduct their own trials and select their own varieties that do best under these particular conditions. Under a compulsory VCU regime, such as in Europe, such independent trials are not taken in consideration, and those farmers who wish to do so have to lobby for separate official trials to be performed under their specific farming conditions. When such voices become strong enough, the official VCU trial system should adapt its methodologies to include coordinated trials under such conditions. In various countries, for example, Canada, Switzerland, France and the Netherlands, such variety trials are currently being performed.[11]

Examples of registering less uniform varieties

Variety registration makes the use of registered, uniform varieties mandatory and, as such, may reduce the use of old populations, old varieties or farmers' varieties. For example, in the United States, genetically diverse varieties may be exchanged and marketed (e.g. through the Seed Savers Exchange, <www.seedsavers.org>) parallel to the certified seed of uniform varieties. However, when registration is compulsory under strict DUS criteria and when, at the same time, the marketing of seed is restricted to registered varieties, it then becomes illegal to exchange or sell seed of landraces or farmers' new varieties (Salazar, Louwaars and Visser, 2007). This situation is common in many countries that have created their seed laws based on the European example. In Europe itself, however, recent changes in the seed laws are creating some room for flexibility.

European countries

The basic rules on variety registration in the EU are very strict and essentially make the exchange and marketing of seed from varieties that are not registered in the common catalogue illegal. However, this does not mean that landraces and other old varieties have completely disappeared in the EU. Breeding companies and farmers are allowed to reproduce seed for themselves, but they are not allowed to sell it. Countries differ widely in their implementation of the term marketing. For example, France (except for old vegetable varieties for gardeners) is quite strict in regulating the market and holds very closely to the rules, while seed regulation authorities in the Netherlands openly allow some level of deviation from the rules. In England, farmers have successfully created clubs through which members – those who belong to the same legal entity – can exchange and even sell seed without violating the letter of the law.

Some European countries have had, for quite some time, special parallel lists for the registration of materials that would otherwise not satisfy the European common catalogue and that would therefore risk falling out of use and disappearing. We briefly discuss those lists that are held for France (national/crop approach), Italy (regional approach), and the Netherlands (sectoral approach).

France

France opened a register for old vegetable varieties for home gardeners. To be included in this list, the variety has to be known – for example, it has to have been presented in an old commercial catalogue or in any other document. This requisite could restrict the inclusion of truly local varieties in the register since some may never have been described and published and it may be complicated to prove that these varieties were already known. To date, there has not been a denial of registration on this basis. As of 2009, the register included 300 varieties for about 30 species, including many varieties of squash, pumpkin, lettuce, tomato and melon. These varieties are registered at a minimal cost (about €100), based on a description provided by the applicant (Zaharia, 2003). The varieties are allowed to be commercialized only among amateur gardeners in France and only in small packets (e.g. a maximum of 2 grams for tomatoes, 15 grams for leeks and 5 grams for cauliflower). They must be labelled 'standard seeds' and marked with the statement 'old variety exclusively reserved for home gardeners,' which assumes that the produce is consumed at home. The seed quality parameters (germination, purity and seed health) are the same as in the regular market. France also opened in 1952 a national list of old fruit tree varieties for home gardeners, with more than 1,000 varieties of apple, pear, plum, hazel and walnut trees. A third special list created in 1993 is for old and well-known potato varieties that have a specific use in France. These varieties had been on the market for a long time and did not need additional VCU tests. The objective was to control the health risks in producing seed of these varieties and to regulate its variety maintenance. This list includes only five varieties.

Italy

A number of regions in Italy have adopted regulations since 1997 to protect and promote traditional farmers' seeds and animal breeds as part of a movement upholding regional competence in agricultural issues according to the Italian Constitution. Each region addresses, in particular, the loss of traditional varieties. Variety registration seems to be just one part of a broader regional initiative to preserve traditional plant, animal and forest resources. The main regional mechanisms for conserving and enhancing traditional local varieties and breeds are:

- the establishment of a regional catalogue where individuals and organizations, on a voluntary basis, may register local traditional varieties and breeds; technical and scientific committees evaluate the proposed accessions to the regional register, which allows the variety seeds to be marketed within the region;

- the creation of a 'network of conservation and enhancement' for both on-farm and *ex situ* conservation;
- the recognition that the heritage of the local and traditional varieties belongs to the local communities (referring to Article 8(j) of the Convention on Biological Diversity [CBD]), and recognition of the regional authority as the body that manages and guarantees the collective and sustainable use of such resources.[12]

The framework for saving and exchanging farmers' seeds provided by regional regulations addresses a very specific situation in which only small numbers of farmers still grow and manage traditional varieties. Nevertheless, the regional regulations do create legal space for protecting and promoting traditional products based on specific varieties that are not recognized by national seed laws and registers and for recognizing the collective rights of the communities over their varieties.

The region of Tuscany is a good illustration of this framework as it currently leads the conservation initiatives in the country. The Tuscan legislation creates a register of conservation varieties (i.e. varieties that are at risk of extinction):

- species, races, varieties, crops, ecotypes and clones from the Tuscany region;
- species, races, varieties, crops, ecotypes and clones from outside Tuscany but that were introduced long ago and integrated in the traditional agriculture;
- species, races, varieties, crops and ecotypes that have been selected;
- species, races, varieties, crops and ecotypes from Tuscany that are no longer grown there and are conserved inside or outside Italy.[13]

In principle, this legislation allows for the commercialization of conservation varieties' seeds, but it specifies under which conditions these seeds can be sold. The register included hundreds of varieties of arboreal and fruit species and dozens of herbaceous species, of which the vast majority have been considered to be at risk of extinction.[14] More information about Italian regional registration systems is available in Chapter 21 of this volume.

The Netherlands

At the request of the ecological (organic) farming sector, the Netherlands developed a special VCU testing regime in order to test new varieties that had been grown without the use of chemical fertilizers and pesticides (Osman and Pauw, 2005). The idea was to develop a 'green variety list' that would list new varieties that would perform well under farming systems with no chemical input (so-called organic or eco-farming). However, there appears to be little difference in the results from the trials of the regular and 'green' farming for most crops. Most varieties grown organically also performed well under 'regular' farming practices. In addition, various initiatives that promote the use of particular (non-registered) varieties for local niche products have been implemented simply by ignoring the existing seed regulations. As long as these varieties have remained quite local, the authorities have not taken action to strictly enforce the seed law.

Outside of Europe

There is a wide diversity in variety registration among developing countries outside of Europe. In most countries, there are no rules other than strict compulsory release and registration systems, and these systems are often not implemented at the local level. Algeria, however, introduced a section in its national variety catalogue for new farmers' varieties.[15] This addition has created an opportunity for the products of participatory plant breeding to be officially registered, even though they may not meet all of the uniformity standards. This list is based on a new interpretation of the existing seed legislation and not on a new seed law itself.

There are other cases where the products of participatory plant breeding have been registered through the regular system. For example, farmers participating in the Local Initiatives for Biodiversity Research and Development network in Nepal have recently released rice and maize varieties that appear to be sufficiently uniform and valuable for a wider use. See Chapters 4 and 18 of this volume for more about these developments in Nepal.

Recent regulations at the EU level: conservation varieties

EU member states have made a commitment towards the implementation of the 1992 Convention on Biological Diversity as well as the 2001 International Treaty on Plant Genetic Resources for Food and Agriculture. At the same time, civil society – in particular, the organic farming community – has requested a more open regime for variety registration. In 2009, some legal openings in European Community (EC) Directive 98/95 on the Marketing of Beet Seed, Fodder Plant Seed, Cereal Seed, Seed Potatoes, Seed of Oil and Fibre Plants and Vegetable Seed and on the Common Catalogue of Varieties of Agricultural Plant Species were used to create conditions to support the *in situ* management of so-called conservation varieties by allowing seed of these varieties to be produced and marketed. The European Commission is developing derogations for conservation varieties. These are varieties that contribute to genetic diversity in the field and that are under threat of genetic erosion. By making this effort, the Commission is expecting that the diversity in farmers' fields will be enhanced and that traditional varieties can be maintained *in situ*. In practice, it means that the production and sale of seed from such varieties is being legalized, thereby enabling the reintroduction of old varieties into the farming system.

The EC directive for field crops was adopted in 2008 after stakeholders debated the issues over a four-year period. Directives for vegetables and seed mixtures, being more complex in economic and technical aspects, followed in 2009. These directives attempted to strike a new balance between the need of farmers for information and transparency in the marketplace and the necessity to support the continued utilization of plant genetic resources, landraces and varieties in actual farm production. Key issues in the debate included: (1) if it was possible to define, identify or measure whether a variety is 'threatened by genetic erosion'; (2) how much flexibility could or should be introduced on DUS and VCU standards to accommodate the varieties that are the intended subject of the directive; and (3) how much flexibility can or should be introduced on quality control, production and marketing of seeds of these varieties.

Varieties 'threatened by erosion'

Ultimately, the directive includes the following definitions:

- 'Conservation *in situ*' means the conservation of genetic material in its natural environment and, in the case of cultivated plant species, in the farmed environment where it has developed its distinctive properties.
- 'Genetic erosion' means the loss of genetic diversity between and within populations or varieties of the same species over time or the reduction of the genetic basis of a species due to human intervention or environmental change.
- 'Landrace' means a set of populations or clones of a plant species that are naturally adapted to the environmental conditions of their region.

Only varieties that can be linked to a particular region can thus be called conservation varieties, including not only genetically diverse landraces but also old varieties that have disappeared from the common catalogue for some years and that have proven to be adapted to a particular region. The seed industry lobbied successfully to exclude the possibility of listing new genetically diverse varieties, such as varieties based on participatory breeding. The commercial seed sector was concerned that allowing new farmers' varieties as conservation varieties would create a back-door opportunity for the registration of new varieties that do not meet the regular standards. Some conservationist and farmers groups, on the other hand, lobbied hard to have a broader definition of the term conservation varieties, particularly with regard to the possibility of using such varieties outside their region of origin and of allowing such varieties to be improved through selection (Chable, 2009).

DUS and VCU criteria

The uniformity criterion is flexible in that it includes the word 'sufficient' (as outlined in the UPOV Convention) and requires that only a limited set of criteria be described and that the applicant can provide these descriptions. VCU criteria may also be included in the description. Thus, certain populations (landraces) may be registered, but these species must be maintained in such a way that they are sufficiently stable – that is, they must remain within their description.

In 2014, the European Commission proposed a new seed regulation, which also includes a class called heterogeneous materials. This might create opportunities not only for the maintenance and use of old varieties, but also for new, but not uniform varieties, for example the result of participatory plant breeding. It proves difficult to balance the guarantees that seed legislation provides to farmers with the desire to open up the rules to allow for a wide diversity of wishes. The proposal was rejected by the European Parliament for procedural reasons and because too many issues were to be specified in 'delegated acts' over which the parliament does not have a say. It is likely that the discussion about regulating the seed sector will continue for several years. In any case, there appears to be a general consensus to open more doors for 'particular' varieties, and to keep the main pillars of EU regulation: compulsory registration of varieties and quality control of seeds under official authority.

Quality control, production and marketing

Some specific rules have been developed with regard to seed production. The conservation variety has to be linked to its 'region of origin' where it has to be maintained, and the seed has to be produced and marketed. Countries have the liberty of defining the regions themselves. For example, the Netherlands is able to consider the whole country as one region. The maximum amount of seed from these conservation varieties has also been defined. Each variety is limited to 0.5 percent of the total seed market (or to the quantity of seed required for 100 hectares when it is superior), and all conservation varieties of a single crop should not exceed 10 percent of the total.

Discussion: variety registration and farmers' varieties

Registration – DUS and biodiversity

Strict compulsory variety release in combination with the uniformity requirement is, by definition, almost contrary to the desire to increase agrobiodiversity. Diversity may be looked at from two viewpoints: diversity within, and diversity among, varieties in a farming system. Obviously, DUS requirements reduce the genetic diversity within varieties. Plant breeders often have to add additional selection rounds that focus on the characteristics that will be taken into account in the DUS tests, once a variety is proven to be sufficiently uniform in its agronomically important characteristics (e.g. maturity period, plant architecture and height). Compulsory release systems for seed marketing (as in Europe) thus make it impossible to release varieties that are genetically diverse for a good agronomic reason. For example, the Netherlands delayed the release of a multi-line wheat variety – appropriately named Tumult (uproar) – which displayed a resistance to rust that was based on the presence of different resistance genes in different components of the variety. The release system could not deal with the obvious lack of uniformity in this single character.

In mechanized agriculture, most diversity is not productive, but breeding for diverse ecologies (e.g. in developing countries) may require the use of genetic diversity to enhance yield stability. Participatory plant breeding that is aimed at a specific adaptation of certain varieties commonly, but not by definition, results in varieties that are less uniform than conventional breeding under more controlled conditions. In situations where DUS testing is compulsory, such breeding strategies may not lead to new varieties in the market.

The situation may be different when diversity is looked at from the point of view of total diversity in a region where many uniform varieties are used. For example, molecular studies show that diversity and allelic richness have decreased in durum wheat in France but have increased in maize, peas and bread wheat in that same country as a result of breeding (FAO, 2010). This means that promoting investments in plant breeding, particularly when it is coupled with incentives to broaden the genetic basis of breeding, is likely to result in higher levels of diversity.

The new rules for conservation varieties create a solution for only part of the issue of diversity. It allows countries to register old varieties and amateur varieties of vegetables, but it does not allow them to improve those varieties that are the result of participatory breeding. Any new variety needs to be registered under the conventional (strict) system.

VCU requirements

The compulsory variety testing creates a wealth of information for both farmers and breeders. Having the latest varieties of competing companies in one trial field allows for an excellent comparison of the competitor's results. These trials may also serve as a demonstration plot for farmers to see how the use of these new varieties can improve their agriculture. Having all of the data analyzed and presented in a national list of recommended varieties, as has been the case in the Netherlands since the early 1920s, greatly supports the farmers in their decision about which varieties to plant.

However, the trials have to be conducted with great care (at the farmers' level of management) and must be done impartially. Farmers' varieties commonly do not pass the official VCU tests, partly because their yield potential is not sufficient (their value often lies in yield stability and special quality aspects) and because their adaptation is commonly quite specific, which means that even though their performance may be excellent in their region of origin, their average performance throughout all testing sites may be insufficient.

Options

Options for countries to allow and promote the use of a wider range of varieties are multiple. In countries with a voluntary registration system, such as the United States, the regulatory framework does not create obstacles for the use and sale of seed from varieties that may not meet the registration standards (DUS or otherwise). In countries with a compulsory variety registration system, there are two options: either create a special class of varieties, to which other registration standards apply (as is currently done in the EU) or carefully demarcate the scope of the seed regulation to the formal seed system, leaving the informal system free. The European example is a great advancement in regard to the possibilities for farmers to use their own farmers' varieties, but since it remains a registration system there is a risk that it cannot easily deal with the plasticity of highly diverse farmers' varieties, which are likely to change as a result of environmental and farmers' selection. It is still too early to assess the effect that the implementation of EC Directive 98/95 will have on different EU countries in this respect.

Developing countries with compulsory variety registration systems will likely find it very difficult to follow the same solution. The large numbers of farmers' varieties that have resulted due to the vast ecological diversity within their countries, in combination with the limited efficiency of variety registration due to insufficient human and technical resources, has made the registration of all farmers' varieties an impracticable solution. On the other hand, turning their registration into a voluntary system may not suit their policies either since this option is based on a very competitive seed market and highly educated farmers. It might be much more suitable for developing countries to carefully design their seed regulatory framework in such a way that the formal sector would be closely regulated and the regulations and implementing institutions would not impinge on the informal sector. This strategy links up with the concept of 'integrated seed systems' (Louwaars, 1994), which has recently been adopted by the African Union and the FAO (FAO, 2006). This concept formally acknowledges that within the same country different seed systems must operate in parallel with the different needs for government support and controls. Translated to seed regulation, this means that a boundary has to be framed between the formal system and the farmers' seed system, leaving the latter unregulated. Since farmers' varieties would be used, exchanged and further developed in the farmers' seed system, such deregulation of this system would provide sufficient opportunities to continue, and to support, the use of farmers' varieties. However, for now, we have no full concrete example of this strategy.

Registering farmers' varieties and the objectives of registration

The original objectives of variety registration were transparency in the market, the provision of information for farmers and increased agricultural production through the use of improved varieties. The inclusion of a fourth objective (biodiversity) should not counter the original ones. EC Directive 2008/62 on the Marketing of Conservation Varieties actually tries to do this very thing. Registration of traditional varieties through relaxed standards and without official agronomic trials allows for a clear naming of varieties. Even when the varieties are not fully stable in all characteristics, farmers will know what to expect from the seed of the named variety. Information on the agronomic performance of the variety is not very important in this case since we typically speak of well-known, old varieties and not of new varieties of which farmers know little. The objective of increasing agricultural output is still valid. However, such output should focus on the creation of monetary value by raising the prices of regional products or through organic farming rather than on increasing yield levels.

Conclusions

Variety registration has developed alongside plant breeding and seed production. In some countries, such as the United States, government involvement is limited. In most other countries, including those in Europe and most developing countries, registration is a compulsory requirement for taking seed to market. Seed certification schemes and plant breeders' rights systems all involve some form of variety registration.

Registration requirements and procedures are being harmonized partly in an attempt to stimulate international trade and partly as a result of international pressure to create globally harmonized intellectual property rights systems. Distinctiveness is a basic requirement of any registration system (a variety needs to be identified as distinct from any other in order to be registered), as is stability (a variety must remain the same in its important characteristics after repeated reproduction). Uniformity is considered to be the most reliable precondition for stability: uniform varieties are least likely to change over time. When more varieties are being developed, the similarities are bound to increase, which means that distinctness and uniformity standards are bound to become more strict over time, leaving less and less room for genetically diverse varieties to enter the market. Agronomic performance testing is often an additional component of the registration process. Ineffective implementation of this system in many developing countries can create a bottleneck for the number of varieties available to the farmer and may thus decreased diversity in the field.

Countries need to balance different policy objectives under the overall goal of promoting agricultural growth. For example, they need to (1) create a transparent market for seeds with a level playing field for competing companies; (2) provide farmers with suitable protection with respect to the identity of varieties and qualities of seed in the market; (3) support the conservation and use of genetic resources allowing farmers to use genetically diverse varieties; and (4) promote seed security, so that developing countries can no longer restrict the informal seed system and use of farmers' varieties. In countries with a voluntary variety registration system and an educated farmer community, such objectives may be pursued jointly. In countries with compulsory variety registration systems, special derogations may need to be specified in the law (as is the case with conservation varieties in Europe), or the scope of the seed regulatory law framework may need to be limited to only the formal seed sector. Different countries have different farming and seed systems and obviously will arrive at different solutions. Whatever the system, countries that subscribe to the CBD and/or the UPOV Convention have a responsibility to promote the conservation and use of biodiversity.

Notes

1. A major part of this chapter is based on a European Union-funded project, Farm Seed Opportunities STREP 044345 of Framework Programme 6 (<www.farmseed.net> (last accessed 15 June 2012)). Niels Louwaars is in the Centre for Genetic Resources, Wageningen University, Wageningen, The Netherlands. François Burgaud is in the GNIS Inter-Professional National Group for Seeds and Plants, Paris, France.
2. Symposium on the Improvement of Plants Continuities and Ruptures, Montpellier, October 2002.
3. The need for this catalogue was explained in a monograph published in 1951 on common wheat in France by a scientist from the INRA, M. Jonard. He indicated that, since 1919, wheat improvement in France had achieved great progress. However, at the same time, he noted that many new marketed types were often incompletely fixed or were a simple copy of old cultivars.
4. Decree on the Fraud Prevention in Wheat Seed Trade, 26 March 1925, published on 29 March 1925 in the Official Gazette.
5. Presidential decree published on the 19 November 1932 in the Official Gazette.
6. Decree no. 81–605, 18 May 1981.
7. Plant Patent Act, 35 U.S.C. §§ 161–164.
8. International Convention for the Protection of New Varieties of Plants, 2 December 1961, online: <www.upov.int/en/publications/conventions/index.html> (last accessed 15 June 2012).
9. Agreement on Trade-Related Aspects of Intellectual Property Rights, Annex 1C of the Marrakech Agreement Establishing the World Trade Organization, 15 April 1994, 33 ILM 15 (1994).
10. See Trade and Agriculture Directorate, online: <www.oecd.org/document/0/0,3343,en_2649_33905_1933504_1_1_1_1,00.html> (last accessed 15 June 2012).
11. For Canada, see <www.npsas.org/ovt.html> (last accessed 15 June 2012); for Switzerland, see <www.fibl.org/en/switzerland/location-ch.html> (last accessed 15 June 2012); for the Netherlands, see <www.louisbolk.org> (last accessed 15 June 2012).
12. Convention on Biological Diversity, 5 June 1992, 31 ILM 818 (1992).
13. Tuscany Regional Law, no. 64, 16 November 2004, Article 2.
14. See Razze e Varietà Locali, online: <http://germoplasma.arsia.toscana.it/Germo/modules.php?op=modload&name=MESI_Menu&file=Manager&act=D_1:@201> (last accessed 15 June 2012).
15. Personal communication with S. Ceccarelli of the International Center for Agricultural Research in Dry Areas, 2010.

References

Bombin, L. (1980). *Seed Legislation,* Food and Agriculture Organization, Rome.

Bos, H. (1920). 'Patentering van nieuwigheden op tuinbouwgebied' (Patenting of novelties in horticulture), *Floralia* [on file with the author].

Bustarret, J. (1944). 'Variétés et variations,' *Annales agronomiques* 14: 336–62.

Ceccarelli, S. (1996). 'Adaptation to low/high input cultivation,' *Euphytica* 9: 203–14.

Chable, V. (2009). *Task 1.1: Definitions of Landraces, Conservation Varieties, Amateur Varieties in Europe (M1–M24),* Interim Report by Farm Seed Opportunities. Available at: www.farmseed.net/scarica.php?id=245 (last accessed 15 June 2012).

Dhar, B. (2000). *Options under TRIPS: A Discussion Paper,* Quaker United Nations Office, Geneva.

Feistritzer, W.P. (1984). 'The FAO Seed Improvement and Development Programme,' *Seed-Science and Technology* 12(1): 31–4.

Food and Agriculture Organization (2006). *African Seeds and Biotechnology Programme*, Twenty-Fourth Regional Conference for Africa, Bamako, Mali, 30 January–3 February 2006. Available at: ftp://ftp.fao.org/unfao/bodies/arc/24arc/J6882E.pdf (last accessed 15 June 2012).

Food and Agriculture Organization (2010). *Second Report on the State of the World's Plant Genetic Resources for Food and Agriculture*, Rome. Available at: www.fao.org/agriculture/crops/core-themes/theme/seeds-pgr/sow/en/ (last accessed 15 June 2012).

Kloppenburg, J.R. (1988). *First the Seed: the Political Economy of Plant Biotechnology*, Cambridge University Press, Cambridge.

Le Buanec, B. (2006). 'Protection of Plant-Related Innovations: Evolution and Current Discussion,' *World Patent Information* 28(1): 50–62.

Leskien, D., and M. Flitner (1997). *Intellectual Property Rights and Plant Genetic Resources: Options for Shaping a Sui Generis System*, Issues in Genetic Resources No. 6, International Plant Genetic Resources Institute, Rome.

Louwaars, N.P. (1994). 'Connecting Formal and Local Seed Systems in Development: Integrated Seed Supply Systems,' in P. Stolp (ed.), *Organisation and Management of National Seed Programmes*, International Center for Agricultural Research in Dry Areas, DSE Feldafing, Germany, pp. 10–17.

Louwaars, N.P. (2002). 'Variety Controls,' *Journal of New Seeds* 4(1/2): 131–42.

Louwaars, N.P. (2005). 'Seed Laws: Biases and Bottlenecks,' *Grain* (July): 3–7. Available at: www.grain.org/seedling_files/seed-05–07–2.pdf (last accessed 15 June 2012).

Louwaars, N.P., and G.A.M. van Marrewijk (1996). *Seed Supply Systems in Developing Countries*, Technical Centre for Agricultural and Rural Cooperation, Wageningen.

Muhhuku, Fred (2002). 'Seed Industry Development and Seed Legislation in Uganda,' *Journal of New Seeds* 4(1/2): 165–76.

Nilsson, N. Hjalmar (1898). *Einige Kurze Notizen über die Schwedische Pflanzen-Veredlung zu Svalöf* [*Some notes on the Swedish plant breeding in Svalöf* (in German)]. Skanska Lithografiska Aktienbolaget, Malmö.

Osman, A., and J.G.M. Pauw (2005). *Zomertarwerassen voor de bioteelt* (in Dutch). Available at: http://library.wur.nl/WebQuery/wurpubs/348362 (last accessed 15 June 2012).

Salazar, R., N. Louwaars and B. Visser (2007). 'Protecting Farmers' New Varieties: New Approaches to Rights on Collective Innovations in Plant Genetic Resources,' *World Development* 35(9): 1015–528.

Schribaux, E. (1928). 'Les meilleures variétés de blé à cultiver ou à essayer,' *La grande revue agricole* 1: 1465–8.

Tripp, R., and N. Louwaars (1998). 'Seed Regulation: Choices on the Road to Reform,' *Food Policy* 22(5): 433–46.

Zaharia, H. (January, 2003). 'La directive européenne 98/95/CE: une avancée législative européenne pour les semences paysannes?' Available at: www.semencespaysannes.org/dossiers/bip/fiche-bip-12.html (last accessed 15 June 2012).

Zeven, A.C., and J.M.J. de Wet (1982). *Dictionary of Cultivated Plants and Their Regions of Diversity*, Centre of Agricultural Publishing and Documentation, Wageningen.

11 Defensive protection of farmers' varieties

Isabel López Noriega

Introduction

In 2006, I was invited to participate in a seminar arranged by the Institute of Agricultural Research in Peru (INIA, Instituto Nacional de Innovación Agraria) on the creation of a national official register for landraces and traditional varieties of potato and maize. This register was to be an officially recognized list that would be supported by the government, where different users could enter and access information about Peruvian traditional varieties of potato and maize. The main purpose of the seminar was to discuss possible objectives of the register with a number of people involved in genetic resources conservation and use within the country, including research institutes, farmers' associations and organizations working with indigenous communities. During the seminar, several participants argued that one of the most important objectives of the register should be to officially recognize traditional farmers as the originators of the huge biodiversity of potatoes and maize that are conserved and cultivated in Peru.

Being educated in Europe and having a legal background, I quickly and naturally assumed that they were somehow talking about granting farmers intellectual property rights over their varieties. I raised my hand and asked if they were actually meaning to confer this idea and, if so, how they were planning to do it. The reaction from the audience was strong and immediate. Participants who were representing the farmers as well as the national research institute stated that neither the farmers nor the public institutions in Peru wanted to be given any monopoly rights over their traditional varieties and landraces but that they simply wanted to be recognized as the developers and conservers. As they explained to me, these varieties have to be freely available for anyone to use, just as they always have been, and no one should claim any intellectual property rights over them or benefit from the use of their genetic resources without recognizing the provenance of the varieties and the efforts of traditional farmers. By documenting the existence of these varieties and their provenance, they said, the register will show that they are not new, thereby preventing enterprises in the developed world from getting patents or plant variety rights over them. In addition, thanks to the information stored in the register for each variety, these enterprises will know with whom they have to share the benefits if they make money off the use of the registered varieties.

Peruvians' fear that their plant genetic heritage could be misappropriated is well founded. The expansion of the scope of intellectual property rights and their protection's standardization through multilateral and bilateral international agreements have made it possible to apply for intellectual property on ideas and inventions; 20 years ago scholars would have agreed these were not eligible for intellectual property protection in countries with a long tradition of little to no recognition or enforcement of intellectual property rights. The number of patent and plant breeders' rights (PBR) applications to get control of the exploitation of plants, plant varieties and their seeds has dramatically increased in the last decade, in particular in developed countries and emerging economies.

In the middle of this intellectual property fever, some intellectual property offices have granted patents or PBRs over varieties and plants that were not new or over products and processes that are actually similar, if not identical, to indigenous or local communities' traditional uses of plants. These cases of misappropriation have increased stakeholders' awareness, in both developing and developed countries, about the importance of protecting the public nature of plant and plant uses traditionally available to anyone, such as landraces and most farmers' varieties. As my Peruvian colleagues argued, one way of ensuring their protection is by documenting the existence of these varieties in public databases so that their lack of novelty makes them ineligible for patents and plant variety protection (PVP). But how should this documentation be completed in order to ensure the effective defensive protection of farmers' varieties? Is an online register enough? What information should be published and how? What are the criteria that effectively define farmers' varieties in defensive protection strategies? What form of publication has the best chance to be taken into consideration by patent and PVP offices? And have these offices enough capacity to access and use this information anyway?

Since the INIA-Peru workshop in 2006, I have followed very closely the creation of the Peruvian national register of landraces and farmers' varieties of potato and maize.[1] I have tried to assess the actual capacity of the Peruvian register and other similar initiatives to protect traditional and farmers' varieties against misappropriation and analyze possible measures to make the defensive protection of farmers' varieties more effective. This chapter is the result of this analysis.

Threats to plant varieties in the public domain

In many countries, the term public domain is widely used to describe public goods such as land or water. However, the usage of this term in the intellectual property world comes from the French term *domaine public*, which was adopted in the language of the Berne Convention for the Protection of Literary and Artistic Work.[2] In this context, the term public domain is used to describe those creative works whose use is not restricted by copyright – that is, all of the original works of art, literature, music and so on whose copyright has expired, that cannot be subject to copyright law or that was created before the existence of copyright. Public domain is also utilized in the context of patents, although patent legal texts do not actually refer to the public domain per se. In the case of patents, the public domain embraces all of the inventions for which the term of patent has expired, those that have been disclosed without patenting them and those that are not eligible for patentability according to the law. Since proprietary rights are founded in national laws, the extension and the boundaries of the public domain differ among countries. An invention can be patented in one country and at the same time be in the public domain of another country. Similarly, in the context of PVP or PBR, the public domain is constituted negatively, without explicit mention in international agreements or national laws, by varieties (and populations) that have not been protected through breeders' rights or whose protection has expired.

The public domain has traditionally been defined in a negative manner, as whatever is not subject, and cannot be subject, to intellectual property rights. However, in recent years, a number of scholars have started to pay attention to the affirmative elements of the public domain with the idea of recognizing its own entity and conferring its own protection.[3] This recent attention to the public domain responds to the increased concern about the extension of property rights over creations, ideas and facts that were not eligible for intellectual property protection a few years ago.

In the field of plant genetic resources, the public domain's usual territories are being threatened not only by recent trends to extend patentability to elements that were not considered patentable before but also by the serious failures of the current intellectual property system. Let us briefly analyze these two factors.

Trends to extend patentability on life forms

Several decisions in the patent offices of the United States and Europe have paved the road towards the patentability of life forms, including plant varieties. Since these countries deal with the majority of patents and PVP in the world, it is worth analyzing with some detail their current patent legislation and its interpretation in the field of life forms. In 1979, the US Court of Customs and Patent Appeals clarified that living organisms modified by human intervention fall under the definition of 'manufacture' or 'composition of matter' according to section 101 in Title 35 of the US Code on Utility Patents.[4] One year later, in the famous case of *Diamond v. Chakrabarty*, the same court ruled that claims were not outside the scope of patentable inventions merely because they dealt with live organisms and stated that a live, human-made microorganism is patentable subject matter.[5] Any remaining issues about the possibility of patenting plants and seeds in the United States were clarified in the case *Ex parte Hibberd*, where the Board for Patent Appeals and Interferences admitted a patent application on maize plant tissues and seeds and stated that patents could be granted on plant inventions despite the fact that they could also be protected by the *sui generis* system under the International Convention for the Protection of New Varieties of Plants (UPOV Convention), which was implemented in the United States through the Plant Variety Protection Act.[6]

In 1990, the Examining Division of the European Patent Office (EPO) initially refused a patent application on a transgenic mouse, among other things, on the grounds that the 1973 Convention on the Grant of European Patents (European Patent Convention) excluded patentability of animals per se.[7] This decision was appealed, and the Board of Appeal held that animal varieties were excluded, in particular, by Article 53 of the European Patent Convention, while animals as such were not excluded from patentability.[8] The Examining Division then granted the patent in 1992. The Board of Appeal confirmed this position some years later, clarifying that claims which do not refer to particular varieties of plants are not excluded from patentability according to Article 53 of the European Patent Convention, even when they may include plant varieties.[9] This has been the approach adopted by EC Directive 98/44 on the Legal Protection of Biotechnological Inventions, which states that 'inventions which concern plants or animals shall be patentable if the technical feasibility of the invention is not confined to a particular plant or animal variety' (Article 4.2). The current guidelines for examination in the EPO state that 'a process claim for the production of a plant variety (or plant varieties) is not *a priori* excluded from patentability merely because the resulting product constitutes or may constitute a plant variety' (European Patent Office, 2013). Bearing in mind that a patent's protection can apply not only to the process but also to the resulting plant, its parts and its seeds, then the exclusion of plant varieties from patentability appears superfluous. On this basis, in the last 15 years the EPO has granted a number of patents on plants, plant seeds and breeding techniques – a number of which have been opposed due to the lax interpretation of patent requirements.[10]

A common topic in the debate about patenting life forms has been the need to distinguish between inventions and discoveries. Only products or processes that are the result of human intervention should be patentable. However, this principle has been interpreted in a lax manner with regard to biological compounds and processes. European legislation has tried to safeguard the inventiveness step that is required for patenting life forms by forbidding the patenting of processes that are essentially biological, but, in practice, the distinction between essentially biological and human-directed processes is very fuzzy if one looks at how this prohibition has been implemented on a case-by-case basis.[11] According to the guidelines for examination in the EPO, 'a process for the production of plants or animals is essentially biological if it consists entirely of natural phenomena such as crossing or selection.'[12] Following this rule, the EPO considers that a method of selecting, crossing or interbreeding is essentially biological and therefore unpatentable. However, the office has granted a number of patents to traditional breeding methods that involve the use of markers and other advanced tools, even if all of the steps in the improvement process are still essentially biological.[13]

Many developing countries have excluded the patentability of animals and plants in their recent patent laws. This restriction minimizes the probability of subjecting plant varieties to patents and guarantees that there is an extended public domain in comparison to developed countries.

Failures of the intellectual property system

In the last few decades, various intellectual property and PVP offices have granted patents and PVP over varieties of plants that were actually in the public domain. It is worth looking at a couple of these examples in detail in order to understand the issues that are involved and how the patenting system has failed. I have selected two famous examples that present different characteristics in terms of the typology of actors involved, the issues concerned, the applicable laws and their resolution. These two cases will be used later to illustrate some of the issues that are related to the defensive publication of plant varieties.

Ayahuasca case

The psychoactive plant *Banisteriopsis caapi* has been traditionally used by indigenous peoples in the Amazon region to prepare a ceremonial drink called Ayahuasca ('soul's wine' in the Quechua language). This drink is used in religious and healing ceremonies. It is also used to diagnose and treat illness, communicate with the spirits and predict the future. A tribe from Ecuador gave some samples of *B. caapi* to Loren Miller in 1974, who cultivated them in Hawaii where, he argued, he managed to develop a stable variety. In 1986, Miller obtained a plant patent on a plant called 'Da Vine,' which he defended to be a new variety of *B. caapi*. Later, he founded a laboratory to study and exploit the plant's properties. The Center for International Environmental Law (CIEL), which represented the Coordinadora de las Organizationes Indígenas de la Cuenca Amazónica (COICA), submitted a request for a reexamination of the patent in 1999, arguing that Da Vine did not meet the novelty requirements since the features described in the claim were typical of the species as a whole and had been described in scientific literature before the patent application was submitted. In addition, the request affirmed that the plant could be found in an uncultivated state and that the patenting of a plant that was sacred to indigenous peoples was against public policy and morality principles. The US Patent and Trademark Office (USPTO) accepted the request because the plant was almost identical to other plants that had been described by the herbarium at Chicago's Field Museum.[14] Miller submitted several briefs to the USPTO requesting the office to reconsider its decision. He argued that the herbarium's description lacked the necessary authority and that their plants demonstrated substantial differences with the flowers, leaves and stem of the Da Vine plant. Due to the regulations controlling patent applications submitted before 1999, the CIEL could not contest Miller's requests, and in 2001 the USPTO issued a notice reversing its earlier rejection of the Da Vine patent claim. They based this decision on some slight variations between the Da Vine plants and those included in the Chicago herbarium, arguing that these variations indicated that the plants described by the herbarium were not the same plant as the Da Vine.[15]

The scope of protection awarded in a US plant patent is to the single germplasm (i.e. the single and particular plant within the species) and its asexually reproduced progeny. In the case of the Da Vine plant, the differences in the leaves were substantial enough for the USPTO to affirm that it was a unique plant within the *B. caapi* species and therefore was worth patenting. It is important to highlight that Miller never compared the Da Vine plant with living specimens of *B. caapi* but only with dry samples from herbariums and written descriptions and drawings in existing literature. No grow test was conducted when the patent was reexamined nor was the molecular characterization examined in order to detect differences between the plants at the genetic level.

The Ayahuasca patent, which is perhaps the most famous case of misappropriation, had far-reaching consequences that went well beyond the CIEL's and the COICA's concerns. It resulted in political conflict between the United States and Ecuador and increased general concern about the reliability of the US patent system. It also raised considerable concern over the moral issue of patenting indigenous' communities' sacred plants and knowledge (Center for International Environmental Law, 1999).

Enola bean case

In 1999, the USPTO and the US PVP office granted a patent and a PVP certificate, respectively, to Larry M. Proctor for a common field bean called Enola. In the patent application, Proctor explained that he had bought some beans in a market in Mexico and, after few years of planting, had developed 'a new field bean variety that produces distinctly colored yellow seed which remain relatively unchanged by season.' Several organizations denounced the Enola patent, including the International Centre for Tropical Agriculture (CIAT), the Food and Agriculture Organization (FAO) and the nongovernmental organization ETC Group. CIAT was able to dispute Proctor's claims by providing evidence of 260 yellow beans among the samples of beans conserved in its gene bank, and it also presented several scientific articles on yellow beans that showed the existence of prior literature. In the course of the patent revision, several studies showed Enola's near to complete identity with preexisting Mexican Peruano-type cultivars commonly grown by Latin American farmers as well as the identity of the yellow seed colour genotype with that of existing yellow bean cultivars documented in scientific literature prior to the patent application (International Centre for Tropical Agriculture, 2002; Pallottini et al., 2004). Azufrado Peruano 87, which was released for the first time by the Mexican Ministry of Agriculture in 1987 and described in an article by Perez Salinas and

Ildefonso Lepiz in 1983, was shown to have an identical genetic fingerprint as the claimed Enola seed. The USPTO issued a preliminary decision in 2003 rejecting all patent claims and gave a final rejection in December 2005. Proctor filed an appeal through the USPTO, and the patent remained in force while the appeal was being considered by the Board of Patent Appeals and Interferences. The board finally rejected all of the patent claims in April 2008, 9 years after Proctor had started to exploit the patent by claiming US$0.6 for every pound of yellow beans sold in the United States.[16] This decision was confirmed by the US Court of Appeal for the Federal Circuit in July 2009.[17] The PVP certificate is still valid.[18]

The Enola bean case has raised serious concerns for the CGIAR, a global partnership that unites 15 centres engaged in agricultural research, as well as for other international organizations involved in the conservation and use of genetic resources. These groups were not only worried about the immediate economic impact of the patent, but they were also particularly alarmed over whether the patent would establish a precedent that would threaten public access to plant germplasm that is held in trust by the CIAT as well as by other international research centres worldwide. As a result, the CGIAR centres have started to recognize the need to adopt preventive actions to avoid future cases of misappropriation (International Centre for Tropical Agriculture, 2008).

In addition to these two well-known cases, there have been various attempts to patent plant varieties in the public domain that have failed in the end. In several cases, this failure was a result of prior art that showed the lack of novelty of the claims. One example is the patent application on a warted pumpkin. In February 2009, the ETC Group denounced the patent that was being claimed by the Siegers Seed Company on a warted pumpkin, which was identified as having at least one wart associated with the outer shell of the body (ETC Group, 2009). The patent application included 25 broad claims covering a range of pumpkins with bumpy surfaces, a range of wart sizes relative to the pumpkin's surface and a range of wart colours.[19] It also included specific varieties and plants, seeds as well as the tissue of warty pumpkins. The patent was rejected for a number of reasons, including a sloppy application, the prevalence of warts on cucurbits historically and the fact that warted pumpkin seeds have already been available from other vendors.

There are also a number of cases where patents have been granted on the use of plants that were almost identical to the traditional uses of those plants, therefore lacking enough inventiveness to deserve patent protection. For example, one well-known case is the turmeric patent. Turmeric is one of the most basic ingredients of Indian food, and its antiseptic properties are widely known. In 1995, two researchers based at the University of Mississippi were awarded a patent on the use of turmeric for healing wounds, which consisted in administering turmeric powder topically or orally. The Indian Council of Scientific and Industrial Research challenged the patent on the ground that the alleged invention was part of the public domain in India. The patent was reexamined, and all former claims were cancelled.

It is difficult to provide an approximate number of cases of uncertain patents and PBRs. We only know those cases where there have been claims against the application or the patent itself. There could be a number of incorrect patents or PVPs that have passed unnoticed because they have escaped the attention of interested groups or because these groups do not have enough resources to enter into the revision procedures. Taking into consideration the fact that PBRs for ornamental crops account for more than half of the total applications granted in both the United States and Europe, and the fact that plant-related utility patents are a recent phenomenon, the number of applications that try to subject landraces and traditional farmers' varieties to patents and PVP should not be very high (Koo, Nottembur and Pardey, 2004). However, the literature produced by nongovernmental organizations (NGOs) and governmental agencies committed to fighting questionable patents argues that numerous patents and PVP certificates should be revised because they refer to plants and plant uses that could be in the public domain before the patents and PVP were requested. For example, the task force established by the Department of Indian Systems of Medicine and Homoeopathy in 2000 estimated that about 2,000 wrong patents concerning Indian systems of medicine were being granted every year. This prompted the Indian authorities to create a public database, the Traditional Knowledge Digital Library (TKDL), which could serve as register of prior art (Gupta, 2011). Since then, the Indian government has signed an agreement with the EPO, the USPTO, the Japan Patent Office (JPO), the German Patent Office, the Patent Office of Australia, the Canada Patent Office and the United Kingdom Trademark and Patent Office to make the TKDL available for patent examination procedures (Indian Council of Scientific and Industrial Research, 2011). As a result, according to the records of TKDL, around 170 patent applications have been withdrawn by their applicants or rejected by the patent offices based on information provided by the TKDL (TKDL, 2014). In Peru, the National Commission Against Biopiracy systematically reviews patents and patent applications that are based on genetic resources for which Peru is a source of diversity. So far, they have identified 13 patents or patent applications (referring to five plant species) that, according to their knowledge, do not meet the novelty requirement since they are plant-based uses in the public domain in Peru. In the last years, the commission has prevented foreign companies from obtaining seven patents related to the use of Peruvian traditional knowledge on plants (Comisión Nacional contra la Biopiratería, 2013). In 1998, the international NGO Rural Advancement Foundation International, together with the Heritage Seed Curators Australia, published a report on irregular plant variety protection grants and applications at the Australian Plant Breeders' Rights Office and a number of other patent and plant breeders' rights in other industrialized countries (Rural Advancement Foundation International, 1998). They identified 147 cases that had significant irregularities, of which 124 were presumed farmers' varieties from at least 43 countries and seven International Agricultural Research Centres (ibid.).

Protecting the public domain through defensive publishing

In view of this situation, defensive strategies aimed at preventing the granting of intellectual property rights on genetic resources and related traditional knowledge in the public domain have gained more and more importance in the last few decades. In contrast to forms of positive protection, these defensive strategies seek to ensure that third parties do not gain or maintain unfounded intellectual property rights (WIPO, 2007).

Defensive strategies are generally well-established intellectual property practice. There are several different strategies, but all of them rely on the same principle: by disclosing an invention it is placed in the public domain and, in this way, renders any identical invention ineligible for patenting or *sui generis* protection because of lack of novelty or uniqueness.

Patents and defensive publishing

Patents are granted to inventions that satisfy the requirements of utility, novelty and nonobviousness. An invention is novel when it does not form part of the state of the art – that is, when it has not been disclosed to the public before the patent application.[20] The nonobviousness requirement states that the invention will be patentable only when it is not an obvious consequence of applying the existing prior art by a person skilled in the subject matter. In order to determine the novelty and nonobviousness of the claims described in a patent application, patent examiners conduct a search of prior art by way of a literature exploration. In doing this search, patent examiners may ask the patent applicant to provide additional information about the invention and point out any prior art in the relevant literature. They may also consult with experts in order to fully understand whether the claimed invention is new and nonobvious, particularly for patent applications involving complex technologies.[21]

The actual scope of prior art differs from country to country. In the United States and Japan, prior art includes everything that is known or used by people in these countries or described in a publication distributed in these countries or in any other country. This means that unpublished knowledge or use in a foreign country may not prevent patentability.[22] Similarly, the Patent Cooperation Treaty limits the scope of prior art to 'everything made available to the public anywhere in the world by means of written disclosure.'[23] In Europe, prior art extends not only to publication but also to oral description, use or any other way of disclosure utilized anywhere.[24]

In reality, practical reasons limit the actual scope of prior art to published documents. Lack of published documentation on the existence of a certain invention may indicate to a patent or PVP examiner that such an invention is indeed new and worthy of intellectual property protection (Adams and Henson-Apollonio, 2002). Thus, if inventors want to ensure that their inventions are considered to be part of prior art, they must be careful to publish their description in specialized literature and not rely on nonpublished knowledge.

Defensive publishing has traditionally been used as a tool for inventors who are not interested in obtaining a legally enforceable monopoly, but rather wish to make sure that their inventions cannot be patented by someone else. Many corporations use defensive publishing as a part of their intellectual property management strategy. For example, in Japan, it is a relatively common practice to apply for patents for inventions that the applicant does not intend to use but that the inventor does not want to fall into the hands of competitors who may reinvent them. A practical solution is to file a patent application, wait for it to be published and then not continue the application process. In this way, the application will fall into the public domain and will necessarily be taken into consideration by patent examiners when assessing the patentability of claims filed by competitors. The USPTO institutionalized this process by developing a system called statutory invention registration, whereby inventions are published in the form of patent applications and made easily available for prior art searches. Some big companies have opted for developing their own defensive publication tools. The *IBM Technical Disclosure Bulletin*, which was published from 1958 to 1998, was a technical journal that was well known for disclosing inventions in order to prevent competitor companies from obtaining patents on them. Nowadays, there are a number of companies that provide expertise in defensive publishing and disclose inventions by publishing them in digital publications where patent examiners can easily search for prior art.[25]

The strategic use of a defensive publication is crucial in countries that have no, or very little, tradition in intellectual property protection. Often in these countries, despite the globalization of intellectual property standards, the majority of innovators are not interested in positively protecting their inventions or are unable to afford the patent application and the maintenance costs.[26] The strategic placement of innovations in the public domain is also important for public research agencies that want to ensure that their research products remain available for everyone.

When defensive publications are successful in protecting inventions in the public domain, the inventor, or any other interested person, does not need to intervene at a later stage, thereby saving the cost involved in the revision of a possible inappropriate patent over the invention. For this reason, it is more convenient for the inventors to adopt precautionary measures.

An important obstacle for the success of defensive strategies is the fact that patent examiners have to deal with an increasing number of patents and have a limited amount of time for prior art searches. The USPTO and the JPO receive around 400,000 applications per year and the EPO receives approximately 140,000. While in Japan the figure has been more or less constant during the last decade, as shown in Table 11.1, the number of patent applications in the United States and Europe has doubled since 1998 – biotechnology being one of the fields where applications have increased most dramatically.

Table 11.1 Number of patent applications and patent grants in EPO and USPTO for selected years

		EPO	*USPTO*
Patent applications	1998	82,087	243,062
	2007	140,763	456,154
	2012	148,229	542,815
Patents grants	1998	36,718	147,520
	2007	58,730	164,291
	2012	65,665	253,155
Plant patent applications	1998	NA	720
	2007	NA	1,049
	2012	NA	1,149
Plant patent grants	1998	NA	561
	2007	NA	1,047
	2012	NA	860
Average annual number of patents granted in the area of biotechnology (1998–2012)		1,335	4,020

Sources: EPO patent and patent application statistics, WIPO IP Statistics Data Center and US Patent Statistics Chart (Calendar Years 1963–2013).

The limited number of examiners in new areas of technology such as biotechnology have been overwhelmed with this current number of applications and there is a risk that they will not be able to dedicate enough time to their prior art searches. In the USPTO, a patent examiner spends an average of 21.2 hours on a full application review (US Patent and Trademark Office, 2007). The Patent Public Advisory Committee, a committee created to advise Congress on the goals and performance of the USPTO, has repeatedly pointed out to the lack of enough patent examiners capable of dealing with the increasing number of patent applications (US Patent and Trademark Office, 2013). Some critics also denounce these examiners' lack of necessary skills in the new technologies, which can be partially explained by the USPTO's inability to compete with corporations and law firms that offer examiners much greater salaries than the US government pays (Jaffe and Lerner, 2004).

An interesting proposal that has been suggested to overcome the current limitations of overwhelmed patent offices is the establishment of an online peer review system to help patent examiners find the right prior art and access those experts who can provide advice on the application (Noveck, 2006). The USPTO has been testing this system since January 2007, through a pilot project called 'Peer to Patent,' which encourages the public to review published patent applications that have been volunteered online and submit technical references and comments on what they believe to be the best prior art to consider during the examination. The Australian and the Japanese patent offices have also started pilot projects to test the convenience of the peer-to-patent system in their countries.[27] If widely adopted, the peer-to-patent system could be an effective channel to make defensive publications available to patent examiners.

PVP and defensive publishing

According to *sui generis* systems inspired by the UPOV Convention, a plant variety can be subject to PBR if it is new, distinct, uniform and stable. The meaning of new is different in patent law and in plant variety protection law. While in patent law novelty refers to the uniqueness of the invention, in the context of PBRs the novelty requirement is limited to commercial novelty; a variety is considered new when it has not been sold or used for more than one year prior to the application date. For this reason, unlike in the context of patents, defensive publishing strategies to prevent unfair PBRs are based not on the novelty requirement but rather on the distinctness requirement.

The UPOV Convention states that 'the variety shall be deemed to be distinct if it is clearly distinguishable from any other variety whose existence is a matter of common knowledge at the time of the filing of the application.'[28] Therefore, the fact that a variety is distinct is what makes it unique, different from the existing varieties and therefore new, in an absolute sense, and worth protecting.

When plant variety protection officers receive a PBR application, they conduct an examination to assess the distinctness, uniformity and stability of the candidate variety, comparing it with similar existing varieties. In the course of the examination, the authority may grow the variety, carry out other necessary tests or take into account the results of growing tests or other trials that have been carried out already. The UPOV General Introduction to the Examination of Distinctness, Uniformity and Stability and the Development of Harmonized Descriptions of New Varieties of Plants states that authorities need to examine distinctness in relation to all varieties of common knowledge (International Union for the Protection of New Varieties of Plants, 2002). However, field trials may not be needed whenever a candidate variety can be distinguished in a reliable way from varieties of common knowledge by comparing documented descriptions. In order to help the examination process, the PVP office requests the breeder to provide certain information about specific distinguishing characteristics, information on the breeding scheme of the candidate variety and any other information that may help to distinguish the variety. The breeder is also requested to identify similar varieties and characteristics by which the candidate variety may be distinguished from the existing ones.

Existing literature on defensive publishing does not address the use of defensive publishing in the context of *sui generis* systems for the protection of plant varieties. This fact may be due not only to the limited scope of PVP as a *sui generis* system but also to the particularities of the US system with regard to PVP. In the United States, new plant varieties can be subject to three different types of intellectual property rights: utility patents, PVP (according to the Plant Variety Protection Act enacted in 1970) and plant patents (according to the Plant Patent Act passed in 1930).[29] Utility patents can be granted on all types of inventions involving plants, including new plant varieties; PVP offers protection to new varieties of sexually reproduced or tuber-propagated plant species; and plant patents are limited to new varieties of asexually reproduced plants (except tubers).

The extent of protection varies, with utility patents being the form that provides the strongest protection for plant varieties. The requirements for obtaining a plant patent are the least strict of the three regimes, and the protection granted by this intellectual property right is the narrowest. The asexual reproduction requirement limits plant patent infringement to the narrow circumstance where the stock from the patentee's original parent plant is obtained and asexually reproduced. Independent breeding of a variety that closely looks like the subject of a plant patent is not considered infringement, nor is seed propagation or sexual crosses of the plant.

The US plant patent shares some of the requirements for PBR. Like plant variety protection, a variety must be new – meaning that the plant variety has not been sold or used for more than one year prior to the application date – and distinct. However, unlike PVP, plant patents do not require that the plant variety be uniform and stable. The process of examining and granting a plant patent is much closer to that of a utility patent. The plant patent examiner compares the description of the claimed plant with the closest available prior art. Unlike in PBR and similarly to utility patents, the process relies almost exclusively on existing literature and does not involve growing trials to compare the candidate variety with existing ones. In general, if the disclosure of the application does not distinguish the claimed plant from those previously known, the claims will be rejected as failing to showing distinctness of the candidate plant.

Several studies have analyzed how companies design their protection strategies in view of the existing parallel systems and have offered interesting conclusions that point towards a clear preference for utility and plant patents and a limited use of PVP.[30] Some of these studies conclude that PVP does not stimulate research and development in the United States, particularly for important crops such as wheat, maize and soya (Alston and Raymond, 2002; Janis and Kesan, 2002). It seems that the main reason the United States maintains the PVP system is because of the advantages that it derives from being a member of UPOV, in particular, the benefits of national treatment and a 12-month right of priority (Janis and Kesan, 2002).

However, defensive publication in the context of PVP is still important in those countries where plant varieties are not eligible for patenting – that is, in most of the countries of the world or in those countries where PVP is already the property right form that is most commonly sought to protect plant varieties.[31] Due to the UPOV Convention's very specific and strict examination trials with respect to the distinctness, uniformity and stability (DUS) conditions of the candidate plant varieties, the possibilities for obtaining a PBR on a plant variety that is already in the public domain are minimal if the examination is conducted properly. However, there have been several cases involving irregular PVP certificates, some of them showing a clear failure of the PVP national system, such as the Enola bean case discussed earlier. Many NGOs have denounced situations in which farmers' varieties have been misappropriated through PBR.[32]

Criteria to describe farmers' varieties in defensive strategies

For practical reasons, the description of farmers' varieties in defensive strategies must take into consideration the definition of plant variety in the patent and PVP systems, bearing in mind that the criteria that are needed to articulate a general definition of plant variety are different and independent from the criteria required by these systems to grant protection on plant varieties.

The UPOV Convention's definition of plant variety, which has been adopted with few changes by most national legislation in regard to both patents and PVP, reads as follows:

> 'Variety' means a plant grouping within a single botanical taxon of the lowest known rank, which grouping, irrespective of whether the conditions for the grant of a breeder's right are fully met, can be
>
> - defined by the expression of the characteristics resulting from a given genotype or combination of genotypes;
> - distinguished from any other plant grouping by the expression of at least one of the said characteristics; and
> - considered as a unit with regard to its suitability for being propagated unchanged.[33]

According to this definition, a plant variety must be distinct – that is, it must be different from all other existing varieties – in order to be considered as such by patent or PBR examiners. Does this mean that defensive publications need to show that farmers' varieties are distinct in an absolute sense? In my opinion, this is not necessary. The objective of the defensive publication is to provide evidence that a farmer's variety exists by making available enough information about the identity of the variety, but it does not need to prove that the variety is unique. The description of a farmer's variety in a defensive publication has to show those characteristics that make it distinguishable from similar varieties found in a limited geographical area, but it does not need to prove that the variety is different from all of the existing varieties worldwide. The fact that other identical varieties could be found, perhaps with different names, in different geographical regions should not disqualify a farmer's variety as a variety of common knowledge or of prior art as long as it has been properly documented in a publication. It will be the responsibility of the patent or PBR applicant to show that the candidate variety is different from the other documented varieties, and the patent or PBR examiner will need to confirm that this is so. After all, those publishing farmers' varieties with defensive objectives are not seeking to establish that their varieties are universally distinct or unique – this is obviously only required if they were seeking to have exclusive property rights. All they are asserting is that this variety exists at this moment in time, so that others may not claim to have developed it themselves de novo or that it is universally distinct or unique.

The use of new methods of varietal identification that allow for pinpoint differences at the genetic level raises questions about how distinctness between a new variety and the existing ones should be measured and what minimum level of difference between varieties should be accepted as being indicative of distinctness. Should differences at the genetic level be taken into consideration when they do not translate into observable morphological differences? Do morphological or agronomical differences that are not recognizable at the genetic level through molecular identification make a plant variety distinct? These questions are particularly relevant when dealing with farmers' varieties since it is not uncommon that the traits that are most important to farmers for distinguishing a variety are not the same ones used by the researcher to distinguish varieties genetically. For example, Busso et al. (2000) found that the traits that Ugandan farmers use to distinguish different varieties of pearl millet do not lead to genetic identity at the molecular level.

The rapid development of different techniques for genetic mapping, their proven efficiency in uncovering the diversity within plant species and identifying the differences between cultivars at the genetic level and the dramatic decrease of some of the costs of these techniques have brought to the forefront the question about what extent they can be used, or should be used, for DUS testing. A number of scholars have defended the benefits of using molecular markers in DUS testing processes when these techniques are shown to be more rapid and cost-effective than the classic comparison of morphological traits between candidate varieties and existing ones (Morrell et al., 1995; Giancola et al., 2002; Noli et al., 2008). On a number of occasions, national courts have requested the application of these techniques to solve cases where the morphological traits were not sufficient to distinguish closely related genotypes (Kumar et al., 2000) or they have accepted scientific studies that provide evidence about genetic differences between similar varieties.[34] However, there is a general consensus among experts that DNA fingerprints must not fully replace morphological traits, given that the distinctness identified by molecular markers may not necessary reflect morphological distinctness. This was the position adopted by the International Seed Federation in 2009 and reiterated in 2012 (International Seed Federation, 2012). The federation strongly endorses the use of DNA-based markers for variety identification purposes (e.g. in the case of enforcement of intellectual property rights) and to help determine genetic similarity between varieties in disputes on essential derivation. Its approach to the use of DNA-based markers in the DUS testing is much more cautious: the federation holds that DNA-based markers can be useful in the DUS testing and examination process whenever DNA-based makers are fully predictive of the expression of phenotypic characteristics, and that their use alone for establishing DUS could significantly decrease the scope of protection and should therefore not be accepted.

This topic has received much attention at UPOV in the last decade. Finally, in 2013, based on the recommendations of the Working Group on Biochemical and Molecular Techniques and DNA Profiling, the UPOV Council adopted UPOV's guidance on the use of biochemical and molecular markers in the examination of distinctness, uniformity and stability (UPOV, 2013). According to this guidance, the use of molecular markers in DUS testing are acceptable in the following two cases:

1 for examining DUS characteristics that satisfy the criteria for characteristics set out in UPOV General Introduction to the Examination of Distinctness, Uniformity and Stability and the Development of Harmonized Descriptions of New Varieties of Plants, if there is a reliable link between the marker and the characteristic;
2 where a combination of phenotypic differences and molecular distances can be used to improve the selection of varieties to be compared in the growing trial provided the molecular distances are sufficiently related to phenotypic differences, and the method increases the risk of not selecting a variety in the variety collection which should be compared to candidate varieties in the DUS growing trial.

The examples of the Ayahuasca plant and the Enola bean demonstrate that molecular techniques may be particularly useful in PVP and patent examination and reexamination procedures where the morphological characterization does not offer definitive responses about a plant's identity. The possibility of using irrefutable evidence provided by genetic information may increase objectivity and certainty in the examination process. The morphological and agronomical description of several varieties of yellow beans in prior art literature was not enough for the USPTO and the PVP office to deny novelty and distinctness of the Enola bean and Da Vine plant respectively. The Enola bean patent was eventually revoked because the Board of Patent Appeals and Interferences of the USPTO considered recent studies showing genetic identity between Enola and several existing varieties of yellow bean by using molecular markers (Pallottini et al., 2004).

In the case of the Ayahuasca plant, the USPTO examiners confirmed the patent on the Da Vine plant because they concluded that the differences in the leaf size showed that the Da Vine variety and *B. caapi* did not share the same germplasm and were therefore genetically different, even if in its previous decision the office had accepted that such morphological differences were very probably due to the plasticity of the plants and not to actual differences at the genetic level.[35] If the PVP examiners had applied molecular techniques to check out the genetic differences between Enola and other varieties of yellow beans, the PVP certificate would have probably never been granted.

Key components in defensive strategies to protect farmers' varieties

A good description of the variety

A detailed description of a farmer's variety is a key condition for it to be considered prior art or a plant variety of common knowledge. Without such a description, there is no possible way to challenge novelty and distinctness requirements in patent and PBR applications. As highlighted earlier, characteristics used to describe a farmer's variety in a defensive publication should be those that allow it to be distinguished from similar ones. They should be easily observable at several different stages of the plant cycle and fairly consistent across generations. This does not mean necessarily that the variety has to be stable according to the requirements for protection under the UPOV Convention, but rather that its key morphological characteristics must be transmittable through genetic heritage and not solely a result of the way a genetically identical plant grows in a different environment. Otherwise, it would be difficult for patent and PVP examiners to recognize different plants that belong to the same variety.

It is important that the morphological and agronomical traits of the variety are described in detail following internationally acceptable descriptors, such as the List of Multi-Crop Passport Descriptors developed by the FAO and the International Plant Genetics Resources Institute, and that characteristics such as the shape of the leaves and stem and the colour of the flower petals, fruits and leaves are presented according to acceptable standards. In addition to providing quality, certainty and transparency and facilitating a clear understanding between plant users, the use of commonly accepted descriptors and standards makes it easier to analyze plant varieties with respect to novelty and distinctness.

It is recommended that the records of the variety also include photos and drawings and that they indicate where living samples of the plant and seeds or other reproductive material can be obtained. This information will not only show that the plant is already publicly available but also will help the examiner find plant samples if necessary (Boettiger and Chi-Ham, 2007).

Defensive strategies for farmers' varieties must take into consideration that, while the only element to be protected by a PBR is a plant variety, patents can also protect single plants and their parts, including tissues, chemical compounds, genes and so on. Breeding processes are also patentable, as long as they are not considered to be biological processes by patent examiners, as well as the uses of plants and plant components. For this reason, a defensive publication that is aimed at preventing patent grants on farmers' varieties should provide not only a detailed description of the variety but also a description of other elements that could be subject to patent claims, such as breeding techniques associated with the plant variety, its agronomical and other uses and, if possible, any useful chemical compounds that can be derived from it. Obviously, there is a limit to what defensive publishing can cover and achieve. A good morphological description of a farmer's variety and of its common uses and associated breeding practices may provide very little protection when faced with a patent application claiming the use of plant chemical processes and compounds that are the result of advanced scientific knowledge and technologies. The novelty of such patent claims can be very difficult to challenge, and the wisdom of questioning such claims questionable when the claims present a novel, not obvious and useful application or use of the plant components.

One of the biggest challenges for defensive publications is language. In general, documents that are not written in English will rarely be taken into account by patent and PBR examiners in the largest intellectual property offices. For this reason, it is highly recommended that the descriptions of farmers' varieties include a summary or abstract in English or, at the very least, a list of English keywords that will help the examiner assess the relevance of the publication as prior art and make a decision about translating it into a language that he or she can understand.

Another issue that concerns defensive publication is whether defensive publications should provide molecular information in order to effectively describe farmers' varieties. In the case of landraces, demonstrating their genetic identity would not only be expensive but also extremely difficult, if not impossible. Farmers' varieties are usually made up of genetically variable populations (Halewood et al., 2005, citing Zeven, 1998), and thus it would prove very challenging to obtain reliable, sufficiently exhaustive molecular-level information to prove the existence of a distinct, yet genetically diverse population. For this reason, plant descriptions in defensive publications should continue to rely primarily on phenotypic characteristics.

Making defensive publications easily reachable by patent and PVP examiners

Since examination procedures are different for patents and PBRs, defensive protection efforts may need to adopt different strategies to ensure that published information concerning farmers' varieties is accessible by patent and PVP offices.

PBRs

The UPOV Convention and the General Introduction to the Examination of Distinctness, Uniformity and Stability both provide key information for determining how a defensive protection strategy has a better chance of ensuring that PVP examiners take farmers' varieties into consideration as varieties of common knowledge. The convention reads:

> The filing of an application for the granting of a breeder's right or for the entering of another variety in an official register of varieties, in any country, shall be deemed to render that other variety a matter of common knowledge from the date of the application, provided that the application

> leads to the granting of a breeder's right or to the entering of the said other variety in the official register of varieties, as the case may be.[36]

Since the text does not specify what type of official register makes a variety a matter of common knowledge, it can be understood that varieties registered in any official register can be considered common knowledge for all UPOV members as long as they respond to the definition of plant variety in the UPOV Convention. Such registers include the official list of protected varieties, the register of commercialized varieties as well as any existing registers of traditional varieties such the ones described later in this chapter.

The General Introduction to the Examination of Distinctness, Uniformity and Stability points out that, in addition to the aspects described earlier, the following elements define, among others, varieties of common knowledge:

- if the propagated or harvested material of the variety has been, or is being, commercialized;
- if there are publications providing a detailed description of the variety;
- if there are samples of the variety in publicly accessible plant collections.

The general introduction also clarifies, in the same way as the UPOV Convention, that in order to be considered a variety of common knowledge the variety does not necessarily need to fulfil the DUS criteria required for granting a PBR under the UPOV Convention. A way to ensure that PVP examiners consider farmers' varieties to be varieties of common knowledge is by: (1) registering them in official registers of traditional varieties or other registers adapted to farmers' varieties; (2) making samples available in public plant collections; and (3) publishing the description of the varieties in journals or catalogues commonly used by plant scientists.

Patents

Patent office search tools are very good at identifying patent literature, but they do not always provide access to all relevant prior art, particularly when it has been published in a foreign country. Defensive publishing depends on the ability of patent and PBR examiners to find publications in nonpatent prior art searches. Stephen Adams and Victoria Henson-Apollonio (2002) provide a comparison between defensive publication mechanisms in the context of patents. Table 11.2 represents their assessment of the accessibility of different mechanisms. The table shows that the current scenario is not very promising for farmers' varieties. A large proportion of plant variety developers, conservers and users do not have the need or the means to systematically publish the necessary information about the plant varieties that they develop and use. This is not only true for farmers – particularly farmers in developing countries – but also for plant collectors and curators of gene banks and botanical gardens, who have not always kept detailed records of the materials they have collected

Table 11.2 Comparison of defensive publication modalities

Self publication			
Institution publicity materials	Institution series	Occasional publications	Gray literature
Moderate to poor	*Moderate*	*Generally poor*	*Generally poor*
Third-party publication			
Commercial research disclosure publications (IP.com and others)	Peer-reviewed journals	Unexamined patent application	Other intellectual property title
Good to very good	*Good*	*Very good*	*Good*

Source: Adapted from Adams and Henson-Apollonio (2002).

or maintained. This situation is very similar for breeders, who usually do not document any information beyond that which is necessary for them to carry out their work. In general, these categories of plant users do not disclose plant varieties and plant uses in a way that makes the varieties easily available in prior art searches, except when they have actually sought patent or PVP protection of the disclosed plant or use. Consequently, a great deal of information is poorly documented and unpublished, particularly in the case of farmers' varieties, and a large portion of the prior art may only be found in nonscientific literature, gene bank databases, herbarium descriptions and nonofficial registers of plants and plant uses. This information is generally beyond the reach of patent examiners conducting prior art searches.

Defensive protection strategies must be aimed at filling these informational gaps with respect to farmers' varieties by publishing information about existing farmers' varieties in a way that allows patent examiners to access the information easily. Some of the mechanisms that are ranked well or very well by Adams and Henson-Apollonio are often too complicated to access or unaffordable for a range of actors that are potentially interested in protecting farmers' varieties through defensive publication, including public agencies in developing countries and small society organizations and farmers. For these actors, a good alternative may be to make their defensive publications known by getting them included in nonpatent literature databases used by the patent offices. The agreement between the International Crop Research Institute (ICRISAT) and the EPO is a good example of how this can be done. In 2005, ICRISAT signed a memorandum of agreement with the EPO allowing ICRISAT to include its publications as part of the EPO's nonpatent literature. Thanks to this agreement, information and knowledge generated by ICRISAT is being provided to European patent examiners for consultation in prior art searches. About 70 documents produced by ICRISAT have been consulted in prior art searches since the agreement was signed. National agricultural research institutes could pursue the same type of agreement with some of the big patent offices in the world, and they could even explore how regional organizations and networks promoting agriculture research and development could provide coordination and advisory services to interested countries.

A close look at biodiversity public registers and databases

In recent years, there has been a tendency among civil society organizations, research institutes and governmental agencies to develop registers and databases that document the biodiversity and associated traditional knowledge of a given area or a whole country. These registers and databases serve many different purposes, but, in most of the cases, these documentation efforts do not seek to create defensive publishing but, rather, arrive at it by default. Consequently, the actual ability of the registers or databases to perform their role as defensive publications is not clear.

Many of the existing registers and databases focus on local communities' traditional knowledge. Their primary need is to preserve this knowledge rather than to put it in the public domain or disseminate it outside the original community. For this reason, they have been purposefully designed so that users outside the community cannot access all of the information in the register or database. In this way, the registers do not disclose the information beyond the circle of viewers that is permitted by customary law.[37] These characteristics therefore limit these types of registers and make them unable to function as defensive publications, which is perfectly fine according to their objectives.[38]

Public local registers initiated by research or civil society organizations to document biodiversity and biodiversity's uses at the local level are becoming more and more popular, particularly in developing countries. The local registers differ very much in terms of their primary objectives, but all of them represent, to a greater or lesser extent, a way of 'memory banking.' This term was coined by Virginia Nazarea-Sandoval (1998) to refer to the collection and documentation of farmers' knowledge for future use and is an analogy to the storage and documentation of germplasm in a gene bank. Memory banking serves to capture and record the cultural dimensions of plant biodiversity, including local names, indigenous technologies and uses associated with different plants and varieties that have been traditionally passed from one generation to another by oral means, for access and management by local communities.

The local registers address a range of objectives, including:

- capturing and recording the cultural dimensions of plant biodiversity;
- recording the present status of biodiversity;
- monitoring changes in ecosystems and genetic erosion;
- documenting the uses of genetic resources;
- protecting genetic resources and traditional knowledge from patenting or PBRs through defensive publication;

- creating a sense of ownership and empowering local communities in regard to local activities oriented to the conservation and sustainable use of genetic resources and related traditional knowledge;
- perpetuating and promoting the development of ecological knowledge of local communities;
- identifying the conservers of traditional crop varieties and associated traditional knowledge with whom an equitable share of the benefits arising from the use of such resources and knowledge should be shared;
- enhancing the collaboration between people working in research and education institutes, government agencies and civil society organizations and farmers, fishermen and traditional healers.

In India, the People's Biodiversity Registers had the original objective of documenting community-based knowledge of medicinal plants and their uses, but, after some initial experiments and some early consultations, the registers' promoters decided to broaden the scope of the exercise to all elements of biodiversity and to record knowledge and perceptions at all levels, from individuals, households and ethnic groups to multiethnic communities (Gadgil, 1996).[39] Currently, these are operative in 12 Indian states. The information recorded in the registers relates to present status, changes over recent years and factors affecting the distribution and abundance of living organisms as well as known uses of biodiversity. They also record the perceptions of local people about ongoing ecological changes, their own development aspirations and their preferences about the management of living resources and habitats (Gadgil et al., 2002).

The main limitations of the People's Biodiversity Registers for serving as defensive instruments to protect farmers' varieties are the fact that many of them:

- embrace miscellaneous information on local biodiversity, making it difficult to identify the actual plant varieties and their specific characteristics;
- do not provide enough information about the farmers' varieties, their uses and the breeding methods;
- do not follow international descriptors;
- even if public, are difficult to access by common people outside the communities and, therefore, are unlikely to be considered in prior art searches.

In Nepal, civil society organizations have promoted the establishment of community biodiversity registers in different regions of the country. Table 11.3 shows the minimum information that is required in registers maintained by communities in the Begnas and Rupa watershed areas.

Compared to the People's Biodiversity Registers, the Nepalese Community Biodiversity Registers offer information about farmers' varieties that is much more focused, complete and detailed, although the descriptions of the varieties are often based on the characteristics that are most important for farmers, which are not always the same as the ones internationally accepted by the

Table 11.3 Minimum information required in registers maintained by communities in the Begnas and Rupa watershed areas

Information on cultivar/breeds/species/varieties	Primarily with consumptive use values. However, it should not mean that the species with nonconsumptive, existence or intrinsic values are undervalued.
Existence history (year)	Since which year the particular bioresource exists in the habitats or ecosystems.
From where the species was introduced	Name of the original place from where the new bioresources were first introduced.
Nature of the species	The nature of the bioresources we defined here as annual, perennial, evergreen, deciduous, herb, shrub, tree and so on.
Mode of reproduction	Means of propagation: seed, clones, sapling, stem and leaf.
Natural habitats	The natural habitats as recognized by farmers.
Extent and distribution	The extent and distribution of bioresources in terms of frequency and area – described as rare (R), medium (M) and widely grown (W).
Local techniques	Processing techniques may vary by species linked with product quality.
Uses	The bioresources are valued in terms of food, clothes, medicine, religion, culture and so on. These values are described by means of their specific purpose and their significance on certain occasions.
Useful parts, stages and times	The sustainable harvest of any local product – each bioresource is defined in terms of harvesting time and cultivation stage.
Life cycle	Time of emergence, growth, regeneration and harvest is recorded.
Information on custodians	An address is provided for the person who has supplied the following information.
Photographs/drawings	Distinguishing characters or useful parts of the recorded bioresource.

Source: Subedi et al. (2005).

scientific community. Like the Indian registers, the Nepalese registers are very likely to be beyond the reach of patent examiners in their prior art searches or of PVP examiners in their analysis of plant varieties of common knowledge.

National registers covering the whole national territory are more rare. As mentioned in the introduction of this chapter, Peru has recently developed an online register of native potato varieties, which is managed by the National Institute of Agriculture Research (Instituto Nacional de Investigación y Tecnología Agraria y Alimentaria (INIA)).[40] Another example is the Portuguese register of autochthonous plant material, which includes local varieties and spontaneously occurring material as well as associated traditional knowledge, which was created by law in 2002.[41] In terms of information provided, the Peruvian register has more opportunity to function as a defensive publication than the databases and registers described earlier. The main reasons for its success include the following:

- It focuses exclusively on traditional varieties of certain crops. This narrow focus allows it to cover each record with much more detail, and it follows the same unique description system for each type of crop.
- It is comprehensive. It includes descriptions of all of the farmers' varieties and landraces known in the country.
- The description of each farmer's variety is very detailed and combines descriptors that are internationally recognized by the scientific community as well as those that are relevant only for farmers, such as taste. The descriptions also include photographs and drawings.
- It is the responsibility of the national public authority that deals with plant genetic resource issues, lending a 'mark of quality' to the database and ensuring that it is permanently available and regularly updated in the long term. It also qualifies as an 'official' register according to the UPOV Convention.

One significant limitation, however, is that access to the database is not open to the public but must be requested to INIA authorities, which makes it difficult for patent and PVP examiners to use it as a source of information of prior art.

In order to be included in the Peruvian register, potato and maize varieties do not need to pass a formal DUS examination. It is understood that the fact they have been used for decades automatically categorizes them as varieties, as long as their distinctive and valuable traits remain stable, at least to a certain degree, across generations. Their inclusion in the register does not provide any right to the variety holders. However, they can benefit from the efforts made by the INIA to promote the conservation of traditional varieties. For example, the INIA is using the register as a means of certifying some traditional varieties by the National Seed Certification Authority, so that their seeds can be sold as commercial varieties (CIP, Centro Internacional de la Papa, 2008). The necessary requirements to catalogue these varieties as commercial are more flexible than they are for varieties outside the register – DUS characteristics are not required, nor is it necessary to assess the agronomical value since these values have largely been proved by the years of cultivation in farmers' fields. In addition, the certification is free of charge and does not have to be renewed every 5 years, as is done for nontraditional commercial varieties.

Efforts to document national and local plant genetic heritage and to prevent its misappropriation are not exclusive to the developing world. In Chile, an initiative engaging NGOs, institutions of the academia and other organizations have created a national catalogue of traditional seed which is available through the websites of the 20 institutions which have participated in the development of the catalogue. The catalogue is a living document which is updated on an ongoing basis. We can find examples of official registers and catalogues of traditional varieties in Portugal (mentioned earlier), France and Italy. In general, the main requirement for introducing a variety in these registers is the provision of a good description of the variety and confirmation that it has been used for a long time and therefore can be considered traditional (see Chapter 10 of this volume). The main purpose of these registers is to maintain information on existing varieties for conservation purposes. Some of them also regulate the commercialization of farmers' seeds. Table 11.4 summarizes the characteristics of some of these selected registers, highlighting those features that facilitate or hinder their role as defensive mechanisms to prevent the misappropriation of traditional crop varieties.

Supporting defensive strategies

According to what we have explained so far, an efficient defensive strategy may often require a capacity and level of resources that goes far beyond what most NGOs and research institutions in developing countries can afford. One way to overcome this limitation is by strengthening the responsibility and support of public agencies. It is essential that those governments that defend the need to conserve and protect their genetic heritage provide sufficient support to registers and databases of plant genetic resources. Indeed, in countries with a long tradition on community-based registers, the registers initially managed by NGOs and community-based organizations at the local level have eventually been integrated in national public supporting programs and received official recognition (Subedi et al., 2013).

The acceptance of developing country databases by foreign intellectual property rights authorities may be easier when governments choose to officially recognize these databases and when the information contained in them is verified in accordance with standard processes. The current patchwork of NGO or community-led biodiversity registers may not be very useful in the context of defensive publication strategies unless they receive official sanction and are subject to standard procedures that ensure their accurateness for defensive strategies.

International agencies can also assist in different ways, and some of them have already adopted measures to support the efforts of developing countries. Following the recommendations of the Inter-Governmental Committee on Intellectual Property, Genetic Resources, Traditional Knowledge and Folklore and as part of its efforts to harmonize intellectual property rights with the protection of genetic resources, traditional knowledge and folklore, the World Intellectual Property Organization (WIPO) has recently been working on the development of a comprehensive and worldwide online database of traditional knowledge and genetic resources that can be used by patent examiners in their prior art searches. This database, which is still in its initial stages, browses information from different online catalogues, including information on traditional knowledge and genetic resources. Currently, this information is not always sufficient or

Table 11.4 Characteristics of selected registers

	Objective	*Scope*	*Does it provide a detailed description of the plant varieties?*	*Does the plant description follow internationally accepted descriptors?*	*Is it an official register or catalogue (official meaning managed by a state agency)?*	*Is it in English, totally or partially?*	*Is it public?*
Community Biodiversity Registers in Nepal	• Documentation for conservation and use purposes. • Monitoring genetic erosion. • Protection against misappropriation.	Each register has its own scope. The most common elements are: • traditional plant varieties • traditional knowledge • forest biodiversity • bioresources in general.	Yes.	No.	No, but for some registers (Mustang, Tanahun and Sunsari) the Ministries of Forest and Soil and Agriculture and Cooperatives have provided training and logistical support.	No. The registers are available in local languages only.	Interested people outside the communities can consult the registers but cannot copy them.
Honey Bee	• Conservation and diffusion of knowledge. • Protection against misappropriation. • Adding value to the traditional knowledge. • Recording information for the purpose of benefit sharing with knowledge holders.	• Mainly traditional knowledge and innovation.	No.	No.	No[a].	Yes. The database is available in English as well as in Hindi, Gujarati, Tamil, Kannada, Telugu, Malayalam and Oriya.	Partially. The database is available online.[b]

(*Continued*)

Table 11.4 (Continued)

	Objective	*Scope*	*Does it provide a detailed description of the plant varieties?*	*Does the plant description follow internationally accepted descriptors?*	*Is it an official register or catalogue (official meaning managed by a state agency)?*	*Is it in English, totally or partially?*	*Is it public?*
People's Biodiversity Registers in India	• Documentation for conservation and use purposes. • Promotion of knowledge-based sustainable management of resources. • Adding value to biodiversity resources • Protection against misappropriation. • Recording information for the purpose of benefit sharing with genetic resources and knowledge holders.	• Agrobiodiversity. • Wild biodiversity. • Urban biodiversity.	Yes.	No.	The National Biodiversity Authority establishes the template for the registers and provides guidance and technical support to the management committees in charge of the registers.	Partially.	Yes, but restrictions may be established.
Register of Native Varieties of Potato in Peru	• Documentation for conservation and use purposes. • Recognition of traditional varieties as Peruvian genetic heritage.	• Native Peruvian potato varieties and associated traditional knowledge.	Yes, including molecular characterization.	Yes.	Yes. The register was established by the Ministry of Agriculture. The National Institute for Agricultural Innovation is in charge of developing and managing the register.	Available in Spanish only.	No. Access to information is subject to permission.

	• Protection against misappropriation. • Recording information for the purpose of benefit sharing with varieties holders.						
Register of Autochthonous Plants of Portugal	• Recognition of national sovereignty over traditional varieties. • Documentation for conservation purposes. • Providing grounds for appellations of origin and geographical indications. • Promotion of secure exchange of plant genetic resources. • Recording information for the purpose of benefit sharing with genetic resources holders. • Protection against misappropriation.	• Autochthonous plant material of current or potential interest to agrarian, agroforest and landscape activity, including local varieties and spontaneously occurring material as well as associated knowledge.	Yes. However, specific criteria and conditions for inclusion in the register have to be established by ministerial decree.	No. The register is not yet developed.	Yes. The catalogue was established by decree-law and is managed by the directorate general for crop protection, under the Ministry of Agriculture, Rural Development and Fisheries.	The register is not yet developed.	The register is not yet developed.

(*Continued*)

Table 11.4 (Continued)

	Objective	*Scope*	*Does it provide a detailed description of the plant varieties?*	*Does the plant description follow internationally accepted descriptors?*	*Is it an official register or catalogue (official meaning managed by a state agency)?*	*Is it in English, totally or partially?*	*Is it public?*
Catalogue of Local Varieties in Tuscany (Italy)	• Conservation and utilization of traditional varieties.	• Traditional plant varieties of Tuscany (Italy)[c].	Yes.	Yes.	Yes. The catalogue was established by regional authorities and is managed by the Regional Agency for Agricultural Development and Innovation.[d]	No. The catalogue is only available in Italian.	Yes, the information is public and available online.[e]
Catalogue of Traditional Seeds of Chile	• Conservation and utilization of traditional varieties.	• Traditional varieties of numerous crops of Chile.	Only for some entries in the catalogue.	No.	No.	No. The catalogue is only in Spanish.	Yes, the information is public and available online.[f]

[a] The network is managed by the following organizations: the Society for Research and Initiatives for Sustainable Technologies and Institutions, the National Innovation Foundation, the Grassroots Innovation Augmentation Network, the SEVA (Madurai), the Pritvi, PEDES (Kerela), the Innovation Club (Orissa), and the Network of Gram Vidyapeethas.

[b] www.sristi.org/wsa/index.htm.

[c] According to applicable legislation, local varieties are genetic resources (including species, varieties, cultivars and populations):

- originated in Tuscany;
- came from outside Tuscany but traditionally integrated in the regional agriculture;
- ωερε derived from the previous ones;
- αρε no longer present in Tuscan agriculture but conserved in botanical gardens or research institutions.

[d] Legge regionale, 16 novembre 2004, no. 64 Tutela e valorizzazione del patrimonio di razze e varietà locali di interesse agrario, zootecnico e forestale; Decreto del Presidente della Giunta Regionale, 1 marzo 2007, no. 12. Regolamento di attuazione della legge regionale, 16 novembre 2004, no. 64.

[e] http://germoplasma.arsia.toscana.it/Germo/.

[f] In several websites, for example: www.terram.cl/images/DOCotros/catalogo-semillas-tradicionales-de-chile.pdf.

available in a user-friendly format due to the fact that it includes records coming from many different sources, presented in different formats and dealing with very different topics. WIPO could play a valuable role in setting standards in the development of biodiversity databases and registers that seek to work as defensive publications as well as in facilitating the negotiation of agreements between intellectual property offices and the authorities in charge of these databases.

The establishment of an information system that supports the objectives of the multilateral system on access and benefit sharing under the International Treaty on Plant Genetic Resources for Food and Agriculture (ITPGRFA) opens the door for another multilateral tool for defensive strategies. Such information system will eventually make information on the accessions of germplasm included in the ITPGRFA's multilateral system publicly available. It may be worth exploring how this huge database can serve the defensive purposes of farmers' varieties and combine the coordinated efforts of the ITPGRFA and WIPO.

Conclusions

By disclosing farmers' varieties to the public through defensive publication, the public nature of such varieties can be protected against misappropriation through patents or PBRs. However, not all publications are created equal, and many publications do not meet the necessary conditions for defensive protection. In order to be effective, defensive protection strategies must rely on some key elements delineated by the legislation on patents and plant variety protection and by patent and PVP examination practices. The core of a defensive protection strategy must be a detailed description of the farmer's variety in a public document. Such a description should always include those characteristics that make the variety distinguishable (but not necessarily distinct) from other similar varieties. Making the publications easily accessible to patent examiners is also a key element of defensive protection strategies. There are numerous ways to get defensive publications included in examiners' nonpatent literature searches, such as by signing agreements with patent offices. The USPTO pilot project 'Peer to Patent,' which is also being tested by the Australian and Japanese patent offices, may also be a promising channel.

Local and national registers of biodiversity could integrate these elements into their usual operations in order to effectively serve as defensive publications and therefore protect farmers' varieties against unfounded intellectual property rights. However, this shift may change their original scope and way of functioning. In order to increase the efficiency of defensive strategies, governments should provide public support for the development and maintenance of registers as well as for the negotiation of possible agreements with intellectual property offices. They should also encourage the development of a monitoring system that ensures that the genetic resources described in the registers are not misappropriated. Intergovernmental agencies and agreements such as WIPO and the ITPGRFA can explore various avenues of operation whereby they can support national efforts to protect their genetic heritage in the public domain.

I would like to finish with a word of caution. Ultimately, defensive protection is only a tool used to protect the public nature of goods available in the public domain and to ensure that information about these goods is made easily available to the public. It is worth noting that defensive protection is not the appropriate tool to deal with issues commonly raised in regard to plant patents and plant-derived product patents, for example:

- the extension of patent protection on plant varieties;
- the lax interpretation of patent requirements in plant innovation;
- the general lack of recognition or compensation for farmers and indigenous communities when new products are based on traditional plant species and varieties or on ancestral knowledge;
- the difficulties that arise in establishing a clear line between common knowledge and novelty in many patents involving the use of traditional knowledge in the use of plants;
- the use of genetic resources and traditional knowledge in patented products without consulting the countries of provenance and the holders of the resources and the knowledge;
- the incapacity of current intellectual property system to adequately protect traditional plants and knowledge.

Defensive strategies may somehow support wider initiatives to deal with these issues, but they themselves do not promote any significant change in current intellectual property schemes. Instead, defensive publication works to reduce the failures, inefficiencies and/or inequities of existing patent and PBR systems.

Acknowledgements

I would like to thank Michael Halewood for his general guidance and useful discussions, Alejandro Mejías for helping me in putting together Table 11.3, the anonymous peer reviewers and Kay Chapman for their thoughtful comments, Rolf Jördens for a nice conversation on the UPOV Convention and DNA fingerprinting and Victoria Henson-Apollonio for helping me navigate the labyrinth of the Enola bean case.

Notes

1 The register was officially recognized by Ministerial Resolution no. 0533–2008-AG, Lima, 1 July 2008, published in *El Peruano*, 3 July 2008.

2 Berne Convention for the Protection of Literary and Artistic Works, 9 September 1886, online: <www.wipo.int/treaties/en/ip/berne/trtdocs_wo001.html> (last accessed 3 June 2014), Article 18:

> (1) This Convention shall apply to all works which, at the moment of its coming into force, have not yet fallen into the public domain in the country of origin through the expiry of the term of protection. (2) If, however, through the expiry of the term of

protection which was previously granted, a work has fallen into the public domain of the country where protection is claimed, that work shall not be protected anew.

3 See, for example, D. Lange (2003), J. Boyle (2003) and E. Samuels (2002).
4 *In re Bergy*, 596 F.2d 952 (C.C.P.A. 1979): 'Whoever invents or discovers any new and useful process, machine, manufacture, or composition of matter, or any new and useful improvement thereof, may obtain a patent therefore, subject to the conditions and requirements of this title.'
5 *Diamond v. Chakrabarty*, 447 U.S. 303 (1980).
6 *Ex parte Hibberd*, 227 USPQ 442 (Bd. Pat. App. & Int. 1985). International Convention for the Protection of New Varieties of Plants, 2 December 1961, online: <www.upov.int/upovlex/en/upov_convention.html> (last accessed 3 June 2014). Plant Variety Protection Act, 1970, 7 U.S.C. §§ 2321–2582.
7 Convention on the Grant of European Patents, 5 October 1973, online: <www.epo.org/law-practice/legal-texts/html/epc/2013/e/ma1.html> (last accessed 3 June 2014). The convention creates a unique patent-granting procedure for all country members (members totalled 35 in May 2009). It entered into force in parallel to the Patent Cooperation Treaty, online: <www.wipo.int/pct/en/texts/articles/atoc.htm> (last accessed 3 June 2014). Applications under the European Patent Convention are processed by the EPO.
8 Case T19/90, *Onco-mouse/HARVARD*, Doc. EP-A169, OJ EPO 1990, 476.
9 Case T1054/96, *Plants/NOVARTIS*, OJ EPO 1998, 511.
10 See, for example, Patent EP 1069819 on a method for the selective increase of anticarcinogenic glucosinolates in Brassica species, which is currently pending decision at the EPO Enlarged Board of Appeal, the highest court at EPO. This patent was granted in 2002 by the EPO to Plant Bioscience, a UK company, on the breeding methods, the broccoli seeds and the edible broccoli plants obtained through these breeding methods. In 2003, the plant breeding companies Limagrain and Syngenta filed oppositions arguing that the patent claims refer to an essentially biological process, which is not patentable under the EC Directive 98/44 on the Legal Protection of Biotechnological Inventions and Article 53(b) of the European Patent Convention. Another similar case is the patent EP 1211926 on the method for breeding tomatoes having reduced water content and the product of such method, which belongs to the Ministry of Agriculture of Israel. This patent is also pending consideration by the EPO Enlarged Board of Appeal after the opposition submitted by the Dutch company Unilever, which is based on the same arguments as the broccoli case. All official documents related to these two patents can be found in the European Patent Register, online: <www.epo.org/searching/free/register.html> (last accessed 3 June 2014).
11 According to the European Patent Convention's guidelines for examination, whether or not a process is essentially biological depends on the level of human intervention in the process and its impact on the results. In Decision T320/87 on Hybrid Plants/LUBRIZOL, Doc. EP-A 4 723, OJ EPO 1990, 71, the EPO Board of Appeal concluded that the claimed processes to produce hybrid plants did not constitute an exemption to patentability. Even though none of the steps implied enough human intervention, the process design did since the steps would not have been combined in such a manner without human intervention.
12 See European Patent Office (2013, section 5.4.2).
13 See, for example, Patent EP 0483514 on the use of molecular markers in tree breeding, which involves one of the most common methods in genetic fingerprinting: the restriction fragment length polymorphism (RFLP) for the selection of trees. Claim 1 reads:

> A method of forest tree breeding wherein RFLP technology is applied to samples of tree material from a plurality of forest trees; the data derived from said RFLP technology is statistically analyzed thereby to cluster genetically similar trees of said plurality of say trees; two of said trees of genetic diversity are selected based on

the statistically analyzed RFLP data; and a further tree is/are derived from the two selected trees.

14 US Patent and Trademark Office, 'Office Action in Reexamination,' Re-examination no. 90/005307, Art Unit 1661 (4 April 2000).
15 US Patent and Trademark Office, 'Notice of Intent to Issue Reexamination Certificate: Statement of Reasons for Patentability and/or Confirmation, Re-examination no. 90/005307, Art Unit 1661 (26 January 2001). The Center for International Environmental Law provides a detailed description of the process and the issues involved in G. Wiser (2001).
16 *Ex Parte POD-NERS*, LLC, 2007–3938 (Bd. Pat. App. & Int. 2008).
17 *In re POD-NERS*, LLC, 2008–1492 Re-examination no. 90/005,892 (US Court of Appeals for the Federal Circuit, 2007).
18 Plant Variety Protection no. 9700027. Information about the variety and the PVP certificate can be found in the National Plant Germplasm System of the US Department of Agriculture, online: <www.ars-grin.gov/cgi-bin/npgs/html/showpvp.l?pvpno=9700027> (last accessed 29 May 2014).
19 Patent US 20080301830A1.
20 Some countries recognize a general so-called grace period that offers a specific period of time in which a patent application may be filed despite a previous disclosure of the invention. In this way, the grace period avoids the consequences of an inconsiderate publication. US, Japanese, Canadian, Russian and Chinese patent laws include a general grace period. In contrast, the European Patent Convention and the patent laws of most European Union member states do not offer a grace period – that is, every publication that makes an invention available to the public before the date of the patent application eliminates its novelty. Although Article 55 of the European Patent Convention does recognize a specific grace period for abusive publications that have been made in spite of confidentiality agreements. Only New Zealand applies local novelty, which means that only publications, uses or sales that have taken place within the national jurisdiction are capable of destroying novelty.
21 The *Manual of Patent Examining Procedure* of the US Patent and Trademark Office (USPTO) (2014) provides a very detailed description of the tasks to be carried out by patent examiners in patent examinations.
22 Citation of Prior Art, 35 U.S.C § 102, <http://uscode.house.gov>; Japan Patent Law, available in English at: <www.wipo.int/clea/en/details.jsp?id=2652>, Article 29 (last accessed 29 May 2014).
23 Patent Cooperation Treaty, supra note 7, Rule 64.1 on Prior Art for International Preliminary Examination.
24 Convention on the Grant of European Patents, supra note 7, Article 54.
25 See, for example, <http://IP.com> (last accessed 29 May 2014).
26 From 2002 to 2006, the patent offices of the United States, Europe and Japan have received an annual average of 75% of the total number of patent applications worldwide. The other countries increased their applications on average by 15% per annum. A large amount of this growth was made by China and South Korea – their combined share went up from 11% in 2002 to 20% of all filings in 2006. In 2009, 85% of the 7.3 million patents in force were valid in the jurisdictions of the European Patent Office, the Japan Patent Office, the Korean Intellectual Property Office and US Patent and Trademark Office. See Trilateral (2007) and (2010), online: <www.trilateral.net/index.html;jsessionid=kj7rdscdhte6> (last accessed 29 May 2014).
27 Peer-to-Patent pilot project, online: <www.peertopatent.org> (last accessed 10 January 2015).
28 UPOV Convention, supra note 6, Article 7.
29 Plant Patent Act, 35 U.S.C. §§ 161–164; Plant Variety Protection Act, 7 U.S.C. §§ 2321–2382 [PVP Act].

30 Plant patents are preferred by the US nursery industry, particularly for fruit trees such as apples, peaches and tangerines as well as by flower breeders, while utility patents have become the mechanisms of choice for patenting plants, especially for high-tech, genetically engineered plants and plant parts. See Rural Advancement Foundation International (1995).

31 In the United States, the number of PVP applications under the PVP Act, supra note 29, has never been higher than 450 per year. In the period 2003–7, the average number of applications per year was around 1,350 in Japan, 1,000 in China, 350 in Australia, 250 in France and 200 in Argentina. The total number of PVP and plant patent applications in the United States for this period was 7,416, while the number of PVP applications in the European Community for the same period was 13,617 (International Union for the Protection of New Varieties of Plants, 2008).

32 See Rural Advancement Foundation International (1998).

33 UPOV Convention, supra note 6, Article 1(vi).

34 See, for example, the Enola bean case described earlier in this chapter.

35 Citation of Prior Art, 35 U.S.C. § 163, states: 'In the case of a plant patent the grant shall be of the right to exclude others from asexually reproducing the plant or selling or using the plant so reproduced.' Therefore, the scope of protection awarded in a US plant patent is to the single germplasm of the subject plant and not to a range of plants having similar characteristics. In *Imazio Nursery v Dania Greenhouses*, 69 F.3d 1560–67 (Fed. Cir. 1995) (a case referred to by the USPTO in the Da Vine process), the court ruled that a plant breeder could sue for plant patent infringement only if the infringing plant was an asexual reproduction of the protected plant – that is, if it shared the same DNA. In this way, the court confirmed that plant patent protection extends only to a single germplasm and its asexually reproduced progeny.

36 UPOV Convention, supra note 6, Article 7.

37 This is a case of the Honeybee Network, which is an example of a database on grassroots innovations and contemporary and traditional innovative practices, mainly from India. Before uploading the information in the digital database, innovators decide the amount of information they want to publicize. Another example is the register of the Potato Park in Peru. Here, the ONG Andes and the indigenous communities have developed a database that has different thresholds of accession depending on the type of information that they want to store.

38 Several authors have highlighted the risks of placing traditional knowledge in the public domain through open-access databases (Berglund, 2005; Argumedo and Pimbert, 2007). The most obvious one is when it is impossible for the knowledge holder to get a positive protection over it. Another consequence is that, once the traditional knowledge is made public without restrictions, its use outside the original community is very difficult to control, limiting the community's ability to apply their own institutional and customary laws and to get compensation from the use of such knowledge.

39 The People's Biodiversity Registers in India were initiated by the Foundation for the Revitalization of Local Health Traditions and the Centre for Ecological Sciences. Later, they were included in the Bioversity Conservation Prioritization Programme of the Worldwide Fund for Nature in India.

40 Online: <www.inia.gob.pe/ente-rector/registro-nacional-de-la-papa-nativa-peruana> (last accessed 2 June 2014).

41 Law Decree 118/2002, 20 April 2002.

References

Adams, S., and V. Henson-Apollonio (2002). *Defensive Publishing: A Strategy for Maintaining Intellectual Property As Public Goods*, International Service for National Agricultural Research Briefing Paper no. 53, September 2002.

Alston, J.M., and J.V. Raymond (2002). *The Effects of the US Plant Variety Protection Act on Wheat Genetic Improvements*, Discussion Paper no. 62, International Food Policy Research Institute, Washington, DC.

Argumedo, A., and M. Pimbert (2007). *Protecting Indigenous Knowledge against Biopiracy in the Andes,* International Institute for Environment and Development, London.

Berglund, M. (2005). 'The Protection of Traditional Knowledge Related to Genetic Resources: The Case for a Modified Patent Application Procedure,' *SCRIPT-ed* 2(2): 206–22.

Boettiger, S., and C. Chi-Ham (2007). 'Defensive Publishing and the Public Domain,' in A. Krattiger, R.T. Mahoney and L. Nelsen (eds.), *Intellectual Property Management in Health and Agricultural Innovation: A Handbook of Best Practices* Oxford University Press, Oxford, UK. Available at: www.ipHandbook.org (last accessed 3 June 2014).

Boyle, J. (2003). 'The Second Enclosure Movement and the Construction of the Public Domain,' *Law and Contemporary Problems* 66: 33–74.

Busso, C.S., K.M. Devos, G. Ross, M. Mortimore, W.M. Adams, M.J. Ambrose, S. Alldrick and M.D. Gale (2000). 'Genetic Diversity within and among Landraces of Pearl Millet (*Pennisetum glaucum*) under Farmer Management in West Africa,' *Genetic Resources and Crop Evolution* 47: 561–8.

Center for International Environmental Law (1999). *Comments on Improving Identification of Prior Art. Recommendations on Traditional Knowledge Relating to Biological Diversity*, submitted to the United States Patent and Trademark Office. Available at: www.ciel.org/Bio/pto rejection.html (last accessed 3 June 2014).

Centro Internacional de la Papa (2008). *Towards Certified Seed of Native Potatoes*, International Potato Center Annual Report 2008, CIP, Lima.

Comisión Nacional contra la Biopiratería Website (2013). Available at: www.biopirateria. gob.pe (last accessed 29 May 2014).

ETC Group (2009). *Message to USPTO: Squash the Patent on Bumpy Pumpkins; There's Plenty of Prior (W)Art*, news release, 2 February 2009. Available at: www.etcgroup.org/fr/node/721 (last accessed 3 June 2014).

European Patent Office (2013). *Guidelines for the Examination in the European Patent Office*, September 2013. Part G – Chapter II-18, Section 5.4.1 'Plant Varieties.' Available at: http://documents.epo.org/projects/babylon/eponet.nsf/0/6c9c0ec38c2d48dfc1257a21 004930f4/$FILE/guidelines_for_examination_2013_en.pdf (last accessed 28 May 2014).

Gadgil, M. (1996). 'Documenting Biodiversity: An Experiment,' *Current Science* 70(1): 36–44.

Gadgil, M., P.R.S. Rao, G. Utkarsh, P. Pramod, A. Chhatre and members of the People's Biodiversity Initiative (2002). 'New Meanings for Old Knowledge: The People's Biodiversity Registers Program,' *Traditional Ecological Knowledge: Ecological Applications* 10(5): 1307–17.

Giancola, S., S. Marcucci Poltri, P. Lacaze, H.E. Hopp (2002). 'Feasibility of Integration of Molecular Markers and Morphological Descriptors in a Real Case Study of a Plant Variety Protection System for Soybean,' *Euphytica* 127(1): 95–113.

Gupta, V.K. (2011). *Protecting Indian Traditional Knowledge from Biopiracy*. Paper Presented at the International Conference on Utilization of the Traditional Knowledge Digital Library (TKDL) as a Model for Protection of Traditional Knowledge, New Delhi, India, March 22 to 24, 2011. Available at: www.wipo.int/export/sites/www/meetings/ en/2011/wipo_tkdl_del_11/pdf/tkdl_gupta.pdf (last accessed 29 May 2014).

Halewood, M., J.J. Cherfas, J.M.M. Engels, T. Hazekamp, T. Hodkin and J. Robinson (2005). 'Farmers, Landraces and Property Rights: Challenges to Allocating Sui Generis Intellectual Property Rights to Communities over Their Varieties,' in S. Biber-Klemm and T. Cottier (eds.), *Rights to Plant Genetic Resources and Traditional Knowledge: Basic Issues and Perspectives*, Commonwealth Agricultural Bureaux, Wallingford, UK, pp. 173–202.

Indian Council of Scientific and Industrial Research (2011). *India Partners with Japan to Protects its Traditional Knowledge and Prevent Bio-Piracy*, Press Release, 20 April 2011.

International Centre for Tropical Agriculture (2002). *Report on the International Network of Ex Situ Collections under the Auspices of FAO: Further Information Provided by the International Centre for Tropical Agriculture, Regarding Its Request for a Re-Examination of US Patent No. 5,894,079*, Commission on Genetic Resources for Food and Agriculture, 9th Regular Session, Doc. CGRFA-9/02/Inf.7, Rome.

International Centre for Tropical Agriculture (2008). *Us Patent Office Rejects Company's Claim for Bean Commonly Grown by Latin American Farmers*, Press Release, 30 April 2008. Available at: www.cgiar.org/web-archives/www-cgiar-org-newsroom-releases-news-asp-idnews-753/ (last accessed 3 June 2014).

International Seed Federation (2013). *International Seed Federation's View on Intellectual Property*, adopted in Rio de Janeiro, Brazil, 28 June 2012. Available at: www.worldseed.org/isf/on_intellectual_property.html (last accessed 2 June 2014).

Jaffe, A.B., and J. Lerner (2004). *Innovations and Its Discontents: How Our Broken Patent System Is Enduring Innovation and Progress, and What to Do About It*, Princeton University Press, Princeton, NJ.

Janis, M.D., and J.P. Kesan (2002). U.S. Plant Variety Protection: Sound and Fury . . . ? *Houston Law Review* 39: 727–78.

Koo, B., C. Nottembur and P.G. Pardey (2004). 'Plants and Intellectual Property: An International Appraisal,' *Science Magazine* 306: 1295–7.

Kumar, L.D., M. Kathirvel, G.V. Rao and J. Nagaruju (2000). 'DNA Profiling of Disputed Chilli Samples (*Capsicum Annum*) Using ISSR-PCR and FISSR-PCR Marker Assays,' *Forensic Science International* 116: 63–8.

Lange, D. (2003). 'Reimagining the Public Domain: The Second Enclosure Movement and the Construction of the Public Domain,' *Law and Contemporary Problems* 66: 463–83.

Morell, M.K., R. Peakall, R. Appels, L.R. Preston and H.L. Lloyd (1995). 'DNA Profiling Techniques for Plant Variety Identification,' *Australian Journal of Experimental Agriculture* 35, 807–19.

Nazarea-Sandoval, V. (1998). *Cultural Memory and Biodiversity*, University of Arizona Press, Tucson, AZ.

Noli, E., M.S. Teriaca, M.C. Sanguineti and S. Conti (2008). Utilization of SSR and AFLP markers for the assessment of distinctness in durum wheat. *Molecular Breeding* 22(2): 301–13.

Noveck, B.S. (2006). "Peer to Patent': Collective Intelligence and Intellectual Property Reform,' *Harvard Journal of Law and Technology* 20: 123–62.

Pallottini, L., E. Garcia, J. Kami, G. Barcaccia and P. Gepts (2004). 'The Genetic Anatomy of a Patented Yellow Bean,' *Crop Science* 44: 968–77.

Rural Advancement Foundation International (1995). *Sixty-Five Years of the US Plant Patent Act*, RAFI communique. Available at: www.etcgroup.org/content/sixty-five-years-us-plant-patent-act-ppa (last accessed 3 June 2014).

Rural Advancement Foundation International (1998). *Plant Breeders Wrongs: An Inquiry into the Potential for Plant Piracy through International Intellectual Property Conventions*. Available at: www.etcgroup.org/content/plant-breeders-wrongs (last accessed 3 June 2014).

Samuels, E. (2002). 'The Public Domain Revisited,' *Los Angeles Law Review* 36: 389.

Subedi, A., R. Devkota, I. Prasad Poudel and S. Subedi (2013). 'Community biodiversity registers in Nepal. Enhancing the capabilities of communities to document, monitor and take control over genetic resources' in W.S. De Boef, A. Subedi, N. Peroni, M. Thijssen and E. O'Keeffe (eds.), Community Biodiversity Management. Promoting resilience and the conservation of plant genetic resources. Earthscan from Routledge, London, pp. 83-90.

Subedi, A., I. Poudel, B. Regmi, K. Baral, R. Suwal, D. Rijal, B. Sthapit and P. Shrestha (2005). 'Strengthening the Community Biodiversity Registers for Agriculture, Forest and Wetland Biodiversity Management: Users and Livelihood Perspectives,' in *Learning from Community Biodiversity Register in Nepal*, Proceedings of the National Workshop, 27–28 October, Khumaltar, Nepal. Available at: https://idl-bnc.idrc.ca/dspace/bitstream/10625/27838/1/122788.pdf (last accessed 3 June 2014).

Traditional Knowledge Digital Library (TKDL) (2013). *Major milestones of TKDL*. Available at: www.tkdl.res.in/tkdl/langdefault/common/AboutTKDL.asp?GL=Eng#History (last accessed 29 May 2014).

Trilateral (2007). Trilateral Statistical Report 2007 Edition. Alexandria, Virginia (USA), The Hague (Netherlands) and Tokyo (Japan).

Trilateral (2010). Four Office Statistic Report 2010. Alexandria, Virginia (USA), Daejeon (Korea), The Hague (Netherlands) and Tokyo (Japan).

UPOV (2002). *General Introduction to the Examination of Distinctness, Uniformity and Stability and the Development of Harmonized Descriptions of New Varieties of Plants*, Doc. TG/1/3. UPOV, Geneva. Available at: www.upov.int/en/publications/tg-rom/tg001/tg_1_3.pdf (last accessed 3 June 2014).

UPOV (2008). *Plant Variety Protection Statistics for the Period 2003–2007*. Available at: www.upov.int/export/sites/upov/en/documents/c/42/c_42_07_rev.pdf (last accessed 3 June 2014).

UPOV (2013). *Guidance on the use of biochemical and molecular markers in the examination of distinctness, uniformity and stability*, adopted by the Council at its forty-seventh ordinary session on October 24, 2013. Document TPG/15. Available at: www.upov.int/tgp/en/ (last accessed 2 June 2014).

US Patent and Trademark Office (2007). *Patent Public Advisory Committee's Annual Report*. Available at: www.uspto.gov/about/advisory/ppac/index.jsp (last accessed 29 May 2014).

US Patent and Trademark Office (2013). *Patent Public Advisory Committee's Annual Report*. Available at: www.uspto.gov/about/advisory/ppac/index.jsp (last accessed 29 May 2014).

US Patent and Trademark Office (2014). *Manual of Patent Examining Procedure,* 9th edition. Available at: www.uspto.gov/web/offices/pac/mpep/index.html (last accessed 29 May 2014).

Wiser, G. (2001). *US Patent and Trademark Office Reinstates Ayahuasca Patent: Flawed Decision Declares Open Season on Resources of Indigenous Communities*. Available at: www.ciel.org/Publications/PTODecisionAnalysis.pdf (last accessed 3 June 2014).

World Intellectual Property Organization (2007). *Overview of Activities and Outcomes of the Intergovernmental Committee*, Document prepared by the Secretariat for the 11th Session of the Committee, Geneva, 3–12 July, Doc. WIPO/GRTKF/IC/11/9.

12 Institutional capacity and implementation issues in farmers' rights

C.S. Srinivasan

Introduction

The previous chapters have extensively examined the case for a system of farmers' rights from the perspective of supporting on-farm conservation and the development of plant genetic resources (PGR). The rationale for farmers' rights rests principally on two arguments, which may be called the 'conservation argument' and the 'equity argument' (Swanson, Pearce and Cervigni, 1994). The conservation argument rests on the premise that treating PGR as a public good to be freely accessed and used by all provides no incentives for its conservation. The lack of incentives for retaining agricultural biodiversity arises because the crucial role played by farmers and farming communities in the conservation and development of PGR generally goes unrewarded. The conservation rationale sees farmers' rights as a framework that enables farmers/farming communities to appropriate a portion of the value of the agricultural biodiversity that they conserve and develop – thereby creating incentives for their sustained conservation and enhancement. The equity argument rests on the premise that the availability of all PGR used in institutional breeding programs today is the result of enormous efforts towards conservation, selection and improvement of PGR by rural communities in different countries that go unrewarded when PGR are freely appropriated by these programs. Intellectual property regimes reward institutional breeders with monopoly rights when they develop new varieties, but they provide no rewards to those responsible for the conservation of PGR, who provide the foundation for the development of new varieties. Farmers' rights provisions represent an attempt to redress the asymmetry in the rewards accruing to institutional breeders for innovations and farming communities that conserve PGR.

Given this rationale, developing countries could have conceived of legislative provision to craft a system of farmers' rights independently of other developments in the international regime governing the exchange of PGR. In practice, however, farmers' rights provisions have been articulated and developed in response to the extension of intellectual property rights (IPR) regimes for plant variety innovations in developing countries, which have been mandated by international agreements. The introduction of a plant variety protection (PVP) system to comply with the Agreement on Trade-Related Aspects of Intellectual Property Rights (TRIPS Agreement) has largely been viewed as an external imposition by developing countries.[1] A wide range of concerns have been raised by farmers and other stakeholders on the potential adverse impacts of an IPR regime for agricultural innovations on farm livelihoods in developing countries. The anticipated adverse impacts include potential restrictions on farmers' ability to save seed from harvests, higher prices for seed, increasing levels of concentration in the seed industry, dependence on foreign sources for the supply of seed and the erosion of genetic diversity. In introducing IPR legislation to comply with the provisions of the TRIPS Agreement, developing countries have had to address these concerns. Farmers' rights provisions have been incorporated as a counterweight to the monopoly rights granted to institutional innovators under PVP or other legislation. Farmers' rights provisions have also been inspired by the provisions in the Convention on Biological Diversity (CBD) that assert the 'sovereign rights' of nations over their biological resources and promote 'equitable benefit sharing' from the exchange and use of these resources.[2]

Developing countries have considered diverse approaches for the realization of farmers' rights. These range from market-based approaches to promote the use of farmers' varieties to 'rights-based' approaches that seek to confer some form of IPR on farming communities over varieties that they have conserved and developed. Legislative provisions in developing countries have tended to focus predominantly on rights-based approaches. Developing countries have brought in legislative provisions on farmers' rights in a variety of ways.[3] In some countries, farmers' rights provisions have been brought in as an integral part of PVP or IPR legislation, while in others they can be found in biodiversity legislation that seeks to regulate the exchange and use of PGR or that develops a separate set of regulations. Efforts by developing countries to create a system of farmers' rights generally involves one or more of the following elements:

- providing incentives for conservation by facilitating the commercialization of farmers' varieties;
- protecting farmers' ability to access and use existing agrobiodiversity and innovations;
- recognizing and rewarding farmers/farming communities for their contributions to innovations produced by institutional breeding programs/innovators;
- providing incentives for farmers/farming communities to sustain on-farm conservation and the enhancement of PGR.[4]

This chapter attempts to identify the institutional capacity and the administrative, technical and legal infrastructure required in developing countries to effectively implement farmers' rights provisions. Our main argument is that developing countries have not systematically examined or provided for the institutional capacity, infrastructure, processes and operational procedures required for giving effect to farmers' rights provisions, and this lack of institutional capacity is likely to undermine their implementation. Further, even if institutional capacity constraints were to be overcome, farmers' rights provisions are unlikely to provide substantial incentives for on-farm conservation or innovation. However, they are likely to dilute the incentives for innovation for the key institutional players who may increasingly resort to non-IPR instruments to appropriate economic returns from their innovations.

Commercialization of farmers' varieties

In developing countries, traditional varieties conserved and developed by farming communities, which are well adapted to local agroclimatic conditions, may be in extensive use in these communities. However, these 'farmers' varieties' are seldom part of the mainstream organized commerce in seed. There is no organized multiplication of the seeds of these varieties. Transactions in the seeds of these varieties may be highly localized and restricted to informal exchange between farmers, which provides few economic rewards for the communities that conserve and develop these varieties. If larger markets could be developed for these varieties and their seeds transacted in commercial quantities, then the resulting economic returns could provide incentives for the continued conservation of these varieties. Further, if varieties enhanced through selection by these farming communities could be commercially marketed, it would provide incentives for sustaining such innovation by farmers. The rationale for the commercialization of farmers' varieties is that it would provide self-sustaining market-based incentives for farmers' conservation efforts.

The most important barrier to the commercialization of farmers' varieties is the seed regulatory system in developing countries, which may, by design or implication, exclude farmers' varieties from organized seed-sector activity. Most developing countries have a seed regulatory system that is designed as a quality control system intended for the protection of farmers as buyers of seed. These systems seek not only to ensure the physical qualities of seed sold to farmers (e.g. assured germination) but also to confirm the varietal identity of seeds purchased by farmers – that is, to assure farmers that the seeds purchased are in fact of the variety they purport to be. Varietal identity is important because each variety is associated with a set of expected performance characteristics. There are different approaches to seed regulation in developing countries. Some countries have a mandatory variety registration system. This system implies that a variety cannot be marketed unless it has been registered or 'inscripted' in a national register of varieties. Other countries have regulations that allow commercially important varieties to be brought under the purview of quality control mechanisms.[5] Varieties not brought under the purview of quality control regulations may be allowed to be marketed under so-called truthful labelling arrangements. Most systems involve three important elements: (1) establishing the identity of a variety; (2) establishing the performance characteristics of a variety; and (3) enforcing quality control on the seeds sold. It should be noted that establishing the identity of a variety is essential because no regulatory system can be enforced unless the 'object' of regulation can be clearly and unambiguously defined.

Variety registration systems generally use the distinctness, uniformity and stability (DUS) criteria to identify a variety. These are generally the same criteria that are used to identify a variety for intellectual property protection (e.g. under a PVP system), except that PVP systems also require a variety to meet the novelty requirement in order to qualify for protection. These criteria have been developed to meet the legal requirement of identifying the object of regulation, but they also impose constraints on the type of material that can be marketed through the organized seed sector. It may be possible to distinguish farmers' varieties with reference to certain important morphological characteristics, but these varieties may generally not be uniform or stable. In fact, what is referred to as a farmers' variety may be a heterogeneous population of plants with certain common characteristics. Farmers' varieties may fail to meet the requirements of existing variety registration systems and, consequently, may be precluded from being marketed commercially.

It has been argued by proponents of farmers' rights that the legal requirement for identification should not be allowed to dictate what can be cultivated commercially. It has been suggested that farmers' varieties should be accommodated within variety registration systems by adopting 'looser' identifiability criteria. The issues relating to the feasibility of implementing alternative identifiability criteria also arise in the context of IPR protection of farmers' varieties and will be discussed in a later section. What is important to note is that the modification of existing variety registration systems is a critical prerequisite for the commercial marketing of farmers' varieties. While many developing countries have expressed an interest in the commercial development of farmers' varieties, few countries have initiated changes to their seed regulatory system for this purpose.

Some countries are attempting a bottom-up enumeration of farmers' varieties to incorporate them in national variety registers (such as in Nepal and Peru). In Peru, for instance, the public research system in consultation with local communities has attempted to identify the diversity of traditional maize and potato varieties using a broad set of descriptors sufficient to distinguish between varieties (GRPI, 2008). This endeavour is expected to pave the way for the inscription of these varieties in the national register (without insisting on strict conformity to uniformity or stability criteria). Changes are being made to the variety registration system to allow it to accept farmers' varieties identified through this process. In such an enumeration-based approach, varieties identified through a systematic 'vetting' process are allowed to be added to the national register, rather than insisting that a set of predetermined criteria be applied to the candidate varieties. Such enumeration may also serve the purpose of 'defensive publication' when these varieties are sought to be used in the development of other new varieties. This approach may be a useful way of overcoming the barriers posed by variety registration systems.

Even if farmers' varieties can be brought within the ambit of variety registration systems, a number of issues will still remain to be addressed if incentives for conservation are to flow from commercial marketing of farmers' varieties. If seed regulatory systems require varieties to be tested for 'value in cultivation and use,' then farmers' varieties will also have to be tested. This may involve the development of protocols for assessing the agronomic advantages of varieties that may not be uniform or stable. More importantly, commercial marketing of farmers' varieties will probably require the services of an agency or firm that can organize the multiplication and distribution of farmers' varieties on a commercial scale and meet the transaction costs of such commercialization. However, any public or private agency that is mandated/licensed to distribute farmers' varieties could end up appropriating a major share of the economic returns – leaving few rewards for farming communities responsible for the conservation and development of traditional varieties.[6] It is possible to envisage decentralized and dispersed production of seeds of farmers' varieties by individual farmers. However, small farmers in developing countries will not be able to participate in organized seed commerce unless there is an institutional mechanism (e.g. a cooperative) to link them to markets. As farmers move away from informal seed exchange based on trust towards transactions in more distant markets, issues relating to quality control and enforcement of standards will inevitably arise. A key issue will be the quality control standards that could (or should) be applied to the decentralized production of seeds of farmers' varieties and how they could be enforced or guaranteed.

Thus, the existing seed regulatory system in developing countries poses formidable barriers to the commercialization of farmers' varieties. Modification of variety registration systems, testing arrangements and quality control systems are all-important prerequisites for commercialization. However, the removal of regulatory barriers alone is unlikely to kick-start the commercial production of farmers' varieties. Institutional investments are also necessary to organize production and to link farmers with the necessary markets. Regulatory constraints have thus far precluded an assessment of whether markets exist for farmers' varieties and the potential profitability of supplying these markets. There is little empirical evidence on the potential size of these markets and on how these varieties will fare in competition with varieties bred through institutional breeding programs. The commercialization of farmers' varieties in developing countries could in many situations be at odds with policies promoting improved 'modern' varieties from the public research system. Developing country policy will have to clearly delineate the areas, crops or agroclimates in which promotion of farmers' varieties is useful. The most important challenge is to ensure that economic rewards through commercialization filter down to those responsible for conservation without being entirely appropriated at higher levels of the seed production and distribution chain. This requires policy to address the issue of 'ownership' or 'rights' over varieties, which we discuss in the following sections.

Use of innovations by farmers

On-farm seed saving

In developing countries, the adoption of intellectual property regimes for plant varieties has been seen as a major threat to farmers' ability to access and use innovations emerging from institutional breeding programs, affecting prospects of productivity enhancement and livelihood improvements. In the context of PVP legislation, the major concern has been the potential restrictions that farmers could face in the use of farm-saved seed of protected varieties. These concerns have been heightened by trends in the United States and Europe, where the evolution and strengthening of PVP law has led to progressively stringent curbs on farmers' ability to save and use farm-saved seed of protected varieties. In developing countries, such restrictions have been seen as being deeply disruptive of the fundamental rights of farmers to save and use the seed from their harvests – enshrined in centuries-old traditional practice all over the world. Varieties protected by IPR are likely to be more expensive than nonprotected varieties, and restrictions on the use of farm-saved seed flowing from plant breeders' rights could force them to buy fresh seed every planting season from seed companies. Such restrictions may exclude poor farmers from the use of new innovations and increase the productivity/income gap between rich and poor farmers. Restrictions on the use of farm-saved seed would also adversely affect on-farm innovation using protected varieties.

In responding to these concerns, developing countries, almost without exception, have provided explicit safeguards in their PVP legislation for farmers' privilege or farmers' rights to use farm-saved seed. Box 12.1 provides a summary of how legislation in different countries has attempted to provide these safeguards. Almost all developing country legislation explicitly recognizes farmers' rights to save, use and exchange seeds of protected varieties without payment of royalties to IPR holders. Some legislation (e.g. India's Protection of Plant Varieties and Farmers' Rights Act) allows farmers to exchange and even sell seeds of protected varieties saved on farm.[7] The only restriction placed on farmers is that they should not use the protected variety denomination or brand name in the course of exchange or other transactions. The intention in most developing countries appears to be to provide virtually unfettered rights to farmers for on-farm seed saving.

Box 12.1 Safeguards for on-farm seed saving in PVP legislation

PVP legislation in most developing countries contains explicit safeguards for protecting farmers' rights to save seeds of protected varieties obtained from the harvest. The scope of protection for on-farm seed saving, however, varies from country to country. A few examples of safeguard provisions are given here.

African Union

AU Model Law on Rights of Local Communities, Farmers, Breeders and Access

Article 26

(1) Farmers' Rights shall, with due regard for gender equity, include the right to: . . .

(f) collectively save, use, multiply and process farm-saved seed of protected varieties.

(2) Notwithstanding sub-paragraphs (c) and (d), the farmer shall not sell farm-saved seed/propagating material of a breeders' protected variety in the seed industry on a commercial scale.

Andean Community

Decision 345 on Common Provisions on the Protection of the Rights of Breeders of New Plant Varieties

Article 26

Anyone who stores and sows for his own use, or sells as a raw material or food, the product of his cultivation of the protected variety shall not be thereby infringing the breeder's right. This Article shall not apply to the commercial use of multiplication, reproductive or propagating material, including whole plants and parts of plants of fruit, ornamental and forest species.

Brazil

Law no. 9,456 Establishing the Plant Variety Protection Law and Enacting Other Measures

Article 10

The breeder's right in the plant variety shall not be deemed infringed by a person who: . . .

(iv) being a small rural producer, multiplies seed, for donation or exchange in dealings exclusively with other small rural producers, under programs of financing or support for small rural producers conducted by public bodies or non-governmental agencies, authorized by the Government.

China

Regulations of the People's Republic of China on the Protection of New Varieties of Plants 1999

Article 10

Without prejudice to other rights of the variety rights holder under these Regulations, the exploitation of the protected variety shall not require authorization from, or payment of royalties to, the variety rights holder for the following purposes: . . .

(ii) the use for propagating purposes by farmers, on their own holdings, of the propagating material of the protected variety harvested on their own holdings.

India

Protection of Plant Varieties and Farmers' Rights Act 2002

Section 39(1) . . .

(iv) a farmer shall be deemed to be entitled to save, use, sow, resow, exchange, share or sell his farm produce including seed of a variety protected under this Act in the same manner as he was entitled before the coming into force of this Act:

Provided that the farmer shall not be entitled to sell branded seed of a variety protected under this Act.

Malaysia

Protection of New Plant Varieties Act 2004

Article 31

(1) The breeder's right shall not extend to: . . .

- (d) any act of propagation by small farmers using the harvested material of the registered plant variety planted on their own holdings;
- (e) any exchange of reasonable amounts of propagating materials among small farmers.

Thailand

Plant Varieties Protection Act, 1999

Section 33

The right holder of a new plant variety has the exclusive right to produce, sell or distribute in any manner, import, export or possess for the purpose of any of the said acts the propagating material of the new plant variety. The provisions of paragraph one shall not apply to the following circumstances: . . .

(4) the cultivation or propagation by a farmer of a protected new plant variety from the propagating material made by himself, provided that in the case where the Minister, with the approval of the Commission, publishes that new plant variety as promoted plant variety, its cultivation or propagation by a farmer may be made in the quantity not exceeding three times the quantity obtained.

Source: PVP legislation, accessed at <www.upov.int> and <www.grain.org>.

Developing country governments have incorporated provisions to safeguard the practice of on-farm seed saving to mitigate the opposition to the introduction of PVP and to reassure farmers and other stakeholders that their interests will not be adversely affected as a consequence. These provisions have also been cited as evidence of how developing countries have attempted to balance the interests of breeders and farmers while putting in place an IPR regime. These provisions underline the political economy of PVP in developing countries, showing how an IPR regime for plant varieties would have been politically unacceptable unless farmers' seed-saving traditions were respected. However, it is possible that the significance of these provisions in defending farmers' rights is being overstated. In most developing countries, where commercial seed accounts for less than 20 percent of seed use, informal seed exchange is dominant and there are millions of farmers dispersed over a very large number of holdings, thus any restriction on the use of farm-saved seed would probably be unenforceable. Monitoring the use of brand names or variety denominations in informal seed exchanges in rural areas may also be an unrealistic proposition. Most seed companies (domestic or international) have no expectation that they will be able to enforce plant breeders' rights against infringement by farmers; they only expect to be able to deter competitors in the organized sector from poaching protected material (Srinivasan, 2004). The safeguard provisions for on-farm seed saving are perhaps nothing more than a pragmatic adaptation to reality – an acknowledgment of the lack of institutional capacity to meaningfully enforce conventional IPRs when there are millions of farmers with highly dispersed small holdings in rural areas.

While it could be argued that safeguards provided for on-farm seed saving are not likely to make a noticeable difference to farmers in developing countries (because any restrictions on the use of farm-saved seed would be unenforceable), the design of these provisions does have important implications for the incentives afforded to institutional breeders through PVP. The lack of a precise definition of what constitutes permissible on-farm seed saving and the exchange/transaction of seeds of protected varieties makes it difficult for breeders to proceed against *any* type of infringement.[8] The safeguard provisions seldom specify who is regarded as a farmer, to what category of farmers they would apply (and whether they would be applicable to very large farmers), the purposes for which they would apply (would it be limited for the purpose of replanting the farmer's own land), how informal seed exchanges are to be distinguished from commercial transactions and what would constitute unacceptable use of brand names or variety denominations in informal or permissible seed exchanges. This lack of clarity provides a large avenue for activities that would be regarded as infringements (under conventional PVP legislation) to be subsumed under the safeguard provisions (e.g. under the Indian PVP legislation a large commercial seed farmer may also seek the benefits of the safeguard provisions). In developing countries, where judicial enforcement of sanctions against infringement may be weak to begin with, the effect of safeguard provisions would reduce considerably the feasibility of challenging infringements. Keeping a potentially large volume of seeds of a protected variety outside the purview of breeders' rights can have only one impact, which is to limit the appropriation of returns by IPR holders from their innovations. Some stakeholder groups may argue that this is precisely the outcome that developing country governments should seek to achieve – but clearly the original objective of providing incentives for innovation in the private sector is then diluted.

Compulsory licensing

Compulsory licensing provisions have generally been a part of every type of IPR legislation and have been incorporated into PVP legislation in both developed and developing countries. These provisions are meant to guard against situations where the dissemination of an IPR-protected innovation is prevented or hindered as a result of the IPR holder refusing to license the use of his/her innovation or demanding such onerous terms that renders the use of the innovation infeasible. The nondissemination of a protected innovation would go against the objectives of IPR legislation, and therefore, compulsory licensing is a device to ensure that protected innovations can be utilized for social benefit. These provisions enable governments to license an innovation on reasonable terms to intending users, where the IPR holder fails to do so. These provisions generally require the IPR holder to be adequately compensated in order to preserve the incentive effects of IPR protection. Compulsory licensing provisions attempt to strike a balance between incentives for innovation by granting monopoly rights to innovators and disseminating innovation to enhance social welfare.

The relevance of compulsory licensing can be readily seen in the area of drugs and pharmaceuticals, where it appears to have been applied most widely by developing countries. It is easy to imagine situations where governments may resort to compulsory licensing to make available a protected drug for combating a major disease (e.g. HIV/AIDS). In the context of PVP in developing countries, compulsory licensing provisions have been presented as a mechanism to preserve small farmers' access to protected new varieties (for instance, when a multinational seed company protects a variety and then refuses to license it for multiplication on reasonable terms). In the United States and the European Community, where PVP has been implemented for more than four decades, there have been no instances of compulsory licensing of protected new plant varieties. This, perhaps, reflects the fact that in the context of plant variety innovations, situations calling for the invocation of compulsory licensing provisions are likely to be rare. Unlike a protected drug, which may be the only available means to combat a major disease, a given new plant variety is unlikely to be critical for sustaining the production of a crop. A new protected variety will generally be in competition with existing varieties, and the value of the incremental benefits that can be derived from the use of the new variety will set a ceiling on the price that can be charged for it. Except in very rare situations where all existing varieties of a crop have become susceptible to a pathogen or disease or where the supplies of all existing varieties have been disrupted due to natural calamities or other shocks, it will be difficult to make a case for compulsory licensing, as the production of a crop is unlikely to be affected by the nonavailability of a single new variety.

It should also be noted that compulsory licensing provisions would not enable a developing country government to force a foreign seed company to make available an improved new variety. The provisions will be relevant only if the company, in anticipation of marketing the variety, has obtained protection for the variety protected under domestic PVP law. A seed company that is apprehensive of compulsory licensing provisions being applied could well choose not to offer the variety for protection. As an instrument for ensuring farmers' access to new varieties, compulsory licensing is likely to be relevant only in a limited number of circumstances.[9] Farmers' access to new varieties is likely to be influenced much more by other aspects of the seed regulatory regime – regulations governing the release or marketing of new varieties, price controls, if any, on the seed sector, participation of private and public sector players in the marketing and distribution of seed and so on – rather than by contingency provisions in IPR/PVP legislation.

Compulsory licensing provisions are generally worded fairly broadly, allowing them to be invoked where warranted in the public interest. However, developing country PVP legislation provides little guidance on the precise circumstances under which these provisions would be invoked. The Indian legislation, for instance, allows for the application of compulsory licensing in situations where reasonable quantities of the seeds of a protected variety are not made available to farmers at a reasonable price.[10] It is not clear how the determination of what constitutes 'reasonable quantity' and 'reasonable price' would be made in the context of a seed market with a large number of different varieties on offer or what would constitute a 'reasonable return' for the IPR holder. Mechanisms for the enforcement of compulsory licensing arrangements have also not been spelled out in PVP legislation. The ability of the concerned administrative authority to enforce compulsory licensing arrangements depends on its ability to provide the licensee(s) with adequate reproductive material of the protected variety for multiplication. Where the IPR holder refuses to cooperate in implementing a compulsory licensing arrangement, the only material available to the administrative authority is the reproductive material deposited by the IPR holder at the time of making the IPR application.[11] The issue of whether this would be adequate or useful for enforcing a compulsory license for rapid multiplication of the protected variety does not appear to have been examined. More generally, the question of institutional capacity to enforce compulsory licensing provisions does not appear to have been addressed in developing countries.

Extant variety protection

Many developing countries (e.g. India and Brazil) have incorporated provisions for the retrospective protection of public sector varieties in their PVP legislation. The Indian PVP legislation provides that extant varieties[12] shall be deemed to be protected from their date of statutory release/notification under the Seeds Act.[13] The rationale for retrospective protection of public sector varieties has never been clearly explained, but it appears to be the result of intense lobbying by the public research system for recognition of its dominant role in plant breeding in developing countries. This provision may be chiefly intended to protect the interests of the public research system and enable it to derive some returns from past investments in plant breeding, but it does have potential implications for farmers' access to publicly bred varieties. The nature of these implications will depend on how the public sector chooses to use the IPRs granted on extant varieties.

The grant of retrospective protection for varieties bred by the public sector can ensure that these varieties remain in public ownership and are not usurped by other entities. Such an event, however, is incidental because in any well-functioning PVP system it should not be possible for any entity to seek IPRs on an existing variety, whose existence is a matter of common knowledge.[14] In most developing countries, the practice has been that publicly bred varieties are made freely available for multiplication and distribution to public and/or private sector seed producers. If this practice is continued, it could be argued that the public sector is using its IPR on extant varieties to facilitate the continued dissemination of its varieties. However, it is also possible that when faced with fiscal constraints, the public research system in developing countries may seek to derive rents from IPRs on extant varieties to generate revenues. Such efforts could be done through royalty-based exclusive or nonexclusive licensing arrangements for the multiplication of seeds of extant varieties.[15] This is already happening in developing countries (particularly in Latin America) where public research organizations have started granting exclusive or semi-exclusive licenses to parastatals (or even private firms) to market the products of their research/breeding programs. If this trend continues, then the availability of publicly bred varieties to farmers could actually become restricted, and their price could rise, just as in the case of varieties protected by the private sector. The public sector will be faced with a conflict between the objective of revenue generation and the maximum dissemination of its innovations.

The more important implication of retrospective protection for extant varieties is likely to arise when these varieties are potentially used for the development of other new varieties (by the private sector). In such a situation, using the 'essential derivation' clause (if available), the public sector may be able to demand rents from the commercialization of derived varieties.[16] Such rents may support the revenue-generation objective of the public sector, but may also reduce the incentives for developing further innovations using public sector varieties.[17] Thus, retrospective protection of publicly bred varieties may support the public research system, but its impact on farmers' access to improved varieties is ambiguous.

Economic reward for contribution to innovation

Legislative provisions designed to provide an economic reward to farmers or farming communities for their contribution to plant variety innovations are called benefit-sharing provisions, reflecting the principles of access and benefit sharing enshrined in the CBD. These provisions may be found in biodiversity legislation that seeks to regulate access to PGR or as an integral part of IPR legislation such as PVP. These provisions may involve payments by institutional breeders for access to PGR conserved by farming communities. In the context of PVP legislation, they are generally designed to force institutional breeders who apply for the protection of new varieties to share their economic returns with farming communities that may have been the source of the PGR used in the development of new varieties. Different models or mechanisms for benefit sharing may be envisaged. Breeders may be required to share a portion of their royalties with the identified farmers or farming communities, which may be facilitated by allowing farmers/farming communities or their representatives to make benefit-sharing claims when an application for protection is made. Alternatively, breeders may be required to contribute a portion of their PVP royalties to a common conservation or a gene fund, which is then used to promote on-farm conservation activities. The implementation of benefit-sharing provisions may be supported by two provisions that relate to:

- The disclosure of the pedigree of a new variety (these provisions require the breeder of a new variety to disclose the pedigree or breeding history of a new variety offered for protection. Breeders may also be required to disclose the source of parental material used in the development of the new variety and confirm that such material has been legally obtained. The disclosure of pedigree could assist in the adjudication of benefit-sharing claims).
- Prior informed consent (some developing country legislation requires breeders to show that they have obtained the prior informed consent of farming communities from where the PGR used in the breeding program may have been sourced. This requirement provides leverage to farming communities to seek economic rewards when material conserved by them is sought to be used in institutional breeding programs).

The benefit-sharing provisions in the Indian PVP legislation are extracted in Box 12.2.

Box 12.2 Benefit-sharing provisions in India's Protection of Plant Varieties and Farmers' Rights Act

Article 26

(1) On receipt of copy of the certificate of registration under sub-section (8) of section 23 or sub-section (2) of section 24, the Authority shall publish such contents of the certificate and invite claims of benefit sharing to the variety registered under such certificate in the manner as may be prescribed.
(2) On invitation of the claims under sub-section (1), any person or group of persons or firm or governmental or nongovernmental organisation shall submit its claim of benefit sharing too such variety in the prescribed form within such period, and accompanies with such fees, as may be prescribed:

Provided that such claim shall only be submitted by any:

(i) person or group of persons, if such person or every person constituting such group is a citizen of India; or
(ii) firm or governmental or non-governmental organisation, if such firm or organisation is formed or established in India.

(3) On receiving a claim under sub-section (2), the Authority shall send a copy of such claim to the breeder of the variety registered under such certificate and the breeder may, on receipt of such copy, submit his opposition to such claim within such period and in such manner as may be prescribed.
(4) The Authority shall, after giving an opportunity of being heard to the parties, dispose of the claim received under sub-section (2).
(5) While disposing of the claim under sub-section (4), the Authority shall explicitly indicate in its order the amount of the benefit sharing, if any, for which the claimant shall be entitled and shall take into consideration the following matters, namely:

a the extent and nature of the use of genetic material of the claimant in the development of the variety relating to which the benefit sharing has been claimed.
b the commercial utility and demand in the market of the variety relating to which the benefit sharing has been claimed.

(6) The amount of benefit sharing to a variety determined under this section shall be deposited by the breeder of such variety in the manner referred to in clause (a) of sub-section 45 in the National Gene Fund.

(7) The amount of benefit sharing determined under this section shall, on a reference made by the Authority in the prescribed manner, be recoverable as an arrear of land revenue by the District Magistrate within whose local limits of jurisdiction the breeder liable for such benefit sharing resides.

Section 41

(1) Any person or group of persons (whether actively engaged in farming or not) or any governmental or nongovernmental organisation may, on behalf of any village or local community in India, file in any centre notified, with the previous approval of the Central Government, by the Authority, in the Official Gazette, any claim attributable to the contribution of the people of that village or local community, as the case may be, in the evolution or any variety for the purpose of staking a claim on behalf of such village or local community.
(2) Where any claim is made under sub-section (1), the centre notified under that sub-section may verify the claim made by such person or group of persons or such governmental or nongovernmental organisation in such manner as it deems fit, and if it is satisfied that such village or local community has contributed significantly to the evolution of the variety which has been registered under this Act, it shall report its findings to the Authority.
(3) When the authority, on a report under sub-section (2) is satisfied, after such inquiry as it may deem fit, that the variety with which the report is related has been registered under the provisions of this Act, it may issue notice in the prescribed manner to the breeder of that variety and after providing opportunity to such breeder to file objection in the prescribed manner and of being heard, it may subject to any limit notified by the Central Government, by order, grant such sum of compensation to be paid to a person or group of persons or governmental or nongovernmental organisation which has made claim under sub-section (1), as it may deem fit.
(4) Any compensation granted under sub-section (3) shall be deposited by the breeder of the variety in the Gene Fund.
(5) The compensation granted under sub-section (3) shall be deemed to be an arrear of land revenue and shall be recoverable by the Authority accordingly.

Source: Protection of Plant Varieties and Farmers' Rights Act, <agricoop.nic.in/seeds/farmersact2001.htm>.

The architecture of benefit-sharing programs is based on the following three elements:

1 that it explicitly recognizes the role of the past and present conservation efforts of farming communities in making available a wide range of PGR used in institutional breeding programs;
2 that it assumes that the geographical origin of such PGR can be identified with a reasonable degree of precision and that groups of farmers or other entities chiefly responsible for its conservation can be identified;
3 that it forces institutional breeding programs to acknowledge the use of such PGR in current breeding programs.

These three elements create a framework within which economic rewards for conservation and enhancement of PGR can flow to identified farming communities. We examine in the following section some of the conceptual, technical and legal issues that developing countries need to address to give effect to benefit-sharing provisions as well as the operational processes that need to be set out for meaningful implementation. We will also examine the institutional capacity required in developing countries to implement farmers' rights – in particular, the information and administrative infrastructure required that goes beyond what is required for conventional PVP systems.

Complex pedigrees

Conventional PVP systems only require the relevant authority to determine if a new variety offered for protection is sufficiently distinct from existing varieties. Implementation of benefit-sharing provisions, however, requires the determination of the contribution made by different ancestral varieties in the development of a new variety. Modern varieties have incredibly complex pedigrees involving tens of ancestral varieties and multiple breeding stages. The pedigree of many modern varieties can be traced back to 10 to 15 generations. The pedigree tree of a highly successful Indian wheat variety is given in Box 12.3. Developing country legislation does not address the principles that will be used to identify the contribution of a single ancestral variety to a modern variety with a highly complex pedigree. This may be a 'technical' issue, but it is one that is fundamental for implementing farmers' rights provisions. While adjudicating benefit-sharing claims, how many generations on the pedigree tree will be reckoned? Where the pedigree of a new variety involves multiple farmers' varieties, it may be necessary to assess the relative contribution of different varieties. There appears to be almost no discussion in the farmers' rights literature on the principles or algorithms that could be used in making this assessment. The application of benefit-sharing claims to the contributions made by farmers' varieties will also depend on the point of time at which the legislative provision becomes effective. Feasibility considerations are likely to allow only the prospective application of benefit-sharing provisions – that is, the provisions can be applied only to material accessed after the law comes into force. In this case, farmers' varieties accessed prior to a law's coming into force will remain outside the scope of benefit-sharing provisions.[18]

Box 12.3 Derivative history of Indian wheat variety Sonalika for five generations

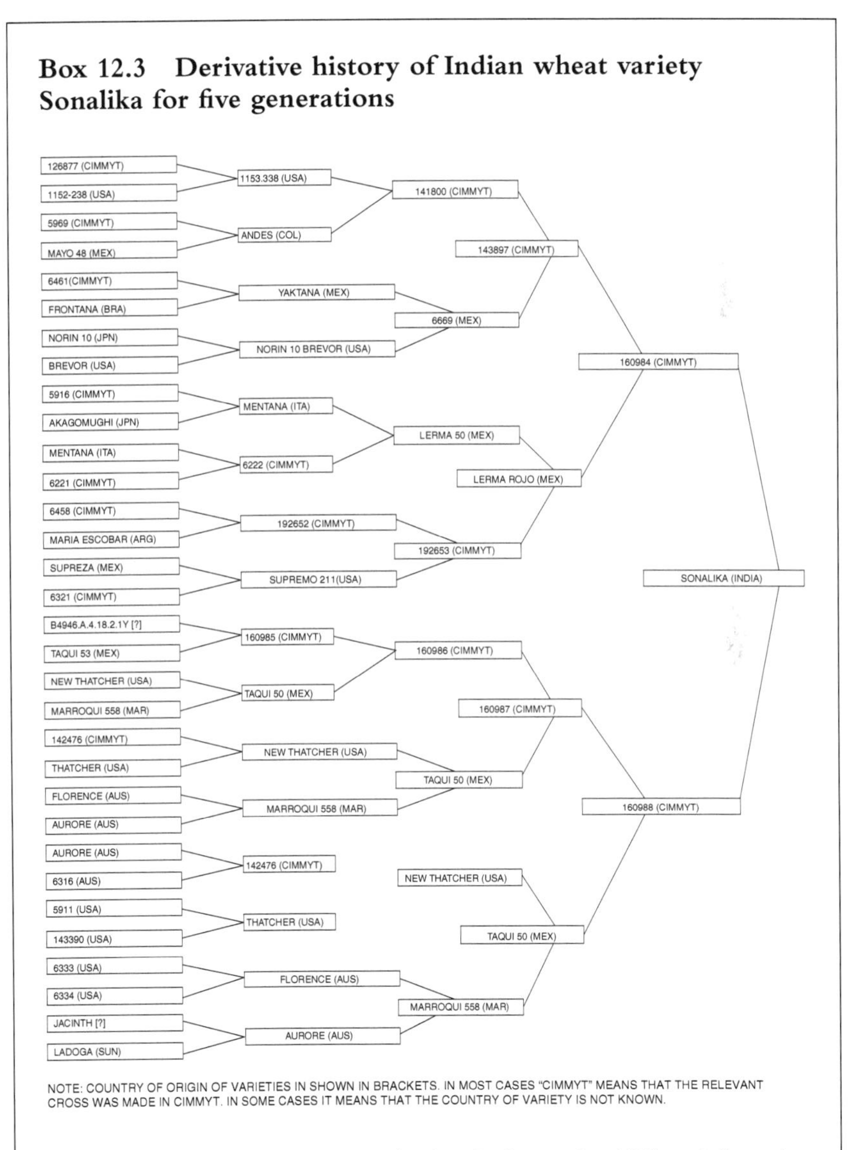

NOTE: COUNTRY OF ORIGIN OF VARIETIES IN SHOWN IN BRACKETS. IN MOST CASES "CIMMYT" MEANS THAT THE RELEVANT CROSS WAS MADE IN CIMMYT. IN SOME CASES IT MEANS THAT THE COUNTRY OF VARIETY IS NOT KNOWN.

Source: The pedigree tree was prepared using the International Wheat Information software and database developed by the International Centre for Maize and Wheat Improvement, Mexico.

Foreign contributions

It should also be noted that many of the new varieties commercialized in developing countries involved the collaboration and exchange of PGR between national agricultural research systems and the International Agricultural Research Centres of the Consultative Group on International Agricultural Research (e.g. in the development of 'green revolution' wheat and rice varieties in South Asian countries). An examination of the pedigree of varieties used in developing countries may reveal substantial contribution from foreign PGR. Legislative provisions on benefit sharing do not specify how the contribution of foreign PGR will be treated – whether they would be acknowledged and whether they could be the subject of benefit-sharing claims.

Attribution of ownership

In conventional PVP systems, the novelty claim of a variety offered for protection is assessed against a reference collection of known varieties, and ownership claims are decided on the basis of the first-to-file (or, in some countries, the first-to-invent) principles.[19] Farmers' varieties (or other PGR conserved by farmers) that are the subject matter of benefit-sharing claims have, by their very nature, been in the public domain for a considerable period of time. They may have evolved through selection/adaptation by farmers over generations and may have involved extensive informal exchange of PGR across farming communities. Verification of claims made under benefit-sharing provisions requires the authentic documentation of the geographical provenance of the PGR that is the subject matter of the claims. It further requires the development of the PGR to be attributed to an identifiable group of farmers or farming community. The 'reference collection' required for adjudication of benefit-sharing claims is a comprehensive database on agrobiodiversity relating to a crop in a country. Even developed countries have not yet managed to comprehensively document all of the agrobiodiversity for important crops within their borders. Developing countries appear to have made no attempt to assess the magnitude of investment required for creating (and updating) databases that would be critical for implementing benefit-sharing provisions.

While it may be possible to broadly identify the geographical provenance of farmers' varieties, the attribution of ownership or other rights over these varieties to identifiable groups of farmers poses formidable conceptual and legal challenges. Where farmers' varieties have evolved through the informal exchange of material over generations, it is likely that the use of these varieties is dispersed over a large area. Attributing rights over these varieties to a particular group of farmers may not be feasible or meaningful – and any determination or attribution of ownership could well be arbitrary and iniquitous (if it ignores the role of other groups of farmers who have also contributed to its development). The use of community biodiversity registers has been suggested as a way of identifying PGR over which local communities can claim rights. However, it is now fairly well established from the long history of PGR that the place of current occurrence of a variety may not be the place where it has been conserved or improved. Developing countries do not appear to have examined the principles or processes by which rights over farmers' varieties can be attributed to a group or a community. They have also not attempted to assess the costs of investigating benefit-sharing claims and set them against the benefits likely to accrue. Given the process by which farmers' varieties emerge, any narrow attribution of rights is likely to be fraught with arbitrariness. It will generally be infeasible to attribute rights over farmers' varieties to individual farmers. However, larger communities of farmers do not exist as legal entities that can exercise the rights granted or make benefit-sharing claims. Developing country legislation does not appear to address the question of defining legal entities that could exercise farmers' rights except to say that farmers may be represented by nongovernmental organizations (NGO) or other organizations. In designing benefit-sharing provisions based on the equity argument, developing countries have ignored the fact that conventional IPR provisions are based on a number of compromises reflecting practical considerations – with a view to keeping the system manageable and limiting transaction costs.[20]

The Indian PVP legislation not only requires the PVP authority to adjudicate benefit-sharing claims but also requires the authority to set the terms for benefit sharing (although it does not specify whether the benefits will be by way of lump sum payments or royalties linked to sales or profits).[21] If benefits are to be specified as lump sum payments, then such efforts will call for the authority to assess the commercial potential of a variety at the time that they are granted protection. This is something that even seed companies or IPR owners cannot do with any degree of precision.[22] Benefits could be related to commercial performance – for instance, royalties linked to the volume of sales – but this assessment will require a mechanism to monitor the commercial sales of seeds at the varietal level throughout the country. Such an endeavour is not something that the seed regulatory system in most developing countries is currently equipped to do.

Many of the issues in the attribution of ownership of farmers' varieties would not arise if benefit-sharing provisions required IPR holders to make only a contribution to a conservation fund for the use of material accessed from farming communities. The conservation fund could then be used to support a variety of on-farm conservation activities. Such an arrangement would, however, break the link between conservation effort and reward and could dilute the incentive effects of benefit-sharing arrangements, which would then depend a great deal on how the conservation fund was applied to different activities.

Disclosure of pedigree and prior informed consent

The disclosure of pedigree of a new variety offered for protection is not an entirely new feature in PVP legislation. In the United States and many other developed countries, information on the pedigree of a new variety is routinely provided in the PVP application (although it may not be compulsory). In conventional PVP systems, this information is used mainly to assess the distinctness of a variety from closely related varieties. In developing countries, the disclosure provision is mainly intended to force breeders to reveal whether any PGR sourced from farming communities has been used in the development of the new variety.[23] Further, the prior informed consent provision places the onus on the breeder to show that PGR used in the development of a new variety has been legitimately acquired. Taken together, these provisions are intended to enable farming communities to identify cases where their PGR is used in institutional breeding programs and to pursue benefit-sharing claims when new varieties are offered protection. The prior informed consent provision may also allow farming communities to garner rents at the development stage of variety.

Many of the issues discussed in the context of attribution of ownership will also affect the prior informed consent provisions. When an institutional breeder accesses PGR from a particular location, it is not clear to which entity (representing the local community with rights over PGR) he or she will apply for obtaining consent. It may be possible to assign the function to local authorities or to other biodiversity regulatory bodies at the regional/local level, but this option has not been spelled out in developing country legislation. It is also not clear how terms for prior informed consent will be set, especially as there are no market values of PGR that can be observed and used as a basis. It may also be necessary to set out the circumstances in which consent can be refused or withheld so that the operations of this provision do not run counter to the objective of *promoting* the use of PGR in generating innovations. Many institutional breeders may secure PGR not from farming communities but, rather, from national or international *ex situ* collections. The application of prior informed consent provisions in such cases (which will be a common occurrence for most institutional breeding programs) has to be carefully thought through. It may be infeasible for breeders to seek and obtain prior informed consent for material sourced from *ex situ* collections. In practice, it may be necessary to delegate authority for granting consent to the *ex situ* collections themselves through national and/or international agreement. The operation of the prior informed consent provisions and the associated time and transaction costs will have a significant influence on the incentives for institutional breeders to develop and offer new varieties for protection.

Protection of farmers' varieties

In many developing countries, providing some form of IPR protection to farmers' traditional varieties and landraces has been proposed as a means of recognizing the contribution of farmers to the conservation and enhancement of agrobiodiversity. While benefit-sharing provisions could be regarded as rewards for past conservation and enhancement efforts, the protection of farmers' varieties can be seen as creating incentives for sustaining conservation efforts in the future. Such recognition could also redress the perceived inequity in conventional PVP systems, which reward only the innovations of institutional players. Indian and Brazilian PVP legislation make explicit provision for the protection of farmers' varieties, while in many other developing countries it is clearly the intention that farmers should be able to seek protection for their innovations on par with institutional breeders.

The protection of farmers' varieties is often presented as being analogous to the protection of new varieties developed by institutional breeders, and it is often seen as bringing equity into the reward for innovation irrespective of the setting in which it takes place. However, the protection of farmers' varieties involves a number of conflicts with the principles of conventional IPR systems, and these conflicts need to be addressed. A fundamental criterion of conventional IPR systems is novelty – that is, they are designed to protect (new) innovations for a limited period of time before they pass into the public domain. They are not designed to push innovations that have long been in the public domain back into some form of private ownership. To circumvent this difficulty, the Indian PVP legislation, for instance, incorporates a provision that novelty shall not be a criterion for the protection of farmers' varieties.[24] More importantly, the protection of farmers' varieties also conflicts with other conventional criteria for protection – namely DUS. It is well recognized that farmers' traditional varieties are unlikely to conform to the DUS criteria, especially to the uniformity and stability criteria. Farmers' rights advocates have suggested that farmers' varieties could be protected using looser identifiability criteria. There has been a great deal of debate over whether less stringent identifiability criteria could be used to identify farmers' varieties and bring them under the purview of protection. The Crucible Group provides an excellent summary of this debate (IDRC, 2001). We will draw on this summary to examine how the identifiability issue affects the feasibility of protecting farmers' varieties.

Those making the case for looser identifiability criteria argue that the legal requirement to identify an innovation for the purpose of intellectual property protection must not determine or dictate the desirable characteristics of the innovation. The DUS criteria essentially serve the purpose of identifying varieties, but their application in intellectual property regimes also serves to exclude varieties (or heterogeneous populations with certain common characteristics) that have definite agronomic advantages but may not strictly meet the uniformity and stability criteria. If such varieties (or heterogeneous populations) offer certain advantages to farmers in certain conditions, then it should be possible to develop alternative mechanisms to identify them and bring them under the purview of protection. Such varieties, for instance, could be identified through 'persons skilled in the art.' This could include farmers who are generally able to unambiguously distinguish between varieties growing in their fields, even when these varieties are not uniform or stable. Such an approach would also have the advantage of not artificially driving the innovation process towards the production of stable and uniform varieties. Farmers' varieties are constantly evolving, and any system of protecting them would have to deal with situations where a variety changes significantly over time in terms of its essential characteristics. However, it is argued that this problem of 'shifting' varieties is no different from the problems faced with open-pollinated varieties bred through institutional breeding programs. Even in the case of the modern open-pollinated varieties, the IPR system has to make a judgment at what point variety has changed sufficiently to be regarded as a new variety, and a similar judgment would need to be made in the context of farmers' varieties brought under the purview of protection. Instead of using stability as a criterion for protection, it could simply be used as a marker to decide when an existing variety has changed sufficiently to be classified as a new variety. A variety that had changed significantly with respect to its essential characteristics would no longer fall under the scope of the right granted to the original variety.

There is certainly a strong conceptual case for making a distinction between criteria required for the legal identification of an innovation and the characteristics of the innovation that make it useful for farmers. However, the implementation of alternative identification criteria is likely to face many technical and administrative challenges. The most important question to be addressed is how to define the boundaries of an innovation for the purpose of protection when its characteristics are changing and are not uniform. Such distinction may necessitate the development of IPRs that are based on an examination of the claims relating to the scope of the innovation (as is done in the case of patents), which would enormously complicate conventional plant variety protection systems. Even if farmers can unambiguously identify varieties that are not uniform or stable, a legal system that offers protection for such varieties would still require explicit descriptors/characteristics against which these varieties could be evaluated, the definition of the degree of variation that would be permissible in regard to each of these characteristics and the type/breadth of claims that would be permissible. In fact, it is the complexity and difficulty of undertaking such an exercise for biologically reproducing material that has led to the emergence of PVP systems (distinct from patent systems) that are based on DUS criteria centred on morphological (rather than genetic) characteristics, without the examination of utility ('value in cultivation and use') and breadth of claims. Developing countries advocating IPR protection for farmers' varieties do not appear to have grasped the complexity of operationalizing alternative identifiability criteria. Moreover, if farmers' varieties that were protected using alternative identifiability criteria did lose their protection when their characteristics changed, then it is not clear what incentives for conservation or on-farm innovation would be provided by the ephemeral protection of such varieties. What is important to note is that developing country legislation, which intends to protect farmers' varieties, generally sticks to the DUS criteria and makes no effort to delineate alternative identifiability criteria that may be necessary for protecting farmers' varieties. The Indian PVP legislation explicitly dispenses with the novelty criterion to allow for the protection of farmers' varieties, but it still retains the DUS criteria. The unintended effect may be that the protection of farmers' varieties remains infeasible, even in a legislation that sees the protection of farmers' rights as its fundamental objective.

The issues discussed in the previous section relating to the establishment of the provenance of varieties that are widely dispersed in the community and the attribution of ownership to individual farmers or farming communities apply with equal force to the protection of farmers' varieties. Developing country legislation provides little guidance on how these issues would be addressed in practice. Even if these issues are addressed and farmers' varieties are protected, the question of how beneficiaries could expect to exercise the rights over traditional varieties would remain. It is unlikely that beneficiaries would be able to exercise any control over the multiplication and distribution of these varieties. The reproductive material of the protected varieties may be widely available through previous informal exchange, and seed multiplication may be taking place largely outside the organized sector. It appears to be quite unrealistic to expect that farmers who have hitherto been freely using these varieties would suddenly start paying royalties on their use simply because ownership over them has now been conferred on a particular farming community. It is also unlikely that the IPR holders will be able to seek rents using the essential derivation clause when the protected farmers' variety is used in the development of other new varieties. Unless the farmers' variety is used directly as a parent (which may rarely be the case given the complex pedigree of modern varieties), the essential derivation criteria are unlikely to be met. Essential derivation criteria are also not likely to be met when a single useful trait is extracted from a farmers' variety and inserted into a new variety. However, the formal protection of farmers' varieties could strengthen the hands of IPR holders in making benefit-sharing claims when new varieties (that use farmers' varieties as parental material) are offered for protection.

The protection of farmers' varieties may provide recognition for farmers' contributions to the conservation and development of agrobiodiversity. However, it is unlikely to provide significant economic rewards for ongoing conservation efforts. Farmers' varieties have evolved over generations through the exchange, adaptation and selection of PGR, and they continue to evolve. Conferring ownership of these varieties, which have so far been in the public domain, on individuals or communities for a limited duration – incurring substantial administrative and transactions costs – may contribute very little to the on-farm conservation of agrobiodiversity.[25] The protection of farmers' varieties could, however, be a useful adjunct to benefit-sharing provisions through defensive publication strategies. Defensive publication of farmers' varieties may not yield economic benefits to farmers, but it can prevent IPRs on farmers' varieties from being appropriated by individuals or institutions. It can also underpin benefit-sharing claims when farmers' varieties are used as parental material in institutional breeding programs. Defensive publication requires comprehensive and authentic documentation of virtually all existing agrobiodiversity, which entails large investments by the public research system or a similar agency.

Implications of institutional capacity and implementation issues

While developing countries have strongly advanced the political economy arguments for farmers' rights, there is little evidence to suggest that institutional capacity and the administrative, technical and legal infrastructure required for implementation have been systematically evaluated. Both market-oriented and rights-based approaches to farmers' rights call for substantial institutional capacity. The commercial development of farmers' varieties appears to offer an attractive self-sustaining route to rewarding and sustaining farmers' innovations and conservation efforts. However, the previous discussion has shown that institutional adaptation required for the commercialization of farmers' varieties involves fairly radical reform of the seed regulatory system, encompassing variety registration and quality control systems. However, the institutional capacity required goes well beyond regulatory reform. Even if the regulatory systems become more accommodative of farmers' varieties, there is still the question of the investment required to set up and maintain a comprehensive and authentic ('official') database of farmers' varieties. It is unlikely that this goal can be accomplished by an evolving patchwork of decentralized, NGO-led community biodiversity registers. The institutional capacity required extends to the ability to assess market potential (to pick winners), creating institutions and linkages for production and marketing and a policy framework that defines a space for farmers' varieties and redirects economic rewards from commercialization to those responsible for conservation. While a few developing countries (such as Nepal and Peru) appear to be initiating regulatory reform, there is no evidence of the other elements of institutional capacity being addressed. In the absence of these elements, regulatory reform can at best provide only a feeble stimulus to commercial development.

In the context of rights-based approaches, it must be noted that even conventional PVP systems that do not address the issues of farmers' rights call for substantial institutional capacity for effective implementation. Many European countries that were pioneers in introducing PVP legislation took considerable time to establish the necessary infrastructure. The first PVP certificates could generally be issued only 4–5 years after the enactment of legislation. PVP requires technical expertise and administrative infrastructure for DUS testing for a range of agricultural/horticultural crops. Even in developing countries with large national agricultural research systems that have built up substantial capabilities in the public sector for variety testing (for instance, in India, Brazil and China), it is still necessary to develop independent and credible arrangements for testing – preferably at an arm's length from the public research system (as the public sector may itself be a large breeder seeking protection for new varieties). Large reference collections need to be established to examine the novelty of varieties submitted for protection. Developing countries also need to establish agreements that will enable them to search reference collections of other countries, which is especially important when the novelty of foreign varieties submitted for protection has to be established. Systems need to be put in place to ensure the security of seeds and other reproductive materials as they pass through the testing system. This is an important concern for private sector entities, which otherwise may remain reluctant to take advantage of PVP. PVP is a private right that has to be defended against infringement by the IPR holder. Effective implementation of PVP requires a judicial system that can provide reasonably quick remedies against infringement without imposing excessive transaction costs on IPR holders.

The institutional capacity required for implementing farmers' rights goes considerably beyond what is required for a conventional PVP system focused on institutional breeders. The discussion in the previous sections has identified several key elements of institutional capacity required in developing countries for the implementation of farmers' rights provisions. Giving effect to farmers' rights requires comprehensive documentation of existing agrobiodiversity in the country and its geographical distribution, which even developed countries are yet to accomplish. It also requires processes by which ownership of traditional varieties or landraces can be attributed to farmers and farming communities – a challenging task in the context of material that has long been in the public domain, is constantly evolving and has been exchanged between communities for generations. The exceedingly complex pedigree of modern varieties (which may include a good proportion of material sourced from foreign countries) makes it very difficult to determine the contribution of specific parental varieties in the development of a new variety. Adjudication of benefit-sharing claims at the stage of grant protection can pose great difficulty since the commercial potential of new varieties cannot be accurately predicted. Monitoring the countrywide sales of protected varieties at the level of individual protected varieties for enforcing benefit-sharing provisions calls for substantial investment in seed industry regulation.

The magnitude of investment required for creating the requisite infrastructure (documentation of existing farmers' varieties, machinery for adjudicating benefit-sharing claims and so on) does not appear to have been assessed at the time of enacting legislation. The discussion in developing countries has largely centred on how farmers' rights could countervail the monopoly rights accorded to institutional players by the newly introduced IPR regimes. The infrastructure required for implementing a conventional PVP regime, and the preparatory work that needs to be done before implementation can commence, have merited very little discussion. Even the build-up of infrastructure for implementing conventional PVP systems has been slow in developing countries. It is mainly for this reason that there is as yet very little evidence of the commencement of implementation of PVP in most developing countries (Tripp, Louwaars and Eaton, 2007). In some developing countries (e.g. India), no PVP certificate has so far been issued even though the legislation has been on the statute books for several years – a testimony to the fact that the operational mechanisms for PVP implementation are not yet in place. In this setting, the prospect of even a beginning being made in the implementation of farmers' rights appears to be a very limited because farmers' rights have been designed as an adjunct to conventional IPR systems rather than as a set of independent measures to support on-farm conservation and innovation.

The institutional capacity for the implementation of farmers' rights has important international dimensions as well. This arises because legislative provision affording protection for farmers' varieties can be enforced only within the respective national jurisdictions.[26] However, much of the potential economic returns from the use of farmers' varieties may arise when they are exchanged and used across national boundaries. Farmers' rights provisions in domestic legislation are not likely to be an effective instrument for capturing economic returns from the international exchange and use of farmers' varieties. For instance, it may be possible to enforce benefit-sharing provisions in regard to a farmers' variety in India, when it is used for the development of a new variety that is sought to be protected in India. However, it may not be possible to enforce benefit-sharing provisions if the Indian farmers' variety is used in the development of a new variety that is sought to be protected in the United States. Such a conundrum is due to two reasons. First, US law may regard the Indian farmers' variety as being in the public domain. IPRs conferred on the farmers' variety in India would be of no consequence unless the variety is also protected in the United States. Further, it may not be possible to seek protection for the variety in the United States because US law may not provide for exemption from the novelty criterion that the Indian law allows (and which enables the farmer's variety to be protected in India). Second, if US law contains no benefit-sharing provisions, then no benefit-sharing claims can be made in respect of that variety even if US law recognizes the Indian farmers' variety as being subject to some form of IPRs in India. The fact that strong farmers' rights provisions in domestic legislation will help very little in benefit sharing in an international context does not appear to be well understood in developing countries. Just as the effectiveness of conventional PVP systems requires the universalization of IPR principles through the International Convention for the Protection of New Varieties of Plants (UPOV Convention), so too the effectiveness of farmers' rights provisions will also call for universal application of farmers' rights principles through some form of international agreement.[27] This means that for the effective implementation of farmers' rights, developing countries will need to work towards an international understanding for the extension of farmers' rights principles to national legislation in developed countries. The international dimension of institutional capacity arises even in the context of defensive publication strategies where the aim may be only to prevent other entities from obtaining IPRs on farmers' varieties rather than the enforcement of benefit-sharing provisions. The authentic documentation of farmers' varieties in India may prevent a breeder or a seed company from seeking protection for the same variety in India, but it will not prevent the variety from being protected in the United States unless the US patent authorities can be persuaded to refer to the India database as evidence of prior art. Therefore, developing countries will not only have to document existing biodiversity, but they will also need to negotiate to make their databases part of the reference material on prior art routinely used by IPR authorities in other countries. The ability to monitor large numbers of IPR applications in different jurisdictions for potential infringement of rights conferred on farming communities is also an important element of institutional capacity that developing countries will need to develop once a framework for international application of farmers' rights principles is developed.

In developing countries, the political economy of farmers' rights appears to have precluded any empirical assessment of the incentives for conservation that can be provided through farmers' rights provisions. There does not appear to have been any effort to assess the magnitude of economic rewards that can potentially accrue to farming communities as a result of benefit-sharing provisions or the resources that can be garnered for conservation funds using farmers' rights provisions. If incentives for on-farm conservation are to flow through a share in the benefits appropriated by IPR holders, it is necessary to examine the magnitude of benefits that accrues to breeders as a result of protection. IPRs are seldom traded in the market, and, therefore, the private returns appropriated by IPR holders are not directly observable. However, empirical studies on large databases of patent cohorts in developed countries using indirect estimation methodologies consistently suggest that (1) the private value distribution of returns from IPRs is highly skewed, with a large proportion of patents yielding little or no economic value to patent holders; and (2) that the average private returns appropriated from patents are exceedingly modest (Pakes, 1986; Schankerman and Pakes, 1986; Schankerman, 1998).[28] These conclusions appear to apply with greater force to PVP, as on-farm seed saving restricts the returns that can be appropriated through PVP (Srinivasan, 2003).[29] The expectation that benefit-sharing provisions can yield large returns is probably based on an inaccurate appreciation of the private value distribution of returns from IPRs. They may be belied by the fact that there are only limited returns to share. This is a much more fundamental issue affecting the effectiveness of IPR provisions than constraints of institutional capacity.

In developed countries, the limited appropriability of returns afforded by PVP has led to it being viewed as a rather weak IPR measure. It has led to a series of changes to PVP to increase the appropriability of returns for breeders, which is reflected in the revisions made to the UPOV Convention in 1991.[30] In many developed countries (such as the United States, Japan and Australia), the weaknesses in protection offered by PVP have led to plant varieties being protected under much stronger forms of protection (e.g. through utility patents).[31] Seed companies have also attempted to introduce contractual arrangements with seed buyers (farmers) that proscribe any on-farm seed saving, overriding the farmers' privilege that flows from PVP legislation. Finally, the seed industry is also exploring technological options to protect innovations that are not reliant on the enforcement of IPRs. The emergence of the so-called terminator technologies can be seen as a technological response to the weaknesses in IPR enforcement. In introducing farmers' rights provisions that limit even the appropriation of returns afforded by conventional PVP, developing countries have chosen to completely disregard the experience of developed countries and the trend towards stronger forms of protection.

While the private returns appropriated from the average PVP certificate may be modest, the cascade of transaction costs imposed on the institutional breeder as a result of farmers' rights provisions will apply to all new varieties offered for protection irrespective of their eventual commercial potential. These include costs that arise from breeders having to keep detailed records of sourcing of parental material used at different stages of the breeding cycle, costs associated with negotiating access and obtaining prior informed consent for the use of farmers' varieties and PGR (where applicable), costs involved in complying with prior informed consent for *ex situ* material, costs involved in contesting benefit-sharing claims (or the level at which they are set) and costs in complying with benefit-sharing arrangements. These are transaction costs over and above the costs that apply in conventional PVP systems. The cost of obtaining protection in developing countries, therefore, is likely to be substantially higher than in developed countries, even as potential returns are constrained by the broadly defined safeguard provisions for on-farm seed saving and the limited capacity and infrastructure that is necessary for IPR enforcement.[32] The inevitable impact of the additional transaction costs imposed by farmers' rights is that incentives for institutional breeders to seek protection for new varieties will be further diluted.

In designing PVP or biodiversity legislation, many developing countries have sought to confer certain advantages on the public research system. The retrospective protection of publicly bred extant varieties and incorporation of the essential derivation principle are two examples of how legislation has attempted to support the public sector in deriving returns from previous investments in plant breeding. The rationale advanced for conferring these advantages is that, unlike the private sector whose research priorities are likely to be influenced by commercial profit, the public research system can be mandated by policy to pursue national objectives (e.g. catering to the needs of resource-poor farmers or areas). It is generally assumed that public sector innovation would be largely unaffected by farmers' rights provisions – a view that most public sector research managers in developing countries appear to share. However, a prima facie examination of developing country legislation suggests that the public sector will be subject to the same transaction costs that are imposed on the public sector by farmers' rights provisions (the public sector will also be subject to the prior informed consent provisions, and the public sector varieties offered for protection can also be subject to benefit-sharing claims). If the effect of farmers' rights provisions is to increase transaction costs and restrict access to farmers' PGR for *all* institutional players, then the development of new varieties by the public sector will also be limited.[33] This result may be unintended, but it will nevertheless be a real consequence of farmers' rights provisions. We have seen how pressures on the public sector to generate revenues using IPRs could restrict farmers' access to publicly bred varieties. This is not a consequence of farmers' rights provisions, but it shows how measures designed to support the public research system could adversely affect farmers' access to public sector varieties.

Conclusions

Armed with the equity and conservation arguments that have a deep resonance with farming communities, developing countries are crafting a range of measures designed to protect farmers' access to innovations, reward their contributions to the conservation and enhancement of PGR and provide incentives for sustained on-farm conservation. These measures range from the commercialization of farmers' varieties to the conferment of a set of legally enforceable rights on farming communities – the exercise of which is expected to provide economic rewards to those responsible for on-farm conservation and innovation. The rights-based approach has been the cornerstone of legislative provision for implementing farmers' rights in most developing countries. In drawing up these measures, developing countries do not appear to have systematically examined or provided for the substantial institutional capacity required for the effective implementation of farmers' rights provisions. The lack of institutional capacity threatens to undermine any prospect of serious implementation of these provisions. More importantly, the expectation that significant incentives for on-farm conservation and innovation will flow from these rights may be based on a flawed understanding of the economics of intellectual property rights. While farmers' rights may provide only limited rewards for conservation, they may still have the effect of diluting the incentives for innovative institutional breeding programs – with the private sector increasingly relying on non-IPR instruments to profit from innovation. The focus on a rights-based approach may also draw attention away from alternative stewardship-based approaches to the realization of farmers' rights objectives.

Notes

1 Agreement on Trade-Related Aspects of Intellectual Property Rights, Annex 1C of the Marrakech Agreement Establishing the World Trade Organization, 15 April 1994, 33 ILM 15 (1994) [TRIPS Agreement].

2 Convention on Biological Diversity, 31 ILM 818 (1992).

3 The literature distinguishes between 'rights-based' approaches and 'stewardship-based' approaches that can be used to recognize the contribution of farming communities and to promote on-farm conservation and innovation. The rights-based approaches rely on the conferment of certain legally enforceable rights on farmers/farming communities, the exercise of which is expected to generate economic rewards providing incentives for conservation. Stewardship approaches focus on providing farming communities with an important role in deciding how the conservation of PGR can be sustained and used for generating productivity-enhancing innovations (for example, through engagement in participatory plant-breeding programs). Developing country legislation to support conservation has focused almost exclusively on rights-based approaches.

4 The analysis in this chapter is based on an overview of developing country legislation that has been enacted or is under consideration for enactment. For a compendium of relevant legislation and regulations, refer to <www.upov.int> and <www.grain.org> (last accessed 10 May 2012). We have focused our attention on common themes in developing country legislation and have not examined the differences in the legislative provision made by different countries. Our illustrations are drawn from the Indian PVP legislation as it appears to have gone the farthest in articulating farmers' rights and includes all three of the approaches mentioned earlier.

Several regional bodies have developed model legislation on plant variety protection and/or on access and benefit sharing in the context of the exchange of biological resources. These include the African Union (Model Law on Rights of Local Communities, Farmers, Breeders and Access), the Association of Southeast Asian Nations (Framework Agreement on Access to Biological and Genetic Resources) and the Andean Community (Decision 391 on the Common Regime on Access to Genetic Resources). Countries that have incorporated elements of the model provisions in their PVP/biodiversity legislation include Argentina, Bolivia, Brazil, Chile, Costa Rica, Ecuador, Ethiopia, Guyana, India, Kenya, Malawi, Pakistan, Panama, Peru, the Philippines, South Africa and Venezuela.

5 For instance, in India, a variety can be 'notified' under the provisions of the Seeds Act. Once a variety is notified, only certified seeds of that variety can be sold. Such a system allows varieties to be brought under the purview of quality control regulations on a selective basis. Varieties developed by the public research system are all invariably notified.

6 This could happen if no marketing or intellectual property rights are conferred on farming communities from which varieties are sourced. The registration of a variety in a national register by itself may not confer intellectual property ownership on the entity registering the variety. However, if the regulations stipulate that only the entity registering a variety is entitled to market it, then variety registration could confer de facto ownership rights on the applicant.

7 Protection of Plant Varieties and Farmers' Rights Act, online: <http://agricoop.nic.in/seeds/farmersact2001.htm> (last accessed 10 May 2012).

8 Under the European Community plant variety legislation, on-farm seed saving of protected varieties without payment of royalty to breeders is allowed only for small farmers and is restricted to certain species (e.g. fodder crops and cereals). Small farmers are defined based on the land area sufficient to produce a certain volume of output (92 tonnes in the case of cereals). In the United States, on-farm seed saving of protected varieties is permitted only for the purpose of replanting the farmer's land. US legislation no longer permits 'brown bagging,' the practice of selling farm-saved seeds of protected varieties without using the brand name or protected variety denomination.

9 Frequent resort to compulsory licensing provisions could be self-defeating since it would go against the fundamental objective of intellectual property rights (IPR) legislation, which is to provide incentives for innovation.

10 See Chapter VII of India's Protection of Plant Varieties and Farmers' Rights Act, supra note 7.

11 PVP legislation generally requires the IPR applicant to deposit a sample of the seeds of the variety for which protection is sought with a specified depository. This is intended for fulfilling the disclosure requirement under IPR law in the case of plant variety innovations.

12 It is not clear whether the provision is also intended to apply to extant varieties developed by the private sector.

13 Seeds Act, 1966, online: <http://agricoop.nic.in/seedsact.htm> (last accessed 10 May 2012). Extant variety protection requires an exception to be made to the novelty

criterion for protection. This is also necessary for the protection of farmers' varieties that have been developed over a long period of time and may be widely in use.

14 Except where this happens through oversight or fraud.

15 The ability of the public research system to derive rents from retrospectively protected extant varieties is doubtful because the varieties may have been in the public domain for a considerable period of time and their reproductive material may be widely available with farmers and seed producers.

16 'Essential derivation' is a concept developed in the 1991 revision of the International Convention for the Protection of New Varieties of Plants, adopted on 2 December 1961, <www.upov.int/en/publications/conventions/index.html> (last accessed 10 May 2012) [UPOV Convention], and is intended to strengthen the protection afforded to holders of PVP rights. It is intended to protect the innovations of first round innovators from being appropriated by second round innovators through minor or (agronomically) unimportant modifications. A variety defined as an 'essentially derived variety' can be protected as a distinct new variety if it conforms to the usual PVP criteria of distinctness, uniformity and stability, but its commercial exploitation requires the consent of the breeder of the original variety from which it was derived. While the UPOV Convention has enunciated the principle of essential derivation, the technical definitions of what constitutes an essentially derived variety are still a matter of considerable debate – they are based on extent to which the genetic compositions of the original and derived varieties are similar. Technical definitions of essential derivation have not been provided in the Indian PVP legislation or in other developing country legislation.

The ability of the public research system to seek rents from the use of retrospectively protected extant varieties in the development of new varieties will depend on whether the extant varieties are used directly as parents of new varieties developed by the private sector breeding programs and the extent to which they have undergone modification in the private sector's own program. In India, while the private sector acknowledges the use of public varieties in its programs, it is often argued that these public varieties have been subject to extensive modifications before they are used as parents in the development of new varieties – the implication being that the essential derivation clause would not be attracted (Srinivasan, 2004).

17 Incentives for follow on innovation depend on the distribution of returns between the first round innovator and the second round innovator. Compared to a situation where the private sector was freely able to appropriate public material in its breeding programs, extant variety protection reduces the incentives for follow-on innovation.

18 Some retroactive application of benefit-sharing provisions may be feasible, but it is unrealistic to assume that this can be stretched back a long distance in time. Much of the expectation that benefit-sharing provisions can yield large rewards for past conservation efforts appears to be premised on the feasibility of retroactive application of these provisions.

19 The novelty of varieties offered for protection may also be checked against reference collections of other countries by reciprocal arrangements. In the case of plant varieties, it is very difficult to establish when a variety was actually bred. Consequently, novelty is decided on the criterion of absence of previous commercial sales.

20 The translation of novelty as 'absence of previous commercial sale' in PVP is one such concession. The attribution of IPRs based on the 'first-to-file principle is another compromise, reflecting the difficulties involved in establishing who was the "first-to-invent."'

21 See section 26 of the Indian PVP legislation, extracted in Box 12.2.

22 The consequences of commercial failure of a new variety for benefit-sharing provisions are not very clear, but the legislation does contain provisions for civil and criminal penalties where breeders fail to make mandatory payments.

23 It is not clear from developing country legislation how disclosure of pedigree can be used to identify the use of farmers' varieties in the development of new varieties. For instance, in the United States, the parental lines of new varieties are often identified using nomenclature assigned to them by breeders. Unless all farmers' varieties are authentically identified and breeders are required to use the nomenclature assigned in the official database, the disclosure provision may not be very effective. It is also not clear how developing countries propose to deal with situations where farmers' varieties are not immediate parents of a new variety but figure further back in the ancestry of parental varieties.

24 See section 15(2) of the Indian Protection of Plant Varieties and Farmers' Rights Act, supra note 7. If rights are to be conferred on material that has been in the public domain, it becomes important to ensure that rights are conferred only on those to whom the development of the innovation can be definitively attributed. The first-to-file principle will not have much relevance to the case of farmers' rights.

25 The intended duration of protection for farmers' varieties is often not clear from PVP legislation in developing countries (e.g. Indian legislation). Some advocates of farmers' rights have even suggested that farmers' varieties should be protected in perpetuity. Perpetual protection for farmers' varieties would imply that innovations would be treated differently depending on the setting in which they emerge. Moreover, using standard economic models of IPR, it is easy to show that perpetual protection of any innovation would be welfare reducing (Nordhaus, 1969).

26 Although international IPR conventions and the TRIPS Agreement, supra note 1, have sought to universalize the application of IPR principles by providing for common standards of protection, national treatment, reciprocity and right of priority, protection conferred by IPR legislation remains national in scope. IPRs on a plant variety have to be secured separately in each country under the respective legislation. Some arrangements such as the European Community's (EC) Community Plant Variety Office do make it possible to protect a variety in all EC countries through a single application.

27 UPOV Convention, supra note 16.

28 These methods attempt to infer the private economic value appropriated from innovations from the economic behaviour of IPR holders. In most countries, IPR holders have to pay an annual renewal fee to keep the protection in force. If it is assumed that IPR holders will pay the renewal fee only if the expected returns from renewing protection exceed the renewal cost, the private value appropriated from holding IPRs can be inferred from the data on the renewal behaviour of IPR holders (the duration for which IPRs are kept in force) and data on costs associated with renewal. See the studies cited in the text.

29 A distinction needs to be made between private returns to holding IPRs on new plant varieties and the returns to the 'complementary assets' that are necessary for realizing economic returns from IPRs (Teece, 1987). Seed companies require complementary assets such as marketing and distribution networks to realize returns from a protected new variety. Seed companies may make large profits from the marketing of new varieties, but much of their profit may be a return to investment in these complementary assets. The empirical studies referred to earlier estimate the 'pure' returns to holding IPRs. Farmers' rights provisions seek a share only of the economic returns to IPRs.

30 These changes that have been gradually adopted in national legislation of countries that are members of the UPOV Convention, supra note 16, include: (1) the extension of protection to all species and genera; (2) an increase in the duration of protection; (3) the provision of farmers' privilege for on-farm seed saving only as an exception to the breeder's rights; (4) the extension of breeders' rights to the harvested material of the protected variety where a breeder has not had an opportunity to exercise the rights on the reproductive material; and (5) the introduction of the concept of essential derivation to prevent breeders' innovations being appropriated by others through minor modifications.

31 This is especially true of new plant varieties created through the application of agricultural biotechnology. Genetically modified varieties in the United States are not protected by PVP – they are protected through utility patents. Farmers' privilege and researchers' exemptions do not apply if a new plant variety is protected through patents.

32 The Indian PVP legislation goes a step further by making the breeder responsible for the 'promised' performance of the protected variety – the breeder is liable for civil and other penalties if a protected variety fails to live up to its promise. This is intended to safeguard farmers' interests, but it creates a substantial financial risk for any breeder seeking to protect a new variety – a somewhat bizarre outcome in a legislation that seeks to provide incentives for innovation (section 39(2) of Protection of Plant Varieties and Farmers' Rights Act, supra note 7).

33 There is as yet very little information on how developing countries plan to implement farmers' rights provisions in the context of public sector plant breeding. It will generally be difficult to make a case for differential treatment of the public sector, if a developing country is also trying to provide a level playing field and encourage private investment in plant breeding. Public sector managers appear to assume that restrictions on access to, and use of, farmers' varieties will apply only to the private sector.

References

Genetic Resource Policy Initiative (GRPI) (2008). *Genetic Resource Policy Initiative – PERU: Final Project Report May 2004 to June 2008,* prepared by Isabel Lapeña, Project Coordinator, GRPI, Peru.

International Development Research Centre (IDRC) (2001). *Seeding Solutions, Volume 2 on Options for National Laws Governing Access to and Control over Genetic Resources,* jointly published by the IDRC, Canada, the International Plant Genetic Resources Institute, Italy, and the Dag Hammarskjöld Foundation, Sweden.

Nordhaus, W. (1969). *Invention, Growth and Welfare: A Theoretical Treatment of Technological Change,* MIT Press, Cambridge, MA.

Pakes, A.S. (1986). 'Patents as Options: Some Estimates of the Value of Holding European Patent Stocks,' *Econometrica* 54: 755–84.

Schankerman, M. (1998). 'How Valuable Is Patent Protection: Estimates by Technology Field,' *Rand Journal of Economics* 29(1): 77–107.

Schankerman, M., and A. Pakes (1986). 'Estimates of the Value of Patent Rights in European Countries during the Post 1950 Period,' *Economic Journal* 96(384): 1052–76.

Srinivasan, C.S. (2003). 'Exploring the Feasibility of Farmers' Rights,' *Development Policy Review* 21: 419–47.

Srinivasan, C.S. (2004). 'Plant Variety Protection in Developing Countries: A View from the Private Seed Industry in India,' *Journal of New Seeds* 6(1): 67–89.

Swanson, Timothy N., D. Pearce and R. Cervigni (1994). *The Appropriation of the Benefits of Plant Genetic Resources for Agriculture: An Economic Analysis of the Alternative Mechanisms for Biodiversity Conservation,* Background Study Paper no. 1, prepared for the Food and Agriculture Organization's (FAO) Commission on Plant Genetic Resources, FAO, Rome.

Teece, D.J. (ed.) (1987). *The Competitive Challenge: Strategies for Industrial Innovation and Renewal,* Ballinger Publishing Company, Cambridge, MA.

Tripp, R., N. Louwaars and D. Eaton (2007). 'Plant Variety Protection in Developing Countries: A Report from the Field,' *Food Policy* 31(3): 354–71.

Part V

Case studies of national laws

13 Commentary on the Indian Protection of Plant Varieties and Farmers' Rights Act 2001

Dwijen Rangnekar

Introduction

This chapter investigates the way in which farmers' varieties are treated pursuant to India's Protection of Plant Varieties and Farmers' Rights Act and the accompanying Protection of Plant Varieties and Farmers' Rights Rules, 2003.[1] Read together, the Act and the Rules represent the first time that farmers' rights have been explicitly recognized and promoted in national law. The Indian law was heralded by M.S. Swaminathan (1998) as unique in the sense that it is the first time anywhere in the world that the rights of both breeders and farmers have received integrated attention. For Olivier de Schutter (2009), the UN special rapporteur on the right to food, India's legislative architecture stands alongside the African Model Law for the Protection of the Rights of Local Communities, Farmers and Breeders, and for the Regulation of Access to Biological Resources[2] as an act of resistance to deepening proprietary claims in plant genetic resources. The drafting history of India's law is testimony to struggles to resist intellectual property rights in plants while seeking to push the canon for the rights of marginalized developers and users of plant genetic resources.

It may surprise some readers that aspirations for plant breeders' rights (PBRs) in India predate the completion of the Uruguay Round. In 1990, the Indian Council for Agricultural Research published a study on introducing PBRs, and, thereafter, a technical mission was commissioned by the Food and Agriculture Organization (FAO) to review the topic (Rangnekar, 1998). Opposition to PBRs came from farmer movements, such as the Beej Satyagraha, and public interest litigation that sought to call the government to account on a range of issues, including plans for seeking membership in the International Union for the Protection of New Varieties of Plants (UPOV). Initial drafts of the legislation failed to explore residual latitude in Article 27.3(b) even while opting for a *sui generis* system for plant variety protection (for a discussion of options, see Leskien and Flitner, 1997; International Plant Genetic Resources Institute, 1999; the Crucible Group, 2001; and Rangnekar, 2002). For that matter, these early drafts also failed to incorporate norms and principles related to farmers' rights as elaborated in countervailing global treaties, such as the FAO's International Treaty on Plant Genetic Resources for Food and Agriculture (ITPGRFA).[3]

Following this introduction, the chapter provides a brief overview of the legal architecture of the Indian Act and Rules and how they are administered. Thereafter, it analyzes the main provisions of the Act and Rules concerning farmers' rights and farmers' varieties, noting in particular the registration criteria, process and the rights conferred therein. In this context, the chapter analyzes the actual performance or implementation of the Act and Rules, considering the numbers of applications and grants of rights over different types of varieties, including farmers' varieties. Finally, it reviews the National Gene Fund and reports on the status of applications and concludes with some apprehensions about the legislative architecture.

Administration

The Act sets out provisions for a Protection of Plant Varieties and Farmers' Rights Authority (Act, Chapter II) with a duty to 'promote, by such measures as it thinks fit, the encouragement for the development of new varieties of plants and to protect the rights of the farmers and breeders' (Act, section 8(1)). This duty includes, among things, ensuring the registration of extant varieties, preparing documentation of the registered varieties, indexing and cataloguing of farmers' varieties, collecting statistics concerning plant varieties, ensuring adequate seed supply of registered varieties and maintaining the Register (Act, section 8(2)). Specific importance is given to the establishment of a Plant Varieties Registry (Act, section 12) and the development of a very detailed National Register of Plant Varieties (Act, section 13; Rules, section 23). Of relevance to the topic is the constitution of two committees: the Extant Variety Registration Committee and a Standing Committee on Farmers' Rights. The Extant Variety Registration Committee, established in 2006, is to advise individuals on the registration of extant varieties, including procedures for registering each species and category of varieties and periodically reviewing applications. The Standing Committee on Farmers' Rights is tasked with translating the provisions for farmers' rights, primarily in Chapter VI of the Act, into practice. Finally, appeals to orders and decisions of the Protection of Plant Varieties and Farmers' Rights Authority will be heard by a Plant Varieties Protection Appellate Tribunal, under whose jurisdiction will also be matters concerning the registration of a variety, claims on benefit sharing, compulsory licensing and payment of compensation, among others (Act, section 56). The decisions of the Tribunal shall be executable as a decree of a civil court (Act, section 57(5)).

Farmers' rights

In the vast literature on the subject, a broad range of rights have been identified as constituting the rights of farmers, including the right to reuse saved seeds of registered varieties, the right to reward and recognize varieties through a

National Gene Fund, the right to benefit sharing, the right to register one's own varieties, the right to information and compensation for crop failure, the right to compensation for undisclosed use of traditional varieties and the right to adequate availability of registered propagating material. Some of these rights are included in the definition of farmers' rights in Article 9 of the ITPGRFA as discussed earlier in this volume. However, this chapter notes that that the Indian Act transcends the ITPGRFA, inasmuch as it states that:

- A farmer who has bred or developed a new variety shall be entitled to registration and treatment (i.e. protection) in a manner akin to a breeder.
- A farmers' variety is entitled to be registered if it fulfils all registration requirements. This has implications for subsequent users of such registered farmers' varieties; they will need to demonstrate obtaining prior informed consent from the farmers concerned if and when they apply for intellectual property rights on new varieties that incorporate a registered farmers' variety.
- A farmer engaged in conservation and improvement of genetic resources shall be entitled to recognition and reward from the National Gene Fund.[4]
- A farmer is entitled to save, use, sow, resow, exchange, share or sell his farm produce, including seed of a protected variety, provided that the farmer does not sell branded seed of the variety.[5]

These provisions for farmers' rights have been among the most difficult hurdles in negotiating India's accession to the 1978 International Convention for the Protection of New Varieties of Plants (UPOV Convention).[6]

Applicants: farmers as breeders

The entitlement for a farmer to be treated as a breeder is a celebrated feature of the Act (Swaminathan, 1998). Quite unlike laws in other jurisdictions, the definition for applicants who may apply to register a new variety under the Act (which confers breeders' rights) include 'any farmer, farmer group or community of farmers' (Act, section 16(1)(d)), which exists alongside the more standard categories of breeders. However, updates in the *Plant Variety Journal* or recent secondary literature (e.g. Nagarajan et al., 2010; Kochupillai, 2011) suggest that as yet no farmer has submitted an application to register a new variety and claim breeders' rights. While this situation may change in the future, the possibilities for farmers to be holders of PBRs could be illusionary. Nothing in either the Act or the Rules provides for differential criteria for the registration of varieties developed by farmers in comparison to other categories of applicants. Farmer applicants seeking to register their varieties for the grant of PBRs have to meet the same distinctness, uniformity and stability (DUS) standards and requirements for novelty that all other applicants need to satisfy (Act, section 15). For that matter, the fees for either registering a new variety or maintaining the registration (and, therefore, the rights conferred) are also the same for farmers as for all other applicants (see Table 13.1). The Act allows some

Table 13.1 Fees for registering a new variety

Test fees	*20,000–50,000*[a]
Annual renewal fees (per year) for:	
Individuals	5,000
Educational institutions	7,000
Commercial enterprises	10,000
Application for registering as agent/licensee	10,000
Registration of essentially derived varieties for:	
Individuals	5,000
Educational institutions	7,000
Commercial enterprises	10,000
Application for variation/cancellation of registration for:	
Individuals	3,000
Educational institutions	5,000
Commercial enterprises	7,000
Notice of opposition	1,500
Application for benefit sharing	5,000

Source: Protection of Plant Varieties and Farmers' Rights Rules, 2003.

Notes:
All monetary values in Indian rupees.
[a] Dependent on the species.

differential treatment for farmers in section 44, wherein farmers and farming communities are exempt from paying fees in proceedings before the Tribunal, the Authority or a High Court.

In this respect, it is useful to recall the negotiating history that led to the UPOV Convention. Discussions in Europe in the 1950s drew attention to how equitable standards for uniformity and stability between varieties developed by breeders and farmers were discriminatory to the breeding practices of farmers, which favoured levels of variability and heterogeneity in the variety. Illustratively, the 1954 Stockholm Conference on the Human Environment, under the auspices of the Organisation for European Economic Co-operation, heard delegates arguing for differential standards of uniformity and stability so as to valorize the work of farmer-breeders (Akerman and Tedin, 1955). These views failed to translate into either national practice or the emergent UPOV system. It is disappointing that India's legal architecture has failed to explore the possibilities of differentiating between different categories of applicants based on their breeding practices by incorporating alternative registration criteria. On the other hand, it is important to note that in 2009, the government passed regulations which specified that, as far as farmers' varieties were concerned, uniformity standards could be relaxed to allow double the number of off-types as otherwise permitted pursuant to the *Plant Variety Journal* of India[7] (see the following discussion).

Independent of these provisions for farmers to register new varieties and be treated like breeders, there are provisions for the registration of farmers' varieties. The latter enable particular provisions associated with farmers' rights such as access and benefit-sharing rights to be triggered. These are discussed in the following section.

The different categories of 'varieties' and conditions for registration

At the heart of India's innovative legislative architecture are the multiple categories of plant varieties, such as farmers' varieties, extant varieties and new varieties, among others (see Table 13.2). In explaining this feature of India's law, it is important to recognize that the Indian effort has either followed existing templates, such as the UPOV Convention's *sui generis* system or sought to open up and travel new avenues. In Article 2, the Act defines variety as a plant grouping, except a microorganism, within a single botanical taxon of the lowest known rank, and it is mapped by the expression of characteristics that are distinguishable from others in the same plant grouping and are stable and uniform. The categories of variety noted in the legislation with possibilities for registration are:

- *Extant varieties:* By definition, extant varieties are those that are already in circulation – thus, those that include 'varieties in common knowledge' (see discussion later in this chapter), farmers' varieties and any other variety in the public domain. Once a species is notified under the Act, a three-year moratorium is provided to allow extant varieties to be registered, which only requires a demonstration of DUS since, by definition, they fall foul of the requirement of commercial novelty.
- *Farmers' varieties:* This is a subcategory of extant varieties, and the Act defines farmers' varieties as those that have been traditionally cultivated and evolved by the farmers in their fields and also includes a wild relative or landrace or a variety about which the farmers possess common knowledge.
- *Varieties in common knowledge:* A subset of extant varieties, the term is not directly defined in the Act. However, while alluded to by the UPOV Convention,[8] it is pronounced a number of times in the Act. A 2009 notice in the *Plant Variety Journal* explains that varieties in common knowledge are those in the public domain and should have been sold or otherwise disposed of in India for at least one year prior to the date of application and less than 13 years. This could include those varieties that are merchandised as 'truthfully labelled.'
- *New varieties:* This category of variety is the subject of registration under the Act for conferring breeders' rights – thus, mutually defined by being (commercially) new and DUS (as discussed later in this chapter). In sum, these varieties will be awarded, upon successful registration, with PBRs.
- *Essentially derived varieties:* These are defined primarily in phenotypical terms with respect to an initial variety from which it is predominately derived while also retaining the expression of essential characteristics related to the

Table 13.2 Categories of plant variety in the Indian plant variety system

	New variety	*Extant variety*	*Farmers' variety*	*Variety in common knowledge*	*Essentially derived variety*
Definition	A variety that meets the conditions for registration, notably commercial novelty, DUS.	A variety already available in India, which is either notified under section 5 of the Seeds Act, 1966; a farmers' variety; a variety about which there is common knowledge or any other variety that is in the public domain.	A variety that has been traditionally cultivated by farmers in their fields or is a wild relative, landrace or a variety about which farmers possess common knowledge.	While not explicitly defined in the Act, this includes varieties in the public domain and should have been sold or otherwise disposed in India for at least one year prior to the date of application and less than 13 years.	With respect to an initial variety, an essentially derived variety is predominantly derived from the initial variety and thus conforms to it in the expression of essential characteristics and is clearly distinguishable from the initial variety too.
Conditions for registration	Commercial novelty: the variety has not been sold or otherwise disposed of in India from the date of application, earlier than one year and elsewhere earlier than 6 years for trees and vines or 4 years for other species.	Since by definition they cannot satisfy commercial novelty, the conditions for registration are limited to DUS. Once a species is notified, there is a three-year moratorium within which extant varieties can be registered.	As a subcategory of extant varieties, there is no requirement for commercial novelty. A June 2009 regulation modifies the standard DUS requirements by tolerating twice the number of off-types for uniformity testing.	The conditions of registration are identical to those of extant varieties.	The conditions for registration are identical to those of new varieties.

	DUS: The variety is clearly distinguishable in its essential characteristics from other varieties, and these features are uniform and stable upon repeated propagation.	
Rights conferred	An exclusive right to 'produce, sell, market, distribute, import or export the variety' (Act, section 28(1)). These rights are conferred for a period of 9 years for trees and vines and 6 years in the case of other crops (Act, section 26(6)), with the possibility of renewal thereafter, up to a maximum of 18 years for trees and vines, 15 years for other crops.	The rights conferred for these categories of varieties remain somewhat ambiguous in the Act and the Rules. However, a careful reading suggests that the rights are quite identical to those granted to new varieties with some differences. In particular, the maximum duration of registration that is allowed for extant varieties is 15 years. Further, some commentaries suggest that registration of farmers' varieties – which otherwise fall within the broader category of extant varieties – may be for longer durations.

genotype of the initial variety. These varieties may also be clearly distinguishable from the initial variety. Interestingly, essentially derived varieties can be registered under the Indian Act, subject to a set of requirements (Act, section 23).

The conditions for registration of new varieties broadly cohere with the UPOV Convention's template of (commercial) novelty, distinctness, uniformity and stability (Act, section 15(1)). The novelty requirement is identical to the UPOV Convention's approach (compare with section 15(3)(a) of the Act). However, there is a different approach for extant varieties – in that, by definition, they fall foul of any construction of novelty and, thus, are exempt from establishing novelty and must only satisfy DUS requirements (Act, section 15(3)).[9]

Even as the conditions for registration cohere with the UPOV Convention, there is an important difference, in that the requirements for DUS (Act, section 15(3)) are pre-fixed in terms of essential characteristics. For example, distinctness requires the variety to be 'clearly distinguishable by *at least one essential* characteristic from any other variety whose existence is a matter of common knowledge' (Act, section 15(3)(b), emphasis added). The Act defines essential characteristics as characteristics that 'contribute to the principal features, performance or value of the plant variety' (Act, section 2). Such an agronomic assessment is a departure from the UPOV Convention's construction of these standards. As evidenced by the amendments that Kenya had to make to complete its accession to the UPOV Convention in 1978, it will prove to be an additional hurdle to India's membership to the UPOV Convention, if and when accession arises (Rangnekar, 2014).

As provisions for registering farmers' varieties are made under the generic class of extant varieties, they also have to meet DUS requirements. Recognizing a paucity of experimental data and also limited understanding of the range of variability tolerated by farming communities, Nagarajan, Yadav and Singh (2008) noted that the species-level DUS standards for farmer varieties should be developed iteratively and carefully. That said, these authors also believe that farmers' varieties have a tendency towards levels of homogeneity and distinctness that is also reflected in their wider acceptance and vernacular classification of folk varieties and wild cultivars. Consequently, they recommended that the parameters and standards of DUS should 'marginally vary' from other categories of varieties (ibid., 710). Subsequently, 19 species were notified under the Act[10] (Nagarajan et al., 2010). A June 2009 regulation passed under the Act relaxed the uniformity requirement for farmers' varieties to allow twice the number of off-types as indicated for variety as provided for under the *Plant Variety Journal* of India.

Rights conferred with respect to different types of registered/protected varieties

What then about the 'rights' that farmers acquire? As far as new varieties are concerned, the Act, at first blush, appears to confer standard 1978/91

UPOV-style PBRs (Act, section 28), including the exclusive right to produce, sell, market, distribute, import or export the variety. As in the case of the UPOV Convention, these rights are subject to various exemptions (e.g. researchers' rights – Act, section 30) and limits (e.g. compulsory licensing – Act, Chapter VII). However, unlike the UPOV Convention, the Indian system introduces particular equity provisions (e.g. benefit sharing – Act, Chapter IV). However, the Act also grants farmers the rights to save seeds to sow and exchange or sell seeds of a protected variety (Act, section 39).

The duration of the rights following the successful registration of a new variety are as follows:[11]

- for trees and vines, initially for 9 years and thereafter they can be reviewed and renewed for a maximum period of 18 years from the date of registration;
- for other species, initially for 6 years and thereafter they can be reviewed and renewed for a maximum period of 15 years from the date of registration.

Unlike rights that result from registering a new variety, the rights that accrue from registering a farmers' variety are largely negative rights – akin to defensive publication to forestall misappropriation and defeat others' novelty claims. These rights are then linked to other sections of the Act. For instance, the documentation that must necessarily support an application for registration and conferment of PBRs (Act, section 18) includes a declaration of prior informed consent from the providers of registered farmers' varieties (Act, section 18(1)(h)). In this sense, India's legal architecture provides an array of measures that are notable for farmers' rights.

Srividhya Nagarajan et al. (2008, 711) insist that the rights here must necessarily be 'notional' as the variety has already been part of the public domain. Consequently, it would appear that the registration would assist in larger struggles against biopiracy while also enabling claims for benefit sharing from the National Gene Fund (compare Act, section 45) or, as Nagarajan et al. explain, to 'negotiate a deal' if and when the variety is used as parental material in breeding a new variety (ibid.).

This brings us to the third possible set of rights that flow from the Act, not exclusive rights such as those discussed earlier with respect to new and farmers' varieties, but, rather, freedom to use rights related to varieties that may be owned or registered by others. Chapter VI of the Act addresses seed saving and benefit sharing from the National Gene Fund. Section 39(1)(iv) of the Act clearly spells out the following:

> A farmer shall be deemed to be entitled to save, use, sow, resow, exchange, share or sell his farm produce including seed of a variety protected under this Act in the same manner as he was entitled before the coming into force of this Act.

With the proviso that the farmer is not entitled to sell branded seed of a variety protected under this Act – a point, as shortly explained, reinforced by seed market regulations. The provision, it could be argued, is quite similar to the practice of 'brown bagging' as elaborated under the US system of plant variety protection. Under the Plant Variety Protection Act in the United States, farmers have provisions that allow for the sale of harvested grain of a protected variety as seed, with the proviso that the variety's name not be used. Hence, the phrase 'brown bagging' is associated with this practice. In 1995, the US Supreme Court in *Asgrow v. Winterboer* decided that the exemption should be understood to limit the amount of seed for sale to the amount that the farmer would need to replant their own farm.

These provisions for saving/selling seeds are, unsurprisingly, contentious. In addition, the Seeds Bill 2004 substantially watered down these (and other) provisions, while ostensibly seeking to promote the provisions of quality seeds. Widespread opposition within and beyond the Indian Parliament ensured the bill's withdrawal and the establishment of a Parliamentary Standing Committee on Agriculture, chaired by Ram Gopal Yadav, tasked with assessing the bill. Their report in November 2006 led to a revised Seeds Bill, 2010.[12] However, state governments have lobbied for further amendments, and this version of the Seeds Bill still awaits parliamentary approval (Singh and Chand, 2011).

Two of the provisions of the Seeds Bill are relevant to the issues discussed in this chapter: (1) protection for farmers' right to grow, save, resow, exchange, share or sell seeds; and (2) the status of farmers' varieties within the National Register of Seeds. With an estimated 70 percent or more of the required domestic seed provided by farmers themselves, it is crucial that the seed market regulations do not erect regulatory barriers to its circulation. The Seeds Bill, 2004, sought to impose quality standards (e.g. germination rates and so on) for all seeds that are transacted, including, therefore, farmers' varieties. Additionally, the bill sought to introduce a mandatory requirement for all varieties to be registered prior to their being transacted (including bartering). For a variety of commentators, these provisions in the Seeds Bill, 2004, not only conflicted with how farmers' rights have been drawn out in the Act but also posed problems for the reality and significance of seed exchange in India (Bala Ravi, 2010). The recommendations from the Parliamentary Standing Committee, were largely – though not entirely – adopted in the Seeds Bill, 2010. For instance, the seeds of farmer's varieties are exempted from a requirement for registration under the National Register of Seeds, which, therefore, removes a possible barrier to this system of seed exchange. Further, the constraints on farmers' rights to exchange, share or sell seeds (including harvested seeds of a registered variety) have been brought into line with the provisions of the Act.

National Gene Fund

Finally, provisions associated with the National Gene Fund contribute to farmers' rights. Reflecting ideas found in the ITPGRFA, India's legislative architecture has made the National Gene Fund a reality (Act, section 45). Different sources of revenues, such as an annual fee and a royalty paid by the breeder, among others, are to constitute the fund (Act, section 35(1)). The fund will then support the benefit-sharing arrangements that the Act prescribes and will also be the financial resource from which farmers and farming communities will be supported in their conservational activities concerning plant genetic resources. A national debate and consultation process was conducted to establish an agreement on the structure of fees and royalty rates for the National Gene Fund, with an agreement being announced in the Gazette in August 2009 (see Nagarajan et al., 2010):

- New varieties: An annual fee Rs 2,000, plus 0.2 percent of the sales value of the seeds during the previous year plus 1 percent of royalty, if any, received during the previous year from the sale proceeds of seeds.
- Extant varieties: For those notified under section 5 of the Seeds Act, 1966, the annual fee shall be Rs 2,000; for other extant varieties, the annual fee shall be Rs 2,000 plus 0.1 percent of the sales value of the seeds during the previous year plus 0.5 percent of the royalty, if any, received during the previous year from the sale proceeds of the seeds.

Naturally, much attention will be placed on how this agreement proceeds to finance the National Gene Fund and how disbursements are made thereafter.

Application status

Applications to register varieties began being received in May 2007, both for new varieties and extant varieties (the latter includes farmers' varieties). In addition, criteria for registering 'varieties in common knowledge' were finalized and published in June 2009, after which point applications were received. The initial applications for registering plant varieties tended to be for extant varieties, and, more recently, there have been applications for new varieties and farmer varieties (Table 13.3). By May 2013, a total of 4,094 applications for the registration of plant varieties – across all categories – were received, of which 37.8 percent were for extant varieties, 33.2 percent were for farmers' varieties and 28.9 percent were for new varieties (compare with Table 13.3). As in other jurisdictions, the examination of applications takes time, and their full assessment involves a number of field trials as well. In addition, the Indian legal system also requires examination, depending on the type of variety and the benefit-sharing and prior informed consent declarations. It is not surprising that applications to register farmers' varieties are overwhelmingly in rice. In 2009–10, three farmers' varieties in rice were successfully registered, establishing the first-ever registration of farmers' varieties. The first new varieties successfully registered in India were in 2009–10, with two bread wheat variety certificates being granted to the Maharashtra Hybrid Seed Company. In addition, there have been numerous certificates of registration granted for extant varieties, totalling over 600 by 2012–13 (compare with Table 13.4).

Table 13.3 Annual applications to register plant varieties (2007–13)

Variety / Year	*2007–8*	*2008–9*	*2009–10*	*2010–11*	*2011–12*	*2012–13*	*Total*
Extant variety	355	260	297	216	177	243	1,548
Farmer variety	2	3	44	30	921	359	1,359
New variety	69	171	227	395	149	176	1,187
Total*	426	460	568	642	1,247	785	4,094

Source: Author's calculations from the annual reports of the Protection of Plant Variety and Farmers' Rights Authority, various issues.

* Note that totals may not add up as all categories of varieties (e.g. essentially derived varieties and varieties in common knowledge) are not included here.

Table 13.4 Annual registrations of plant varieties (2008–13)

	2008–9	*2009–10*	*2010–11*	*2011–12*	*2012–13*
New varieties		2		20	34
Extant varieties	40	123	131	99	216
Farmers' varieties		3			3
Essentially derived varieties					1

Source: Author's calculations from the annual reports of the Protection of Plant Variety and Farmers' Rights Authority, various issues.

Conclusion

The Protection of Plant Varieties and Farmers' Rights Act and associated regulatory interventions provide a complex legal architecture for farmers' rights and the registration of farmers' varieties. While many features of this legal architecture have been celebrated, some of them are possibly illusionary. In particular, provisions to treat farmers as applicants for new varieties may be illusionary as there is nothing in either the Act or the Rules that would ensure differential treatment of the plant material that reflects the different crop improvement, breeding or conservation objectives and practices of farmers. At present, there is very little accommodation for the fact that the materials farmers have developed through informal innovation systems may not satisfy the DUS requirements that are found in most countries' plant variety protection laws and the UPOV Convention from which they draw their inspiration. For that matter, neither will they be differentially treated in terms of payment of fees. Alternatively, the Act does not require commercial novelty as a registration requirement for farmer varieties as they are considered a subcategory of extant varieties.

Although farmers' varieties are held to such high standards for registration, they are not protected in the same way as new varieties under the Act. Indeed, the only positive right associated with farmers' varieties is that parties must show proof of having gained prior informed consent from the owners of the registered farmers' varieties if those varieties are incorporated in new varieties over which those parties are seeking intellectual property rights protection. It appears there is no other instance or checkpoint where proof of the farmers' variety registrant's prior informed consent is necessary.

Despite these limitations, there have been a number of applications to register farmers' varieties under the Act, totalling 1,359 in 2012–13, with three successful registrations in the same period. However, in contrast, as far as this author can verify, there have been no applications submitted by farmers to register a new variety – akin to the status of a breeder. In closing, it is also necessary to flag concerns about the manner in which other regulatory instruments, such as the Seeds Bill, may diminish some of these achievements in farmers' rights, such as the right to sell saved seeds. In addition, concerns remain as to the manner in which securing rights in farmers' varieties impacts farming communities and their cultural and social practices concerning seeds. In this respect, it is important that the laws do not end up disrupting long stabilized cultural practices.

Notes

1 Protection of Plant Varieties and Farmers' Rights Act, online: <http://agricoop.nic.in/seeds/farmersact2001.htm> (last accessed 10 May 2012); Protection of Plant Varieties and Farmers' Rights Rules, online: <www.plantauthority.gov.in/pdf/PPVFRA_RULES_2003.pdf> (last accessed 4 March 2015).
2 African Model Law for the Protection of the Rights of Local Communities, Farmers and Breeders, and for the Regulation of Access to Biological Resources, online: <https://www.cbd.int/doc/measures/abs/msr-abs-oau-en.pdf> (last accessed 4 March 2014).
3 International Treaty on Plant Genetic Resources for Food and Agriculture, 29 June 2004, online: <www.planttreaty.org/content/texts-treaty-official-versions> (last accessed 4 March 2015).
4 Additionally, section 41 of the Act makes available 'rights of communities,' which allows for the filing of claims for benefit sharing and, thereby, seeking benefit sharing from the National Gene Fund.
5 Other provisions in the Act can be seen to buffer farmers' rights. Thus, for example, section 42 allows for 'protection of innocent infringement,' wherein a farmer may avoid infringement on establishing that 'at the time of such infringement [the farmer] was not aware of the existence of such right' (Act, section 42(i)).
6 International Convention for the Protection of New Varieties of Plants, 2 December 1961, revised 1978, online: <www.upov.int/en/publications/conventions/index.html> (last accessed 10 May 2012). In 1997, the UPOV Convention decided to allow accession to the 1978 Act, despite it being closed to those countries that had sought its advice on conformity prior to the entry into force of the 1991 Act. This special provision was open until 24 April 1999. However, at its thirty-third ordinary session in October 1999, it decided to make further special provisions for allowing accession to the 1978 Act for India, Nicaragua and Zimbabwe. Alongside these extraordinary efforts, the negotiations have occurred under a cloud of secrecy and warranted a public interest litigation in October 2002 by a nongovernmental organization, Gene Campaign. Responding, the government of India denied it was 'pursuing' the offer to accede to the 1978 UPOV Convention.

7 Protection of Plant Varieties and Farmers Rights (Criteria for Distinctiveness, Uniformity, and Stability for Registration) Regulation, 2009, GSR 452(E), 29 June 2009, online: <www.plantauthority.gov.in/pdf/gnotifi376.pdf> (last accessed 4 March 2015).

8 The 1978 and 1991 Acts of the UPOV Convention refer to 'varieties in common knowledge.' For instance, distinctness requires the variety to be 'distinguishable by one or more important characteristic from any other variety whose existence is a matter of common knowledge.'

9 As noted earlier, once a species is notified under the Act, a three-year moratorium is provided for the registration of extant varieties. For farmers' varieties, an October 2009 notification extended this period to 5 years from the date of a species being notified.

10 The 19 species are bread wheat, rice, maize, sorghum, pearl millet, pigeon peas, chick-peas, lentils, black grams, green grams, field peas, kidney beans, diploid cotton (two species), tetraploid cotton (two species), jute (two species) and sugarcane.

11 Bearing in mind that there are fees to be paid for renewing the registration (see Table 13.1) and that the rights holder is obliged to maintain the variety true to type throughout the duration of the registration.

12 Bearing in the mind that an earlier version, the Seeds Bill, 2008, lapsed in 2009, thus making this the third iteration (see Bala Ravi, 2010, for a discussion).

References

Akerman, A., and O. Tedin (1955). 'Testing of New Varieties from the Point of View of the Plant Breeder,' in *Development of Seed Production and the Seed Trade in Europe: Proceedings of an International Conference held at Stockholm*, July 1954, Project no. 214. Organisation for European Economic Cooperation and European Productivity Agency, Paris, France.

Bala Ravi, S. (2010). 'Fault Lines in the 2010 Seeds Bill,' *Economic and Political Weekly*, 45(32): 12–16.

Crucible Group II (2001). *Seeding Solutions*, Volume 2: *Options for National Laws Governing Control over Genetic Resources and Biological Innovations*, International Development Research Centre, Ottawa, International Plant Genetic Resources Institute, Rome and Dag Hammarskjold Foundation, Uppsala.

de Schutter, O. (2009). *Seed Policies and the Right to Food: Enhancing Agrobiodiversity and Encouraging Innovation*, Report presented to the UN General Assembly, 64th Session, UN Doc. A/64/170.

International Plant Genetic Resources Institute (1999). *Key Questions for Decision Makers: Protection of Plant Varieties under the WTO Agreement on Trade-Related Aspects of Intellectual Property Rights – Decision Tools*, International Plant Genetic Resources Institute, Rome, Italy.

Kochupillai, M. (2011). 'India's Plant Variety Protection Law: Historical and Implementation Perspectives,' *Journal of Intellectual Property Rights* 16(2): 88–101.

Leskien, D., and M. Flitner (1997). *Intellectual Property Rights and Plant Genetic Resources: Options for a Sui Generis System*, Issues in Genetic Resources no. 6, International Plant Genetic Resources Institute, Rome.

Nagarajan, S., R.K. Trivedi, D.S. Raj Ganesh and A.K. Singh (2010). 'India Registers Plant Varieties under PPV&FR Act, 2001,' *Current Science* 99(6): 723–5.

Nagarajan, S., S.O. Yadav and A.K. Singh (2008). 'Farmers' Variety in the Context of Protection of Plant Variety and Farmers' Right Act, 2001,' *Current Science* 94(6): 709–13.

Rangnekar, D. (1998). '"Trip'ping in Front of UPOV: Plant Variety Protection in India,' *Social Action* 48(4): 432–51.

Rangnekar, D. (2002). *Access to Genetic Resources, Gene-based Inventions and Agriculture*, Commission on Intellectual Property Rights of the UK Government, London.

Rangnekar, D. (2014). 'Geneva Rhetoric, National Reality: The Political Economy of Introducing Plant Breeders' Rights in Kenya,' *New Political Economy* 19(3): 259–83.

Singh, H., and R. Chand (2011). 'The Seeds Bill, 2011: Some Reflections,' *Economic and Political Weekly* 46(51): 22–5.

Swaminathan, M.S. (1998). 'Farmers' Rights and Plant Genetic Resources,' *Biotechnology and Development Monitor* 36: 6–9.

14 Commentary on the Malaysian Protection of New Plant Varieties Act 2004

Lim Eng Siang

Novel conditions of protection for new plant varieties bred by farmers, local communities or indigenous peoples

The Malaysian Protection of New Plant Varieties Act 2004 includes two alternative sets of conditions for granting plant breeders' rights (PBR).[1] First, the act sets out a more familiar set of conditions, found in many countries' PBR laws, that mandate that, to be eligible for protection, varieties must be new, distinct, uniform and stable (Article 14(1)). A second, alternative set of conditions applies if the application for protection is for a variety that has been 'bred, discovered and developed by a farmer, a local community or indigenous people.' These varieties must be 'new,' 'distinct' and 'identifiable' to qualify for protection (Article 14(2)). The act did not come into force until 2007, and the regulations pursuant to the act came into force in 2008.

The definitions for new, distinct, uniform and stable are very similar to those definitions contained in other countries' PBR laws and have been discussed elsewhere in this book, so for this reason they are not reproduced in this chapter. Identifiability, on the other hand, is a relatively novel concept, and, thus, it will be considered in more detail. The act states that a plant variety is identifiable when:

1 it can be distinguished from any other plant grouping by the expression of one characteristic and that characteristic is identifiable within individual plants or within and across a group of plants;
2 such characteristics can be identified by any person skilled in the relevant article (Article 14(e)).

These conditions were derived from the US Plant Patent Act in 1930.[2] Under this system of protection, whoever invents or discovers and asexually reproduces any distinct and new variety of plant, including cultivated spores, mutants, hybrids and newly found seedlings, other than a tuber-propagated plant or a plant found in an uncultivated state, may obtain a patent.[3] According to the US Patent and Trademark Office, asexual reproduction is used to establish the stability of the plant. Acceptable modes of asexual reproduction could include rooting, cutting, apomictic seeds, division, layering, runners, tissue culture, grafting and budding, bulbs, slips, rhizomes, corms and nucellar embryos. Confined as it is to asexually producing plants, it was not necessary to explicitly include the stability requirement in the American plant patent law – stability is biologically built-in to these particular varieties. In addition, the invention would have to be nonobvious – that is, not obvious to one who is skilled in the art at the time of invention by the applicant (US Patent and Trademark Office 2010).

The Malaysian Protection of New Plant Varieties Act 2004 is not explicitly limited in application to asexually reproducing plants, which reflects the fact that the drafters and law makers were interested in extending the conditions of protection inspired by the US Plant Patent Act to sexually propagated cereals and vegetables. However, it is noteworthy in this context that the act specifies that varieties satisfying the new, distinct and identifiable criteria '*may* be registered as a new plant variety, and granted a breeder's right' (Article 14(2); emphasis added). The act does not preserve the same discretion on the part of the Plant Varieties Board with respect to varieties that are new, distinct, uniform and stable, and it does state that such varieties '*shall* be registered as a new plant variety and granted a breeder's right' (Article 14(1); emphasis added). The drafters of the act fully appreciated that there was no precedent for the operation of these newly combined conditions of new, distinct and identifiable for sexually reproducing plants. They included the word 'may,' therefore, in order to provide flexibility to respond to unknowns that could arise when it came to the nuts and bolts of implementation. In the years since the act was passed into law, the Plant Variety Board has considered limiting applications from farmers, local communities or indigenous peoples for asexually reproducing plants.

These alternative criteria – new, distinct and identifiable – respond to the situation on the ground in Malaysia, at least as far as tropical fruit trees are concerned. There are many tropical plant varieties in Malaysia that have been discovered in farmers' fields and that have been propagated by grafting, including, in particular, tropical fruit trees such as durian, mango, guava and rambutan. These fruit trees have long gestation periods, and it would have placed a heavy burden on the farmer (in terms of time and delayed benefits) to ask him to establish their intergenerational stability. Under these circumstances, it is more practical for examiners to determine the novelty of the claimed plant and to ascertain whether its characteristics are distinct and nonobvious to a person skilled in the relevant art. The advantage of this system – at least in theory – is that it cuts down on testing time (for stability and uniformity) and allows farmers the ability to gain protection and to go to the market with their new varieties almost immediately following the discovery in their fields.

Under the Malaysian law, the rights conferred on the registered owners of the new varieties are identical to those contained in the PBR laws of other countries with one exception.[4] Taking effect from the filing date of the application for registration, the breeders' rights shall subsist for a period of: '(a) twenty years for a registered plant variety that is new, distinct, uniform, and stable; or (b) fifteen years for a registered plant variety that is new, distinct, and identifiable' (Article 32(1)). The shorter period of protection for these latter varieties is justified due to the fact that there is no need for intergenerational testing of stability and therefore there is likely to be less investment and less delay in granting farmers their rights.

The act defines a local community as 'as a group of individuals having settled together who are continuously inheriting production processes and culture, or a group settled in a village or area in a eco-cultural system.' A farmer is a person who:

- cultivates crops by cultivating the land himself;
- cultivates crops by directly supervising the cultivation of land through any other person; or
- conserves and preserves, severally or jointly, with any person, any traditional variety of crops or adds value to the traditional variety through the selection and identification of their useful properties.

The act does not include a definition of indigenous peoples other than to note that they are aborigines and/or natives as defined in the federal constitution. A breeder is defined as 'a person who has bred or has discovered and developed any plant variety.' To 'discover and develop,' in turn, is defined as 'activities which lead to the desired phenotypic expression and affect the crop genotype and which may or may not entail deliberate or artificial creation of genetic variability' (Article 2).

Application for plant breeders' rights: proof of prior informed consent to use 'traditional varieties'

In addition to adapting intellectual property rights protection for varieties bred by a farmer, local community or indigenous peoples, the Malaysian law also creates a form of access and benefit-sharing protection for local communities and indigenous peoples vis-à-vis plant varieties. Applications for breeders' rights under the act must be 'accompanied with the prior written consent of the authority representing the local community or the indigenous peoples in cases where the plant variety is develop from traditional varieties' (Article 12(1)(f)). Here again, the law engages the issue of how to define varieties developed by farming communities and indigenous peoples. Traditional varieties are not defined in the act per se. However, the intention behind the use of the term in Article 12(1)(f) is meant to refer to a definite subgroup of material and not to extend to all materials grown by farmers or on farmers' lands (which is another possible way of approaching the issue with an access and benefit–style protection of this nature).

Ultimately, the Plant Varieties Board will have to provide guidance on what is meant by a traditional variety, as it was intended in the act's overall scheme, including the conditions that presumably need to be satisfied before a variety can be recognized as such. In addition to providing such a definition, the Plant Varieties Board will also need to provide guidance on how to identify those individuals who have the authority to represent local communities or indigenous peoples for the purposes of seeking their prior informed consent. The Plant Varieties Board should also provide a standard format for a legally binding document expressing prior informed consent. Ultimately, it may be helpful if the board also considers creating a national list of all such traditional varieties in order to reduce the transactions between developers using those varieties and the communities and peoples whose rights need to be protected. Such a list would be created pursuant to the prior informed consent requirement of Article 12(3)(f).

If the Plant Varieties Board did create such a list, the definition of traditional varieties could include elements that have been proposed under the revised draft provisions for the protection of traditional knowledge published by the World Intellectual Property Organization (WIPO) in 2006 (WIPO 2006). These elements define traditional varieties as those varieties that have been

- developed, conserved and used in a traditional and intergenerational context;
- distinctively associated with a local or indigenous community that conserves and uses the varieties between generations;
- integral to the cultural identity of an indigenous or local community that is recognized as holding the varieties through a form of custodianship, guardianship, stewardship, collective ownership or cultural responsibility;
- identified as having distinctive functional traits such as taste, aroma, cooking quality, colour and medicinal values that are associated with the culture of the local communities.

The act's definition of local community and indigenous peoples, coupled with WIPO's criteria for traditional varieties, could provide the legal certainty that is necessary to make Article 12(3)(f) operational. An example of how this could be applied can be found in the Bario rice varieties of the Kelabit highlands of Sarawak. A formal breeder using the Bario rice varieties of the Kelabit community for the development of a new plant variety would require the prior informed consent of the authority representing the

Kelabit community if the breeder wanted to apply for the new plant variety registration.

Policy rationale

The policy rationale for the Protection of New Plant Varieties Act 2004 is to provide not only for the protection of breeders' rights for new plant varieties but also for the recognition and protection of the contributions made by farmers, local communities and indigenous peoples in creating new plant varieties as well as in encouraging investment in breeding new plant varieties in both the public and private sectors. Given that the Department of Agriculture in Malaysia has been registering fruit clones since the early 1930s and that most of these clones have been discovered and selected in farmers' fields, the act will continue to support such efforts by farmers and to provide them with an easy legal and administrative mechanism under which to register their new plant varieties and acquire breeders' rights. The act also takes note of the fact that there are still about 600,000 smallhold farmers engaged in agriculture in Malaysia. It is therefore necessary to provide an opportunity under the act for these smallhold farmers to also be able to apply for breeders' rights. Thus, the act must not limit propagation by smallhold farmers who have used harvested material of the registered plant variety planted on their own holding, exchanged reasonable amounts of propagating materials among smallhold farmers and sold farm-saved seeds in situations where nonusage is beyond the control of the farmer.

State of implementation

The Protection of New Plant Varieties Act 2004 came into force in January 2007.[5] The Protection of New Plant Varieties Regulations 2008 entered into force in October 2008. Some crop-specific guidelines for testing for distinctness, uniformity and stability have been developed. So far, no such guidelines have been developed for identifiability under the act. The application form that has been developed for registering new plant varieties includes the requirement of a letter of consent from the authority representing the local community or indigenous peoples that is making the application in cases where the plant variety is developed from traditional varieties (Malaysian Department of Agriculture 2010).

As of April 2010, 43 applications for the registration of new plant varieties and request for breeders' rights have been accepted by the Crop Quality Control Division in the Plant Variety Protection Registration Office at the Department of Agriculture in Malaysia. The office is currently conducting substantive examinations with respect to these applications. To date, there have yet to be any applications for protection from farmers, local communities, and indigenous peoples.

Conclusion

The Malaysian Protection of New Plant Varieties Act 2004 represents an attempt – in the context of a plant variety protection law – to provide support for farmers, local communities and indigenous peoples as the conservers and innovators of plant genetic resources. The administrative guidelines to implement the novel aspects of this law are slowly being worked out. Ongoing experiences in the early days of implementation will provide lessons and insights for further improvement and refinement.

Notes

1 Protection of New Plant Varieties Act 2004, Act 634, online: <www.grain.org/brl/?docid=657&lawid=1404> (last accessed 15 June 2012).
2 Plant Patent Act, 1930, 35 U.S.C. §§ 161–164.
3 To be patentable, the following are also required:

- that the plant was invented or discovered and, if discovered, that the discovery was made in a cultivated area;
- that the plant is not a plant which is excluded by statute, where the part of the plant used for asexual reproduction is not a tuber food part, as with potato or Jerusalem artichoke;
- that the person or persons filing the application are those who actually invented the claimed plant – that is, discovered or developed and identified or isolated the plant and asexually reproduced the plant;
- that the plant has not been sold or released in the United States of America more than one year prior to the date of the application;
- that the plant has not been enabled to the public – that is, by description in a printed publication in this country more than one year before the application for patent with an offer to sale or by release or sale of the plant more than one year prior to application for patent;
- that the plant be shown to differ from known, related plants by at least one distinguishing characteristic, which is more than a difference caused by growing conditions or fertility levels, and so on;
- the invention would not have been obvious to one skilled in the art at the time of invention by the applicant (US Patent and Trademark Office 2010).

4 The rights conferred are very similar to those in the plant variety protection laws of other countries, so are not described in this chapter. It is interesting to note, however, that the law creates exemptions for 'small farmers' to plant back – in other words, exchange 'reasonable amounts' of propagating materials of protected varieties (Article 31(d) and (e)). It also allows small farmers to sell farm-saved seeds in situations where they cannot make use of those seeds as a result of a natural disaster or other circumstances beyond their control, as long as the amount sold is not more than what the farmers would need for his own holdings (Article 31(1)(f)). The act defines a small farmer as one 'whose farming operations do not exceed the size as prescribed by the Minister of Agriculture of Malaysia' (Article 2).
5 Protection of New Plant Varieties Act 2004, supra note 1.

References

Malaysian Department of Agriculture (2010). Malaysian Plant Variety Protection. Available at: http://pvpbkkt.doa.gov.my/ (last accessed 15 June 2012).

US Patent and Trademark Office (2010). General Information about the *Plant Patent Act*. Available at: www.uspto.gov/web/offices/pac/plant/index.html (last accessed 15 June 2012).

World Intellectual Property Organization (2006). *Intellectual Property and Genetic Resources: Traditional Knowledge and Traditional Cultural Expressions/Folklore,* Information Resources, Doc.WIPO/GRTKF/INF/1.

15 Commentary on Egypt's plant variety protection regime

Eid M. A. Megeed

Introduction

Plant variety protection in the Arab Republic of Egypt was created by Law 82/2002 on the Protection of Intellectual Property Rights, which was adopted in 2002, specifically Book 4 of this law entitled 'Plant Varieties.'[1] Some aspects of the law closely echo the standards set out in the International Convention for the Protection of New Varieties of Plants (UPOV Convention).[2] For example, to be protectable, a variety must be novel, distinct, uniform and stable.[3] It also requires that newly protected materials represent advanced values for cultivation and use.[4]

However, some aspects of the law introduced requirements and processes that are not present in the UPOV conventions. Until 2015, the law included Article 200, which created an obligation to disclose 'the genetic resource relied on to develop the new plant variety.'[5] It also required that 'the breeder has acquired that resource by legitimate means under the Egyptian law,' noting that this requirement 'extends to traditional knowledge and experience accumulated among local communities the breeder could have relied on in his efforts to develop the new plant variety.'[6] The law also stated that 'a register shall be established in the Ministry of Agriculture and Land Reclamation to include the Egyptian plant genetic resources, both wild and domesticated.'[7] The intention of this register was to facilitate the process of breeders both identifying potentially useful materials to use in their breeding efforts and to identify from whom they needed consent to use the materials. The idea was that, eventually, all wild or domesticated Egyptian plant genetic resources would be registered in order to fully encompass all of the traditional knowledge linked to genetic resources in Egypt.

In 2015, Presidential Decree 82/2002 dismissed article 200 from Book 4 of Law 82/2002. It also modified Article 192 of Law 82/2002 to include the requirement of a breeder's consent prior to the production, sale, export or import of varieties that are essentially derived from protected breeders' varieties. These changes were made to satisfy the requirement of the UPOV Council to accept Egypt's full membership in UPOV. The Presidential Decree also reaffirmed the right of farmers to save and plant on their own holdings seed that they harvested from protected varieties, without the permission of the holder of the breeders' right.

Background

Negotiations for an association agreement with the European Union (EU) were set in motion in 1995 and concluded in June 2001. The agreement entered into force on 1 June 2004, after ratification by the Egyptian People's Assembly and all EU member states. The association agreement states that Egypt should be a member of the UPOV Convention as part of the overall agreement. The association agreement, together with the obligation under Article 27(3)(b) of the World Trade Organization's Agreement on Trade-Related Aspects of Intellectual Property Rights (TRIPS Agreement), prompted Egypt to develop an intellectual property rights protection law including a chapter (Chapter 4) related to new plant variety protection that is compliant with the 1991 UPOV Convention.[8]

The government of Egypt submitted its law to the UPOV Council in 1999 for consideration vis-à-vis its conformity with the UPOV Convention.[9] The UPOV Council indicated that the sections in the Egyptian law requiring declarations of origin of materials used by breeders and agreements with authorities pursuant to national law are not consistent with the UPOV Convention. In addition, they indicated that it was necessary to include treatment of essentially derived varieties (which the law did not recognize). International trade requirements were not the only influences on the development of the Egyptian law. Inspired partly by the Convention on Biological Diversity and the Indian Protection of Plant Varieties and Farmers' Rights Act, the Ministry of Higher Education and Scientific Research wanted to include additional text linking plant breeders' rights to farmers' rights.[10]

Status of implementation

Despite some efforts by the government to establish an official registration list of farmers' varieties as part of the country's plant genetic resources as anticipated in the law, it was not successful. In 2007, a prime ministerial decree suspended Article 200 of Law 82/2002 and the obligation for applicants to provide details about the sources of genetic material used to develop a new plant variety. The same decree suspended creation of the register of Egyptian plant genetic resources, including farmers' varieties. For a while, it was not clear if or when this suspension will be lifted. However, it was confirmed with Presidential Decree 26/2015 referred to above. This decree, along with some other changes to the law, has made it possible for Egypt to be accepted as a member of the UPOV Convention.

Meanwhile, there is still interest among a range of stakeholders in the country to go ahead with exploring a range of mechanisms to advance farmers' rights. The idea came from a study managed by the Genetic Resources Policy Initiative (GRPI)-Egypt project to list all eligible rights related to farmers that include not only registration of their innovative varieties but also all other tabulated rights (land ownership, access to extension services, access to certified seeds, benefit sharing of using their indigenous knowledge, access to fertilizers, free irrigation water and so on). In light of recent developments, it seems that these efforts will need to take place within the existing legal structures – that is to say, without any new laws in support of such initiatives. One possible element would be to further develop a register of farmers' varieties by a farmers' independent association as a means of taking stock of what materials exist and where they are located as well as to 'defensively publish' a list of them. Conditions for registration on a joint farmers/nongovernmental organization-operated database of this nature would likely have to be different from the standard criteria of distinctness, uniformity and stability, perhaps focusing mostly on distinctiveness, with considerably looser conditions of uniformity and stability. The Agricultural Research Centre coordinated a study in 2007–8, surveying hot spots of farmers' varieties throughout the country, identifying 31 varieties of field crops (including cereals, forages and legumes) as well as vegetables, fruit and medicinal plants.

Notes

1 Law 82/2002 on the Protection of Intellectual Property Rights, published in the Egyptian National Gazette, July 2002, online: <www.wipo.int/wipolex/en/details.jsp?id=7309> (last accessed 10 May 2012).
2 International Convention for the Protection of New Varieties of Plants, adopted on 2 December 1961, online: <www.upov.int/en/publications/conventions/index.html> (last accessed 10 May 2012).
3 Law 82/2002, supra note 1, Article 192.
4 Ibid.
5 Ibid., Article 200.
6 Ibid.
7 Ibid.
8 Agreement on Trade-Related Aspects of Intellectual Property Rights, Annex 1C of the Marrakech Agreement Establishing the World Trade Organization, 15 April 1994, 33 ILM 15 (1994).
9 UPOV Council, 33rd Ordinary Session, Geneva, Doc. C/33/18 (20 October 1999), para 17, online: <www.upov.int/export/sites/upov/en/documents/c/33/c_33_18.pdf> (last accessed 10 May 2012).
10 Convention on Biological Diversity, 31 ILM 818 (1992). Protection of Plant Varieties and Farmers' Rights Act, online: <http://agricoop.nic.in/seeds/farmersact2001.htm> (last accessed 10 May 2012).

16 Commentary on Thailand's Plant Varieties Protection Act

Gabrielle Gagné and Chutima Ratanasatien

The Thai Plant Varieties Protection Act, 1999, (PVPA)[1] is a *sui generis* system that contains three types of protections for plant varieties: (1) intellectual property protections for new plant varieties that are novel, distinct, uniform and stable; (2) intellectual property protections for local domestic varieties which are distinct, uniform and stable (DUS), but not necessarily novel; and (3) access and benefit sharing–style protections for general domestic plant varieties and wild plant varieties. Interestingly, while wild plant varieties do not have to be uniform, the Act stipulates that they must be stable and distinct.[2]

Protection of new plant varieties

The conditions for protection of new plant varieties is very similar in some respect to those included in the UPOV Convention (International Convention for the Protection of New Varieties of Plants) system. In addition to having to be DUS,[3] new plant varieties must not have been distributed in or outside the Kingdom by the breeder or with the breeder's consent for more than one year prior to the date of application. This condition is, of course, roughly equivalent to the concept of commercial novelty included in the UPOV Conventions and many countries' plant variety protection laws.[4] The rights conferred with respect to new plant varieties are also roughly equivalent to those provided for under UPOV 1991,[5] although the limits to the scope of protection differ in several manners,[6] and the protection periods are shorter[7] than those established by UPOV 1991.[8]

Going beyond the UPOV Conventions, the Thai law requires applications for new plant variety protection to include details about the origin of the genetic material used for breeding,[9] as well as a proof of a profit-sharing agreement when general domestic or wild plant varieties have been used for breeding of the variety.[10] Accepted varieties are included in a national register of protected varieties.

Protection of local domestic plant varieties

'Local domestic plant varieties' under the PVPA must also be DUS.[11] They do not, however, have to satisfy the novelty requirement. Instead, the PVPA requires that the plant variety 'exists only in a particular locality within the Kingdom and which has never been registered as a new plant variety.'[12] Any person with full legal capacity may apply. The law also provides for a form of collective ownership of local domestic plant varieties stating that,

> A *sui juris* person, residing and commonly inheriting and passing over culture continually, who takes part in the conservation or development of the plant variety which is of the descriptions specified in Section 43 may register as a community under this Act. For this purpose, there shall be appointed a representative who shall submit an application in writing to the *Changwad* Governor of the locality.[13]

The community concerned must have conserved or developed exclusively the plant variety.[14]

The application must include the method of conservation or development of the variety, the members of the community concerned, and details about 'the landscape with a concise map showing the boundary of the community and adjacent areas.'[15]

Once registered, the community has the exclusive rights to develop, study, experiment on, research, produce, sell, export or distribute the propagating material of the plant.[16] The scope of these rights is circumscribed in much the same way as the new plant varieties' protection,[17] with the exception of acts related to education and experimentation.[18] The PVPA states that a profit-sharing agreement must be concluded with a legal representative of the community whose local domestic plant variety is being collected, procured or gathered for purposes of development, education, experimentation or research for commercial outcomes.[19] The PVPA further specifies that this agreement has to be approved by the Plant Variety Protection Commission. Most importantly, the profits go directly to the community providing the resource.[20] The profits generated are separated between the individuals having conserved or developed the plant variety (20 percent), the community having registered the variety (60 percent) and the local organization making the agreement in the name of the community (20 percent).[21] The Act anticipates creating regulations that will further determine the 'profit-sharing among the persons who conserve or develop the plant variety' (section 49 art. 2).

Another particularity of local domestic plant variety protection is the possibility of renewing the term of the protection for an additional 10 years[22] when the variety can still be considered as a local domestic variety and the community still respects the criteria of Sections 44 and 45.[23]

Protection of general domestic plant varieties and wild plant varieties

The PVPA defines wild plant varieties as 'a plant variety which currently exists or used to exist in the natural habitat and has not been cultivated.'[24] As noted earlier, wild plant varieties do not have to be uniform, but they must be distinct and stable.

General domestic plant varieties, on the other hand, do have to meet the DUS criteria. In addition, a general domestic plant variety must be 'a plant variety originating or existing in the country and commonly exploited and shall include a plant variety which is not a new plant variety, a local domestic plant variety or a wild plant variety.'[25] The main differences between local domestic and general domestic varieties are that the former exist only in one well-defined area of Thailand and are conserved by a community. A general domestic variety is commonly exploited throughout the country; it cannot be associated with any particular location or community.

There is no provision for registering general domestic and wild plant varieties under the PVPA. However, would-be users of these materials for 'variety development, education, experiment or research for commercial interest' must first obtain permission from 'the competent official and make a profit-sharing agreement under which income accruing therefrom shall be remitted to the Plant Varieties Protection Fund.'[26] A number of details which must be included in the profit-sharing agreement are listed in the PVPA, with the indication that further rules, procedures and conditions will be provided in a subsequent regulation.[27] The PVPA does not mention benefit sharing with the maintainers or developers of general domestic plant varieties, nor does it establish the percentages of benefit distribution (as it does with local domestic plant varieties) because the varieties are, as the name suggests, generally available, distributed and used across the country.[28] The Plant Varieties Protection Fund's administration is handled by a Fund Committee.[29] This Fund will support communities in conservation, research and development of plant varieties.[30] A Ministerial Regulation adopted by the Council of State of Thailand in January 2011 clarifies that benefit sharing may take many forms, including technology transfer and capacity building (and need not necessarily include monetary benefit sharing). It is understood that the benefits to be included in agreements will be subject to negotiation on a case-by-case basis. The absence of this regulation was a disincentive for some would-be applicants for protection under the Act, and prohibited progress in the processing of their applications (if they made them). Approval for protection required 'a profit-sharing agreement in the case *where a general domestic plant variety or a wild plant variety or any part thereof has been used in the breeding of the variety for a commercial purpose*' (emphasis added).[31] So in these cases, in the absence of the 2011 Regulation, applicants could not complete their application, even though the competent authority would be willing to carry out the DUS standards' test. The varieties that were not affected by this problem were those which were actually finished being developed before the Act entered into force or the ones developed from foreign countries, and therefore before the requirement of providing a benefit-sharing agreement did not apply to them.

As far as noncommercial uses of general domestic plant varieties or wild varieties are concerned, no benefit-sharing obligations are triggered. The Act states that the conditions of use will be set out in a subsequent regulation,[32] which was passed in February 2004: the Plant Variety Protection Commission Regulation in Study, Experiment, or Research on General Domestic and Wild Plant Varieties.[33] The regulation requires that the Department of Agriculture's director general be notified of such uses as well as the details of the project proposal submitted. The results of the study, experiment or research may also be submitted to the Department of Agriculture (DoA) for distribution.

It is widely understood that the protection of general domestic plant varieties and wild plant varieties does not cover the farmers' use of germplasm. The benefit-sharing obligation hence does not apply to farmers who directly exploit plant varieties for example; farmers can collect the varieties for selling or exploitation without owing any profit-sharing.

The treatment of both general domestic plant varieties and wild varieties under the Act is more closely aligned to access and benefit–sharing laws than intellectual property rights. It does not establish a fixed set of rights that vest in an 'owner' that are enforceable against all potential users in the country. Instead, it creates a requirement to get the consent from or notify the competent authorities to use these two kinds of varieties.

Incentives behind the adoption of a *sui generis* system

As a World Trade Organization (WTO) member, Thailand had an international obligation, created by the Agreement on Trade-Related Aspects of Intellectual Property Rights,[34] to develop a plant variety protection by the year 2000. Given uncertainties about which department had the competence to develop the law, both the Thai DoA and Department of Intellectual Property (DIP) drafted versions of a Plant Variety Protection Act between 1995 to 1998. The drafts, both similar in content strictly contained new plant variety protection, although the DIP version was modelled very closely on the UPOV 1978 Convention.

The Thai Council of Ministers considered that the DoA, which has the mandate to conduct research on plant varieties, should lead the drafting process of the plant variety protection law.[35] The cabinet of Thailand demonstrated its preference towards the DoA's draft, electing to adopt it.[36] The DoA conducted a series of public hearings regarding the draft, upon which occasions it became clear there was opposition. Nongovernmental organizations (NGOs), national companies (small and large) and the general public were not in favour of becoming members of UPOV.[37] Moreover, some of these protesting groups asserted the need to protect extant varieties, an approach that appeared to have been inspired by access and benefit–sharing concepts set out in the Convention on Biological Diversity (CBD), and the draft International Treaty on Plant Genetic Resources for Food and Agriculture (ITPGRFA, under negotiation at that time) including text regarding access and benefit sharing, the creation of an international benefit-sharing fund, and farmers' rights. All these inputs were taken into consideration by the Committee for Plant Variety Protection Bill Drafting, which was created especially by the Thai government for the purpose of writing the final draft.[38]

The process for elaborating the PVPA was enriched by much input by the Thai people through the NGOs, academics and civil society groups. Before they voiced their opinions, the draft PVPA contained only protection for new plant varieties. The powerful NGOs influenced the public opinion, thereby applying considerable pressure on the government. The Assembly of the Poor (AOP),[39] with the concert of academics and NGOs, was able to sway the drafting process to extend protection further and to include protection for local domestic plant varieties, as well as general domestic and wild plant varieties. The aim was to include all plant varieties in the sovereign domain of Thailand within the scope of the Act. The felt need to achieve this was motivated largely by concerns about recurring cases of alleged biopiracy and misappropriation of Thai resources, including the very high profile case of 'Jasmati' which the Thai people discovered only in 1998. Such cases made the Thai people very sensitive to these issues. The name Jasmati was registered as a trademark in the US by the American company RiceTec, Inc. in 1993. The company claimed Jasmati was the Texas-grown copy of Thai jasmine rice, when in fact the variety is not derived from any Thai rice variety.[40]

In 1999, the Council of State finally passed the PVPA, which also came into effect that same year.[41]

The implementation and its difficulties

One outstanding regulation

In order to go forward with the implementation of the Act, one further important regulation needed to be adopted concerning the process applications involving local domestic plant varieties.[42]

As of 9 September 2011, there have been only 94[43] registrations since the adoption of the PVPA more than 10 years prior. However, there have been 773 applications: 169 are in the process of consideration the breeding process, 457 in the process of DUS testing, and 34 in the process of reporting testing results.

The most significant factor contributing to the low numbers of registrations was the absence of the regulation concerning access and benefit sharing, described earlier.

Defining locality

While the absence of a regulation concerning local domestic varieties application has created some challenges, this actually has not been the core problem until now. Instead, the more fundamental problem turns out to be that it is difficult to find plant varieties that correspond to the requirement of locality. Under the Act, for a variety to be local, the geographic region where the variety is found must be limited to a well-defined, geographically limited locality. The problem with this criterion is that varieties are often taken from one community to another with cross-community marriages. The entire Kingdom of Thailand cannot be considered a locality, as this would rather qualify the plant variety as a 'general domestic' variety. There has not been any variety registered under the local domestic category. Hopes of finding a local variety were raised when a certain rice variety was found only in a specific area. In 2007, Mr. Somchai Asaiboon applied for registration for the Homhuang Chaiya rice variety that was planted on his premises in the Thoong subdistrict of the Suratthani province. The authority, after investigation, found that the Asaiboons were the only family cultivating this rice. As a result, the authority rejected the application, as it did not qualify under Section 44 of the PVPA[44] because the Asaiaboons were a family, and not a community. The number of registered local plant varieties is still at zero.[45]

Plant Varieties Protection Fund

Another challenge to the full implementation of the Thai PVP law concerns the Plant Varieties Protection Fund. Ministerial regulations creating the fund were adopted[46] in 2007, but there is still no money in the fund. A total of 94 registrations have been granted, which have generated a modest income for the fund. Subsidies from the government have not yet been provided, and donations from other sources have not materialized.

Conclusion

Although NGOs, large and small national companies, and the Thai public were generally not in favour of Thailand becoming a member of the UPOV Convention some years ago, there are signs that opinions are starting to shift due to higher levels of awareness of related issues. Some members of the public and most local companies are in favour of signing the UPOV Convention. Partly as a result of the implementation setbacks described in this chapter, there is talk within the Plant Variety Protection Division, and within some groups of stakeholders, of modifying the legislation so as to render it similar to UPOV 1991.

The government of Thailand unofficially submitted its law to UPOV for verification and the UPOV Council conducted a review of the law, which is summarized in a document entitled 'Comments on the Plant Varieties Protection Act, B.E. 2542 (1999) of the Kingdom of Thailand in Relation to the 1991 Act of the UPOV Convention' in 2006.

The UPOV Office abstained from evaluating the *sui generis* segments of the Thai Act by claiming that 'the Office of the Union is not in a position to provide comments on the protection of general domestic plant varieties, wild plant varieties and local domestic plant varieties.'[47] However, the UPOV Office did opine that the provisions for the protections of local domestic, general domestic and wild varieties should be separate from the new plant variety protection law. In other words, to be in compliance with UPOV, the Thai government should keep the *sui generis* protections for another law. To this day, Thailand is not yet a member of UPOV.

Notes

1 Plant Varieties Protection Act, B.E. 2542 (1999), online: WIPO [www.wipo.int/clea/en/text_pdf.jsp?lang=EN&id=3816].
2 PVPA, S.11.
3 The distinctness criterion exclusively for the new plant variety protection includes the notion that such distinctness must be 'related to the feature beneficial to the cultivation, consumption, pharmacy, production or transformation.' This specification equivalent to the value for conservation and use (VCU) standards is absent in the 'plant variety' definition and was later added in the descriptors of the new plant variety. Therefore, the VCU standards were dropped for the local and general domestic plant varieties.
4 UPOV has longer grace periods in S.6(1), UPOV 1991, which are 4 years before the application date for all varieties, or in the case of trees or vines, earlier than 6 years. The PVPA is therefore not in compliance with UPOV standard delays.
5 The terminology differs in a few aspects, although the idea conveyed is identical.
6 The protection against parallel imports which is detailed in Article 14(2) of UPOV, is missing. Also, the concept of essentially derived varieties found in Article 14(4) UPOV is absent in the PVPA. Moreover, the Thai law incorporates two more exceptions to the protection of the rights holder (33(1) and (4) PVPA).
7 The protection is for 12 years for plants giving fruits within a period of not over two years of the cultivation, 17 years for plants giving fruits after more than two years of cultivation, and 27 years for tree-based plants giving fruits after two years or more of cultivation; PVPA, s.31.
8 UPOV Council, 'The Plant Varieties Protection Act, B.E. 2542 (1999) of the Kingdom of Thailand in Relation to the 1991 Act of the UPOV Convention,' Documents prepared by the Office of the Union for presentation under Topic 5 of the National Workshop on the Protection of New Varieties of Plants under the UPOV Convention, May 8 2006, Bangkok, May 4 2006.
9 PVPA, S.19(3).
10 Ibid., S.19(5).
11 The requirement that they also be distinct uniform and stable is incorporated through the definition of plant variety. See supra note 1 and discussion.
12 PVPA, S.3.
13 Ibid., S.44.
14 Ibid., S.45.
15 Ibid., S.44(3).
16 Ibid., S.47(1).
17 See supra, note 6.
18 PVPA, S.33 & 47.
19 Ibid., S.48
20 Ibid., S.48.
21 Ibid., S.49(1).
22 The initial periods of time attributed for the protection are the same as those for new plant varieties.
23 PVPA, S.50.
24 Ibid., S.3.
25 Ibid.
26 Ibid., S.52.
27 Anon. (2011). 'Ministerial regulation on defining of permission procedures and conditions for collecting, procuring or gathering general domestic plant varieties, wild plant varieties for the purposes of variety development, education, experiment or research for commercial interest and a profit-sharing agreement B.E. 2553', *Royal Gazette*, 128, section 5.
28 There is also no direct benefit sharing with 'suppliers' of wild varieties.

29 The Fund Committee is composed of: 'Permanent Secretary of the Ministry of Agriculture and Co-operatives as the Chairman and not less than seven other members appointed by the Commission and the Director-General of the Department of Agriculture shall be the secretary and a member.'; PVPA, S.56.
30 Ibid., S.55.
31 Anon. (2011). 'A profit-sharing agreement from utilization of general domestic plant varieties, wild plant varieties or any part of such plant varieties according to Plant Variety Protection Act B.E. 2542'. This agreement is an attachment to the 'Ministerial regulation on defining of permission procedures and conditions for collecting, procuring or gathering general domestic plant varieties, wild plant varieties for the purposes of variety development, education, experiment or research for commercial interest and a profit-sharing agreement B.E. 2553'.
32 Ibid., S.53.
33 *Plant Variety Protection Commission Regulation in Study, Experiment, or Research on General Domestic and Wild Plant Varieties*, Royal Gazette of Thailand, No 12, special issue 16 Ngo, 13 February 2004.
34 Agreement on Trade-Related Aspects of Intellectual Property Rights, Annex 1C of the Marrakesh Agreement Establishing the World Trade Organization, art.27.3(b), online: WTO [www.wto.org/english/tratop_e/trips_e/t_agm0_e.htm].
35 Cabinet Resolution of 14 July 1998 on 'Draft on Plant Variety Protection Act B.E.'
36 Robinson, Daniel. 'Governance and Micropolitics of Traditional Knowledge, Biodiversity and Intellectual Property in Thailand,' at p. 62, Table 5.
37 There was a clear division within the country between the multinational and bid national companies in the field of agriculture and the important body of NGOs, which swayed the public, concerning the ratification of the UPOV Convention.
38 Compeerapap, J. 'The Thai Debate on Biotechnology and Regulations,' September 1997, *Biotechnology and Development Monitor*, No. 32, pp. 13–16.
39 The Assembly of the Poor is an NGO in Thailand which was established on International Human Rights Day in 1995. The AOP is a grassroots people's movement consisting of the following social networks: rural poor, farmers, urban poor, workers, indigenous peoples and NGOs [http://blog.world-citizenship.org/wp-archive/427].
40 Lightbourne, M. 'The JASMATI trademark affair,' February 1999, Asia Law & Practice, Hong Kong, online: Grain: [www.grain.org/bio-ipr/?id=299]; Roggemann, Ellen. 'Fair Trade Thai Jasmine Rice: Social Change and Alternative Food Strategies Across Borders,' August 2005, Educational Network for Global Grassroots Exchange, online: [departments.oxy.edu/uepi/uep/studentwork/05comps/roggemann.pdf].
41 The Council of State also passed the Thai Traditional Medicines Act in the same period which further demonstrated the desire of the government to protect local culture and stop biopiracy.
42 See PVPA, S.44, which states that 'The submission of the application and the consideration and approval thereof shall be in accordance with the rules and procedure prescribed in the Ministerial Regulation.'
43 In January 2010, there are still only 56 plant variety registrations, of which 18 are orchids.
44 Plant Variety Protection Division, 'Result of Investigation of Rice Variety Homhuang Chaiya,' 2007, Plant Variety Protection Division, Bangkok, p. 2.
45 As of January 2010.
46 The Ministerial Regulation on determining procedure and allocation rate of PVP fund to the local government organization, promulgated in the Royal Gazette of Thailand, No. 134, issue 39 Kor, 3 August 2007.
47 Anon. (2006). 'Comments on the Plant Varieties Protection Act, B.E. 2542 (1999) of the Kingdom of Thailand in relation to the 1991 Act of the UPOV Convention.' The comments were prepared by the Office of the Union for the presentation of Topic 5 of the National Workshop on the Protection of New Plant Varieties under the UPOV Convention (Bangkok, 8 May 2006).

Bibliography

Legislation and regulations

Agreement on Trade-Related Aspects of Intellectual Property Rights, Annex 1C of the Marrakesh Agreement Establishing the World Trade Organization, art.27.3(b). Available at: www.wto.org/english/tratop_e/trips_e/t_agm0_e.htm.

Anon (2011). 'A profit-sharing agreement from utilization of general domestic plant varieties, wild plant varieties or any part of such plant varieties according to Plant Variety Protection Act B.E. 2542'. 'Ministerial regulation on defining of permission procedures and conditions for collecting, procuring or gathering general domestic plant varieties, wild plant varieties for the purposes of variety development, education, experiment or research for commercial interest and a profit-sharing agreement B.E. 2553'.

Plant Varieties Protection Act, B.E. 2542 (1999). Available at: www.wipo.int/clea/en/text_pdf.jsp?lang=EN&id=3816.

Plant Variety Protection Commission Regulation in Study, Experiment, or Research on General Domestic and Wild Plant Varieties, Royal Gazette of Thailand, No 12, special issue 16 Ngo, 13 February 2004.

Official documents

Cabinet Resolution of 14 July 1998 on 'Draft on Plant Variety Protection Act B.E.'

Plant Variety Protection Division (2007). *Result of Investigation of Rice Variety Homhuang Chaiya*, Plant Variety Protection Division, Bangkok.

UPOV Council, 'The Plant Varieties Protection Act, B.E. 2542 (1999) of the Kingdom of Thailand in Relation to the 1991 Act of the UPOV Convention,' Documents prepared by the Office of the Union for presentation under Topic 5 of the National Workshop on the Protection of New Varieties of Plants under the UPOV Convention, May 8 2006, Bangkok, May 4 2006.

Literary works

Compeerapap, J. (September, 1997). 'The Thai Debate on Biotechnology and Regulations,' *Biotechnology and Development Monitor* 32: 13–15.

Lightbourne, M. (February, 1999). *The JASMATI Trademark Affair*, Asia Law & Practice, Hong Kong. Available at: Grain [www.grain.org/bio-ipr/?id=299].

Robinson, Daniel F. (2006). *Governance and Micropolitics of Traditional Knowledge, Biodiversity and Intellectual Property in Thailand*, Bangkok, University of New South Wales and University of Sydney, February 2006. Available at: www.iprsonline.org/resources/docs/Final%20HRC%20Micropolitics%20Report%20Mar%202005.pdf.

Roggemann, E. (August, 2005). *Fair Trade Thai Jasmine Rice: Social Change and Alternative Food Strategies Across Borders*, Educational Network for Global Grassroots Exchange. Available at: http://departments.oxy.edu/uepi/uep/studentwork/05comps/roggemann.pdf.

17 Commentary on the Zambian Plant Breeder's Rights Act

Godfrey Mwila

The Zambian Plant Breeder's Rights Act, 2007, requires that, to be protectable, varieties must be distinct, uniform and stable.[1] It also requires that they have demonstrable value for cultivation and use. The act does not have any specific provisions on farmers' varieties. This chapter provides an account of the very significant efforts made by a number of actors to include clauses in the act that would have created *sui generis* intellectual property protections for farmers' varieties. Ultimately, these efforts were not successful. It is important, nonetheless, to learn from them.

The emergence of private sector interests and capacities

Prior to its independence, Zambia was not home to any plant breeding work, and hence no locally improved crop varieties were being produced. The small amount of maize seed that was produced locally consisted of one hybrid variety, SR-52, which had been brought in from Zimbabwe (then Northern Rhodesia). Otherwise, most of the seed requirement, especially for maize, was met using imports from Zimbabwe. Immediately after independence, a maize-breeding program was launched, which first focused on maintaining parent lines for SR-52 and increasing SR-52 breeder seed to start seed production in the country. Over the years, the breeding program has expanded, resulting in a number of locally produced maize hybrid and composite varieties. Gradually, this breeding work has extended to other crops, with new varieties of sorghum, groundnuts, pearl millet and cassava being developed throughout the 1990s. All of these efforts were made by breeders working within the public research institution, with no private sector involvement other than the School of Agricultural Sciences at the University of Zambia, which was involved in some collaborative work in breeding. Seed production and maintenance breeding for these varieties was at the time a responsibility of the Seed Services Section (now the Seed Control and Certification Institute [SCCI]), which then fell under the research branch of the ministry responsible for agriculture.

As the capacity for variety development grew, more varieties were made available for commercial use. Then, as the demand for seed increased, it became apparent that the institutional arrangement that was in place was going to prove to be inadequate. This realization led to the creation of the first national seed company in 1981 – the Zambia Seed Company (Zamseed) – which became responsible for the production and marketing of all types of seed, with the exception of cotton and tobacco. Inherent in the establishment of this company was an agreement that provided Zamseed with exclusive rights to produce and market the seed of varieties developed by the Zambia Agriculture Research Institute (ZARI), which was then known as the Soils and Crops Research Branch. This exclusive relationship was maintained until around 1991 when other players in the formal seed sector began to appear.

Partly as a consequence of the overall economic and agricultural policy changes, crop breeding and improvement research underwent some major changes with the establishment of several institutions initiated by the private sector. No longer were crop breeding and research responsibilities the sole domain of the public research institutions. Two of these were agricultural research trusts created as public/private sector partnerships, while four were new seed companies, with active varietal development and improvement programs. Seed production and marketing was transformed from a monopolistic to a competitive market with the arrival of these new companies, leading to the broadening of the seed industry stakeholders covering both the formal and informal sectors.

Development of plant variety protection laws

In time, demands were made by the private sector players – mainly the seed companies – to protect plant breeders' rights in order to encourage private initiative in these areas. In 1998, the government responded to these demands and began to initiate the process. The rationale for plant breeders' rights was to ensure that the efforts of breeders were rewarded adequately in order to encourage further investment in the development of more varieties. The initial efforts made between 1999 and 2001 led to a draft plant breeders' rights bill that was based on the 1991 model of the International Convention for the Protection of New Varieties of Plants (UPOV Convention).[2] This bill, however, could not be approved by the government, reportedly because it did not take into account the interests of small-scale farmers. The decision was also influenced by issues arising from debates in the ongoing negotiations of the International Treaty on Plant Genetic Resources for Food and Agriculture (under the Food and Agriculture Organization's Commission on Genetic Resources for Food and Agriculture) (ITPGRFA) and the African Model Law for the Protection of the Rights of Local Communities, Farmers and Breeders and for the Regulation of Access to Biological Resources (African Model Law) (under the Organization of African Unity).[3] Further influence may have come from changes in the national agricultural policy (Ministry of Agriculture and Cooperatives in 2004), which provided the need to simultaneously recognize and reward plant breeders, farmers and farming communities for their contribution to variety development, although these changes made no specific reference to farmers' varieties. As a result, the Ministry of Agriculture and Cooperatives (MACO) sent the draft bill back to the SCCI for redrafting in order to incorporate farmers' rights issues.

Once the bill had been referred back to the SCCI for revision, nongovernmental organizations (NGOs) got involved and put still more pressure on the government to incorporate *sui generis* intellectual property protections for farmers' varieties in the bill. Technocrats attempting to follow up on these demands considered themselves to be implementing farmers' rights as provided for under the ITPGRFA.

The process of preparing a new draft bill incorporating plant breeders' rights and farmers' rights began in 2002 with the constitution of a working group composed of technocrats from the SCCI, ZARI, the Ministry of Legal Affairs, which represents the public sector, and the Zambia Seed Traders Association, which represents the private sector. The SCCI, being the regulatory agency of the seed industry in the country, provided the secretariat. It is important to note that this process was preceded by the formulation of the National Seed Policy, which was expected to feed into the overall agricultural policy. The working group began the task of developing a new draft bill during a period of increased debate at the regional, subregional, and country levels over issues of intellectual property rights, biopiracy, farmers' rights, community rights, and finally access and benefit sharing, which were arising from the introduction of the Convention on Biological Diversity (CBD), the Agreement on Trade-Related Aspects of Intellectual Property Rights (TRIPS Agreement) and the ITPGRFA.[4]

As part of the process of coming up with the draft bill, the working group reviewed a number of related efforts to develop intellectual property rights protections at the international, regional and national levels. The working group agreed to come up with a broad draft based on a *sui generis* system, combining both plant breeders' rights and farmers' rights. Both the UPOV Convention model and the African Model Law were used as a basis for the new draft. The process attracted a lot of unsolicited external interest from foreign seed companies operating in the country, such as PANNAR Seed and from the regional seed regulatory bodies such as ARIPO (African Regional Intellectual Property Organisation) and the International Union for the Protection of New Varieties of Plants, before the new draft could go through the official review and approval process. The external pressure, which was particularly directed at the SCCI as the coordinator of the process, advocated for the separation of farmers' rights from plant breeders' rights. Added to this debate was the internal pressure from the relatively more powerful private-sector stakeholders in the seed industry, who urged that plant breeders' rights legislation was more urgent than farmers' rights and needed to be dealt with separately in order to hasten the process.

In 2003, a layman's draft legislation based on a *sui generis* system that combined plant breeders' rights and farmers' rights was submitted to the Ministry of Legal Affairs to be drafted into a bill, before being circulated to other relevant ministries for their comments. As a result of the pressures cited earlier, the SCCI withdrew this draft bill before it was completed and circulated to the ministries for comment. The follow-up preparation of a draft plant breeders' rights bill was then based only on the UPOV Convention. The new draft bill, which dealt solely with plant breeder's rights, was prepared and submitted for consideration and approval by the government and by the Cabinet before it went to Parliament for enactment. The bill was presented to Parliament in 2006 and was enacted as the Plant Breeder's Rights Act in 2007.[5]

The process of trying to come up with legislation that reflected a *sui generis* system revealed a number of challenges, the major one being the difficulty of harmonizing farmers' rights with the existing seed and plant variety protection laws. Technical challenges included the decision of how to choose the alternative conditions for protecting farmers' varieties. The drafting committee also considered whether the ownership of farmers' varieties should be individual or communal, and if it would be appropriate and beneficial to treat farmers' varieties in the same way as other improved varieties in terms of the procedures for registration, certification and protection. In looking at these issues, the individuals concerned examined some of the other national plant variety protection laws that had incorporated farmers' varieties, with particular emphasis on the Indian Protection of Plant Varieties and Farmers' Rights Act (see Chapter 12).

Follow-up developments regarding *sui generis* intellectual property protections for farmers' varieties

There are no clear steps to follow when developing policy and law specifically addressing farmers' varieties. What is generally considered to be the most successful avenue to follow involves the overall policy measures necessary to domesticate and nationalize the ITPGRFA. In particular, it is expected that farmers' or local traditional varieties could be addressed within the context of implementing the farmers' rights article under the treaty. In Zambia, it is understood that farmers' rights, which may incorporate the protection of farmers' varieties – including ownership recognition – would be dealt with in a separate piece of legislation. Such a process would be spearheaded by a different government agency, namely ZARI, which is the focal institution with the overall responsibility of implementing the ITPGRFA.

In 2008, a key stakeholder's workshop was held, entitled 'Awareness Creation on the Treaty and Farmers' Rights,' which was coordinated by the National Committee on Plant Genetic Resources. The participants of this workshop recommended that a working group be formed that would be dedicated to reviewing policy and legislation with the aim of realizing farmers' rights in Zambia by including measures to promote farmers' varieties as well as the protection of farmers' rights. Membership in this working group will include representatives of farmers and farmer organizations. It will facilitate capacity building and awareness creation among farmers and farmer organizations by organizing a national forum on farmers' rights. The key strategy will be to create a farmer-driven initiative to revive and encourage the process of developing policy and legislation on farmers' rights. It is expected that these farmers and farmer groups will define farmers' rights based on their own perceptions and will demand that the authorities formulate and enact appropriate legislation. It is unlikely, given past experiences, that there will be room for developing *sui generis* intellectual property rights for farmers' rights. It is more likely that efforts will focus on the creation of legal space to promote farmers' varieties and practices.

The developments outlined earlier illustrate the challenges that Zambia has faced in its efforts to integrate breeders' and farmers' rights into one piece of legislation. Similar challenges may be at play in other African countries and, indeed, in other developing countries trying to undertake similar initiatives. It is clear from this examination that in order to create the environment for enacting laws for the protection of farmers' rights as a counterbalance to breeders' rights, there has to be effective demand emanating from the actual beneficiaries – the farmers and farming communities. It is therefore no surprise that breeders' rights legislation, because of the pressure from the formal seed sector players who are beneficiaries of these rights, was prioritized for enactment into law over farmers' rights. These same formal sector players also create obstacles for the development of farmers' rights legislation mainly due to their desire for self-preservation. Given these factors, although there is a desire and some kind of plan towards implementing farmers' rights into national law in Zambia, and perhaps in other African countries, the prospects for achieving this goal may not be bright. The reality of the situation can be attested to by the fact that to date little or no progress has been made in implementing the process that was agreed upon during the stakeholder consultation workshop alluded to earlier in this section.

Notes

1 Plant Breeder's Rights Act, 31 August 2007, Chapter 239 of the Laws of Zambia.
2 International Convention for the Protection of New Varieties of Plants, 2 December 1961, available at <www.upov.int/en/publications/conventions/index.html> (last accessed 10 May 2012).
3 International Treaty on Plant Genetic Resources for Food and Agriculture, 29 June 2004, <www.planttreaty.org/texts_en.htm> (last accessed 10 May 2012). African Model Law for the Protection of the Rights of Local Communities, Farmers and Breeders and for the Regulation of Access to Biological Resources, online: <www.cbd.int/doc/measures/abs/msr-abs-oau-en.pdf> (last accessed 10 May 2012).
4 Convention on Biological Diversity, 31 ILM 818 (1992). Agreement on Trade-Related Aspects of Intellectual Property Rights, Annex 1C of the Marrakech Agreement Establishing the World Trade Organization, 15 April 1994, 33 ILM 15 (1994).
5 Plant Breeder's Rights Act, supra note 1.

18 Commentary on the Nepalese Seeds Act and the Seeds Regulation

Pratap Kumar Shrestha

Background

In Nepal, the Plant Variety Protection and Farmers' Rights Bill (PVP&FR Bill) has been drafted and is undergoing review and finalization for approval from the government.[1] The PVP&FR Bill will ultimately regulate the registration of plant varieties and grant intellectual property rights protection for these varieties. Plant varieties that are protected in this way will be new plant varieties – both new farmers' varieties as well as varieties developed by formal breeders. The registration of, and ownership over, traditionally grown, local farmers' plant varieties and landraces falls within the remit of the Access to and Benefit Sharing from the Use of Genetic Resources Bill (ABS Bill), which has also been drafted and is awaiting for approval from the government.[2] Until these bills are implemented, the registration and granting of ownership or intellectual property rights to plant varieties in Nepal is currently partly regulated by the Seeds Act 1988 (first amendment, 2012) and the Seeds Regulation 2013 (after first amendment of Seeds Regulation 1997).[3] The National Seed Policy was formulated and brought into force in 1999, which provided further policy guidelines to promote the production and marketing of quality plant seeds in the country.

Seeds Act, 1988 (first amendment, 2012)

Legal provisions

The Seeds Act was enacted in Nepal in 1988 and amended in 2012. The main objective of the Seeds Act was to promote and regulate the increased production and distribution of high-quality plant seeds and to ensure the interest of seed entrepreneurs and farming communities (the consumers of such seeds). There are very limited provisions included in the Seeds Act that are relevant to the protection of intellectual property rights on new plant varieties. The two elements that are key to ensuring the protection of intellectual property rights, namely seeds and breeders, are explicitly defined in the Seeds Act. According to these definitions:

- 'seed' means matured ovules having embryonic plant, food materials and protective covers or seeds that are reproduced sexually or by vegetative

means and that can be used to produce crop by sowing or planting (subsection 2.1.1);
- 'crops' comprise fruits, food grains, vegetables, cash crops and forage crops (subsection 2.1.2);
- 'breeder' means a person, organization or body that brings into use any variety of the crops by producing or selecting it for the first time (subsection 2.1.8).

Thus, the seeds include all planting materials that are used for crop production and the reproduction of seeds. The definition of breeder focuses mainly on the act of developing a new plant variety and recognizes, without any conditions, a person, organization or authority developing such variety as a breeder. By this definition, a farmer developing a new plant variety also qualifies as a breeder.

The Seeds Act also made provision in section 3 for the constitution of the National Seed Board (NSB), with authority and responsibility to formulate and implement seed-related policies and to give necessary advice on seed-related matters to the government. Of the various functions and rights of the NSB, the following two are relevant to intellectual property rights on plant varieties:

- approve, release and register the seeds of new varieties as prescribed (subsection 5.5);
- grant the right of ownership to the breeder as prescribed after testing the seeds of new varieties for distinctness, uniformity and stability (subsection 5.6).

The marketing of the seeds is an important function that is directly related to the intellectual property rights. There are a number of provisions in the Seeds Act that regulate the marketing of seed, including the following:

- If deemed necessary to regulate and control the quality of the seed of any kind or variety used for agricultural production, the government of Nepal may, in consultation with the NSB, prescribe seeds by publishing a notice in the Nepal Gazette (such seeds are called notified seeds), and, while so prescribing, it may also prescribe the kind or variety of the seeds appropriate for different regions (section 11).
- Any person or organization that is willing to engage in the marketing of seed shall have to obtain a permission letter in the format specified by submitting an application to the concerned authority and by paying the specified fee (subsection 11.a.1).
- Except for the purpose of agricultural research, no person shall market seeds that are not notified by the Seeds Act (subsection 11.b.1).
- The kind or variety of seed determined by the breeder as being appropriate for a specific area shall not be sold or caused to be sold in the areas other than prescribed (subsection 13.2).
- Provisions have been made for the punishment of persons or organizations not abiding by the Seeds Act (section 19). The first amendment of the Seed Act has also made a provision for granting ownership rights on the local

plant varieties with the addition of a new section 18.a. It states, "There shall be ownership right on the seeds of traditionally used plant varieties in Nepal as prescribed (page 6)". However, the definition and scope of such a right is not described.

Issues and suggestions

The provisions in the Seeds Act for the protection of plant varieties are not adequate for effective intellectual property rights protection. Although the NSB is authorized to register and grant ownership rights on new plant varieties and a new provision is made establishing ownership rights over the traditionally used local plant varieties, no specific legal provisions have been made for such rights in the Seeds Act. As a result, to date, none of the registrees (breeders) have applied for ownership rights. While it is not explicitly mentioned, these functions appear to have been left to be handled by the Seeds Regulation. The provisions that regulate the marketing of plant seeds are not linked strictly with the ownership rights of the seed breeder. The NSB can authorize more than one person or organization to market a single variety.

The eligibility conditions included in section 11 and subsections 11.a.1 and 11.b.1 of the amended Seeds Act for the marketing of seeds favour a formal seed sector that involves organized seed entrepreneurs and the marketing of new plant varieties. These provisions completely ignore any informal seed systems, where farmers meet most of their seed needs through farmer-to-farmer exchange and by buying and selling such seeds. These eligibility conditions, if implemented strictly, will not only disrupt and dismantle the informal seed systems but also pose a big threat to the conservation of agricultural biodiversity and undermine farmers' customary rights over seeds which they have been cultivating and managing for generations. However, due to a lack of adequate institutional capacity and limited human and financial resources, adequate law enforcement and monitoring is currently not taking place. The size of the commercial seed market for cereal crops is also small, and therefore the pressure and incentives for strict law enforcement are low.

The Seeds Act also restricts marketing of notified seeds in geographical areas that are not prescribed for such seeds. The author is of the opinion that this restriction is an unnecessary and potentially harmful aspect of the Seeds Act. Such a restriction will only limit access to, and the supply of, seeds in resource-poor countries such as Nepal, where market infrastructures are inadequate and marketing networks are poorly developed. As long as there is information about the suitability of the seed for a particular geographical area on the seed packet and the seed users are not misinformed, the marketing of seeds should not be restricted to any particular geographical location. For example, seed entrepreneurs in Nepalganj should be allowed to sell seed prescribed for Jumla. Such restrictions are also detrimental for local innovation and the promotion of agricultural biodiversity. Local Initiatives for Biodiversity, Research and Development's (LI-BIRD's) experience with participatory variety selection demonstrates that some of the new plant varieties released and prescribed for the Terai (the southern plains of Nepal) also performed well and were preferred by the hill farmers. Farmers also experiment with new seeds. They often try out new seeds in their own production environments and management conditions – a process of adaptation and domestication which has made an enormous contribution to the promotion and conservation of agricultural biodiversity.

The scope of the Seeds Act, in terms of the kinds of plants and plant varieties to be included, is not explicitly defined. While the statement of the preamble appears to define the act as covering the seeds of different 'crop,' the definition of the term crop opens up the scope to all kinds of crops – that is, all cultivated plants, both food crops and nonfood crops. For example, 'cash crop' may include nonfood crops such as medicinal plants, ornamental plants, timber plantations and so on.

Similarly, the institutional base of the NSB is also not explicitly mentioned. However, in practice, it operates under the MoAC, possibly because the secretary of the MoAC is the ex-officio chairperson of the NSB. In order to effectively regulate seed production and marketing in the country, the scope of the Seeds Act should be enlarged to include seeds of all kinds of plants and plant varieties, and the NSB should be made an autonomous body directly reporting to the executive head of the government. Such changes would avoid the need for interministerial coordination, which is always difficult to manage.

Seeds Regulation, 2013 (after first amendment of Seeds Regulation 1997)

Legal provisions

The Seeds Regulation was formulated in 1997 and amended in 2013, within the provisions of the Seeds Act 1988 (first amendment, 2012), to define rules and to regulate the production and marketing of quality seeds in the country. Based on the provisions of Rule 4 of the Seeds Regulation, the NSB has constituted a Variety Approval, Release and Registration Sub-Committee (VARRSC). This subcommittee has been authorized to regulate the functions of approval, release and registration of the new plant varieties (Rule 5).

The process of approval, release and registration of new plant varieties is detailed in Rule 11 of the Seeds Regulation. The breeder has to submit an application for approval, release and registration of the new plant variety to the VARRSC in the prescribed format. The new variety has to meet three criteria: it has to be (1) distinct, (2) uniform and (3) stable – these are known as the DUS criteria. The requirement of the new criterion is not explicitly mentioned, but it is implicit in the fact that the provision for approval, release and registration is only for new plant varieties. These criteria, however, are not defined in the Seeds Regulation. The provision for the registration of local plant varieties has also been made in Rule 12.2 of the amended Seed Regulations 2013.

The Seeds Regulation has also made clear provision for the right of ownership to new plant varieties outlined in Rule 13. The breeder willing to acquire such rights has to submit an application to the NSB in the prescribed format.

The right of ownership so obtained is conditional – that is, it remains valid as long as the variety remains in the list of notified seeds. The right of ownership of the variety of de-notified seeds (those that have been removed from the list of notified seeds) is deemed to have been terminated ipso facto after two crop years from the publication of de-notification in the Nepal Gazette. The scope of the right of ownership granted to the breeder is not defined in the Seeds Regulation, and there is no mention about the implications of such rights to the production and marketing of the seeds of right-protected plant varieties. The person or organization willing to market the seeds of registered plant varieties, domestically or internationally, does not have to receive permission from the holder of the right of ownership to such plant varieties. The Seeds Regulation has the capacity to grant authority to more than one applicant. However, this provision has not yet been exercised (personal communication, Madan Thapa, Seed Quality Control Centre, 11 April 2012). Similarly, there is no mention about the period of time for which the plant is protected under the right of ownership on new plant varieties. Likewise, there are no rules included in the amended Seeds Regulation 2013 to establish ownership rights on the traditionally used local plant varieties as provisioned in the amended Seed Act 2012.

Issues and suggestions

The first step towards the realization of intellectual property rights protection on new plant varieties is registration and obtaining a right of ownership on such varieties. Since the DUS criteria required for approval, release and registration are not defined in the Seeds Regulation, these criteria could be subject to interpretation and a possible source of conflict. The 'new' criterion, which is the fundamental requirement for claiming any intellectual property rights protection, should also be explicitly included as a necessary criterion along with the DUS requirements.

Solely granting the right of ownership is not adequate to effectively implement intellectual property rights protection, unless the scope of such ownership is specifically defined. The Seeds Regulation as well as the Seeds Act do not explicitly mention the types of rights the owner of the plant varieties can exercise in relation to the production and marketing of their seeds. One reason for this could be that variety development and registration was initially entirely the responsibility/business of the Nepal Agricultural Research Council, and it was a nonprofit public sector organization. The ownership issue, therefore, may not have received proper attention and thought. There may simply have not been adequate demand to justify developing the appropriate legal provisions. Such a hypothesis is reflected in the fact that more than 574 new plant varieties have been released and registered in Nepal until 28 July 2014, but not a single application for ownership rights has been filed.[4]

It can be argued that the granting of ownership rights should not be subject to a second application and approval procedure, but rather should automatically come into effect after the successful registration of the plant variety. The ipso facto termination of the right of ownership on plant varieties after the de-notification of seeds of such varieties is not logical. Although it is not mentioned, the de-notification may be temporary, and de-notified seeds in one region or country may qualify for notification in another region or country. The right of ownership, therefore, should be made independent of the de-notification of the seeds. Similarly, the Seed Regulation should define the meaning and scope of ownership rights on the local varieties used traditionally in Nepal as provisioned in the amended Seed Act.

The analysis presented in this chapter is for new plant varieties (those with intellectual property rights implications) that are protected by the right of ownership for a specified period of time. Although the protection period is not specified in the current Seeds Regulation, I am arguing that de-notification that is done for reasons other than the completion of the protection period – for example, the degeneration of the genetic quality, a lack of supply in the market, and so on – should not automatically terminate the right of ownership.

National Seed Policy, 1999

Provisions relevant to intellectual property rights protection

The National Seed Policy came into force in Nepal in 1999. The main objective of the National Seed Policy is to provide a policy framework and guidelines to ensure the production and distribution of quality seeds and to conserve and protect rights over seeds of local crop varieties that have distinctive genetic traits.

The National Seed Policy has explicitly mentioned that the variety development, which has so far been carried out by the government sector, will also be done through private organizations and NGOs as well as the private sector (subsection 3.1.1). These organizations will, however, only be given permission to engage in the variety development program after they pre-inform the authority about their infrastructure facilities and the rationale behind developing such variety (subsection 3.1.2). To implement this policy guideline, the NSB made a decision on the requirement for various infrastructure facilities and conditions in a meeting held on 5 September 2003, and these requirements were put on public notice on 20 November 2003. According to this decision, private and nongovernment organizations must have the following infrastructure facilities:

- own or lease land for purposes of research;
- employ a plant breeder with a minimum level of education of a master of science degree;
- employ a seed technologist with a master of science degree;
- employ other staff, each with a bachelor of science degree (one for each crop);
- employ a multidisciplinary team of other individuals as required;
- have a seed store to store the required capacity;
- maintain other equipment that is necessary for plant breeding and seed production.

These organizations must fulfil the following minimum conditions:

- they must submit a plan for variety development to the NSB;
- they must implement the directives/suggestions that are passed down to them from the concerned authorities as necessary.

The National Seed Policy has also made policy provisions for the new varieties that have been developed from agricultural research, which first need to be approved, released and registered with the NSB before being marketed and/or distributed to the farmers (subsection 3.1.3). This policy dictates that the seeds of new varieties are restricted for distribution until they are formally approved, released and registered. The National Seed Policy has also declared that the responsibility of maintaining the quality and production in a required amount of nucleus and breeder seed lie with the breeder.

The seed certification system is a standard certification system adopted by NSB of Nepal. The Quality Standards Determination and Management Sub-Committee of the NSB has set minimum standards for certified seed production, such as isolation, purity, germination percentage, moisture content and so on, and these criteria are published in the *Seed Production Guidelines*. The seed inspector appointed by the Seed Quality Control Centre of the NSB checks compliance with the standards before certifying the seeds and provides a certification tag once the seed has met these standards. The Seed Certification Agency is responsible for ensuring that there is compliance with this quality standard. On the other hand, the truthful label system is a softer and more cost-effective system of producing quality seed. This is a kind of self-certification system where seed producers themselves declare the minimum standards of their seed as set out by the Quality Standard Declaration Sub-Committee of the NSB. Seed inspection is not required for this category of seeds. The seed producers themselves are fully responsible for the quality of the seeds specified.

The National Seed Policy has also introduced the Quality Declared Seed system, which falls somewhere between the seed certification system and the truthful label seed system in terms of quality assurance and quality checks. The seed producers agree to produce quality declared seed as specified by the Quality Standards Determination and Management Sub-Committee of the NSB. The appointed seed inspector may do random testing of, at least, 10 percent of the seed at any stage from production to packaging and retail. The seed producer is responsible for the quality of the seed. To enable NGOs and private-sector seed entrepreneurs to comply with the minimum quality standards, the truthful label seed system does not involve the high cost and bureaucracy of seed inspection and testing for specified minimum standards. As a result, it is much easier and more economical for farmers and farming groups in the decentralized seed production system to meet the high demand for new seed in the domestic market. The quality declared seed system guarantees higher quality compared to the truthful label seed system and is suitable for commercially branded seed production by private companies and NGOs, but it is still less expensive and more straightforward during the certification process.

Issues and suggestions

According to the definition of a 'breeder' in the Seeds Act, any person or organization is able to develop and maintain new plant varieties. In this spirit, the Seeds Act also recognizes farmers as breeders and, therefore, considers them to be eligible to develop new plant varieties. However, it is almost impossible for ordinary farmers and farmers' groups to fulfil the conditions imposed for the production of breeder seeds. Farmers and/or farmers' groups, therefore, will not be able to register new plant varieties since it requires them to submit their breeder seeds to the variety registration authority. The conditionality attached with the production of breeder seeds directly conflicts with the definition of a breeder, and disqualifies farmers and individual breeders from developing new plant varieties.

These conditions further create an artificial barrier for NGOs and small seed entrepreneurs in engaging in the development and maintenance of new plant varieties since they can hardly afford to maintain the specified infrastructures. Such endeavours are also bound to be costly, making variety development a nonviable enterprise. These conditions, therefore, should be removed immediately.

The seeds of new plant varieties are required to be distributed to the farmers for on-farm testing before these varieties are approved, released and registered for further cultivation. However, the policy provision, as stated under subsection 3.1.3, restricts such distribution. Changes to this policy provision should be made to allow distribution of seeds of varieties that are under development and require on-farm adaptation testing as part of the variety development process.

The policy provision that dictates that the responsibility for production and maintenance of the nucleus and breeder seeds lies with the breeder is a supportive provision for the realization of the intellectual property rights protection for new plant varieties. Similarly, the policy emphasis on the conservation of agricultural biodiversity and the protection of rights over local crop varieties needs to be clearly set out in the appropriate provisions of the Seeds Act and the Seeds Regulation. At the moment, both of these legal instruments are silent on both of these matters.

Notes

1 Plant Variety Protection and Farmers' Rights Bill, Ministry of Agricultural Development, Nepal.
2 Access to and Benefit Sharing from the Use of Genetic Resources Bill, Ministry of Forest and Soil Conservation, Nepal.
3 Seeds Act, 1988 (first amendment, 2012), National Seed Board, Ministry of Forest and Soil Conservation, Nepal, online: <www.sqcc.gov.np> (last accessed 18 January 2016). Seeds Regulation, National Seed Board, Ministry of Forest and Soil Conservation, Nepal, online: <www.sqcc.gov.np> (last accessed 18 January 2016).
4 'List of notified crops and varieties until 28 July, 2014', Seed Quality Control Centre (SQCC), Ministry of Agriculture. Available at: <www.sqcc.gov.np> (last accessed 18 January 2016).

19 Commentary on plant variety regulation in the United States of America

Richard J. Blaustein

The system for plant variety regulation in the United States is significantly different than in other countries and regions, especially when compared to its high volume trading partner, the European Union (EU). Unlike in the EU, there is no mandatory registration for varieties as a prerequisite for commercialization in the United States. Also unlike in the EU, there are no explicit requirements that varieties must be 'distinct, uniform and stable'; or that they must explicitly embody improved values for cultivation and use before they can be marketed. However, there are U.S. legal standards for representing a plant as a variety that are comparable to the classic 'distinct, uniform and stable' criterion and that clearly infer improvements. Specifically, the Federal Seed Act (Sec. 101 (12)) specifies that 'The term "variety" means a subdivision of a kind which is characterized by growth, plant, fruit, seed or other characters by which it can be differentiated from other sorts of the same kind.'[1] Overall, a strong commercial ethos of facilitating both market access for new agricultural products and buyer choice accounts for this choice of introducing varieties without registration in the United States.

The Federal Seed Act is the law in the United States that most significantly bears on varieties, and it focuses on the honest representation and labelling of agricultural seeds for commerce.[2] While the Federal Seed Act comprehensively applies to plant seed varieties, it does not direct or mandate the registering of varieties. The US Department of Agriculture's Seed Regulatory and Testing Division, which administers the Federal Seed Act, maintains a varieties names list database, the Variety Name Database (www.ams.usda.gov/services/seed-testing/variety-name-list), which is a voluntary listing that helps variety developers give notice of their name selection and avoid choosing a duplicate name.

Other laws, norms and professional associations have great bearing on how varieties are developed, introduced and publicized in the United States. In terms of development in the United States, it is not the Federal Seed Act but, rather, the active application of intellectual property protections – such as plant patents, utility patents and Plant Variety Protection Act (PVPA) certificates – to agricultural innovations that currently has the greatest influence on the breeding and distribution of new agricultural varieties.[3] This vigorous intellectual property activity is quite different than varieties registration. However, those varieties given a PVPA certificate do have to be registered in the Variety Name Database, if those varieties are vegetable and agriculture species, although varieties such as potatoes, trees and flowers are not required to be registered in the list. With the absence of a mandatory varieties registration system, breeders in the United States are able to seek, and significantly utilize, the nongovernmental, voluntary registering of varieties, particularly with the *Journal of Plant Registrations*, to publish their innovations and learn about the work of other breeders.

Federal Seed Act

The Federal Seed Act is best characterized as what is commonly referred to as a 'truth in labelling' law. The Federal Seed Act specifies its purpose: '(T)o regulate interstate and foreign commerce in seeds; to require labeling and to prevent misrepresentation of seeds in interstate commerce: to require certain standards with respect to certain imported seeds; and for other purposes.'[4] First approved in 1939 and revised in each successive decade in the twentieth century, the Federal Seed Act highlights its direct application to varieties by authoritatively defining a variety as 'a subdivision of a kind of which is characterized by growth, plant, fruit, seed or other characters by which it can be differentiated from other sorts of the same kind, for example, Marquis wheat, Flat Dutch Cabbage, Manchu soybeans, Oxheart carrot, and so forth.'[5] The PVPA offers slightly more specification as to its definition for varieties, but this focus applies to the PVPA's focus on awarding intellectual property protections for breeders' varieties work.[6]

The Federal Seed Act is quite clear about its mandating labels for commerce in varieties. The act states:

> It shall be unlawful for any person to transport or deliver for transportation in interstate commerce . . . [a]ny agricultural seeds or any mixture of agricultural seeds for seeding purposes, unless each container bears a label giving the following information in accordance with rules and regulations prescribed under section 402 of this Act.[7]

Immediately following this foundational principle for labelling, the Federal Seed Act states that the labels must have '(1) The name of the kind or kind and variety for each agricultural seed component present in excess of 5 per centum of the whole and the percentage by weight of each.' Other requirements follow, and the implementing regulations of the Federal Seed Act reiterate and further explain the act's labelling requirements, including:

> percentage by weight of seed;[8] percentage by weight of weed[9] or other crop seed or inert matter;[10] kinds and rate of occurrence of noxious weeds seeds, specified to not exceed regulatory allowance[11] and other ingredients and characterizing requirements.

In short, the Federal Seed Act facilitates commerce by securing for buyers and sellers alike a context of honest representation. The United States' individual states are required to follow the Federal Seed Act and can have supplemental requirements for varieties. In the US commerce-oriented agricultural system, by mandating proper identification and enforcing against misrepresentation, the Federal Seed Act thereby promotes market dynamics and informed consumer choice for new varieties.

The US Department of Agriculture's Seed Regulatory and Testing Division enforces the Federal Seed Act, which includes investigating charges of misrepresentation and scientific testing for seed and instituting penalties and corrective measures. Moreover, the Seed Regulatory and Testing Division maintains the Variety Name Database, which facilitates variety-name seekers under the Federal Seed Act to avoid choosing a duplicate name and also to make public their claim to a variety name. No Federal Seed Act provision or regulation was responsible for the establishment of the varieties names list database. It is administratively maintained by the Seed Regulatory and Testing Division to facilitate its work with the Federal Seed Act. Listing a variety in this database under the Federal Seed Act is voluntary, although varieties under PVPA must be entered into this database.

A few of the stipulations that control variety naming under both the Federal Seed Act and its database include:

- a variety can only have one name;
- the same name cannot be given to more than one variety of the same kind or a closely related kind;
- closely related kinds that are known to intercross, such as wheat and triticale, cannot have varieties with the same name;
- a variety cannot be misleading such as a name that is similar to an existing name but differs only in spelling or punctuation.[12]

It appears that with respect to the Federal Seed Act individual farmers generally have not entered a variety name into the database. Importantly, the naming guidelines for varieties is not the same as the varieties registration system – numerous varieties are not entered into the database. The database, like the Federal Seed Act, primarily supports the American system of facilitating market entry for honestly represented varieties and agricultural products.

Voluntary registration

The absence of a system for variety registration in the United States does create some challenges for variety breeders that must somehow be met. For example, breeders in government and universities often have publishing pressures or requirements as part of their careers. Some mode of recognition that is identical or similar to registration is necessary for documenting the breeding work that they describe in their publications.

The American Society for Horticultural Science and the Crop Science Society of America offer two voluntary registration/listing variety systems in the United States, neither of which is referred to in US law. The American Society for Horticultural Science has published in the past voluntary lists of varieties from time to time in its journal *HortScience,* and these lists pertain to the different editorial focus of the publication. For example, in the May 2010 issue, *HortScience* published 'List 45: Register of New Fruit and Nut Cultivars', and in the April 2014 issue, *HortScience* published 'List 47: Register of New Fruit and Nut Cultivars'.

The Crop Science Society of America's *Journal of Plant Registrations* is the central agronomy-focused volunteer variety register in the United States. The *Journal of Plant Registrations* succeeded the Crop Science Society's central publication, *Crop Science*, in registering varieties. The *Journal of Plant Registrations* states that its goal is 'publish[ing] cultivar, germplasm, parental line, genetic stock, and mapping population registration manuscripts, keeping breeders informed about new advances in the genetic diversity of crops.'[13] In order to register with the *Journal of Plant Registrations*, the breeder must deposit a variety's germplasm with the National Plant Germplasm System, which is administered by the US Department of Agriculture, if that variety is a sexually reproduced crop. Nonsexually reproduced crops (e.g. clones and varieties such as sugarcane varieties) do not need to be deposited. As of 2012, the National Plant Germplasm System maintains a collection of almost 550,000 accessions (germplasm samples) of thousands of plant species. The National Plant Germplasm System also distributes species and germplasm, including patented or protected species, once their intellectual property deadline has expired.

Registering with the *Journal of Plant Registrations* does not require the meeting of the standards of distinction, uniformity and stability (DUS criteria), which is necessary in Europe. However, registration in the *Journal of Plant Registrations* is connected with manuscript submissions for publication, and these manuscripts are vetted to reviewers in the profession. The review system looks seriously and scientifically at the newness and strengths of the new variety. It appears that few farmers or hobbyists have registered with the *Journal of Plant Registrations*, but they are welcome to do so.

Policy rationale

Variety and seed regulation policy in the United States differs from systems in places such as the EU in its fundamental rationale. For example, the EU law depends on state-empowered technocrats to decide if a new variety has better qualities or other significant trait improvements and whether it should therefore be made available for farmers to buy. The US system, on the other hand, is structured to let the market decide what it wants. A variety with no clear superior quality or value can be brought to market in the United States. The lack of registration requirements for varieties reflects a different US approach to the government's interaction with the agricultural sector than in other countries – with a central operating principle of minimum noninterference at the marketing level. Thus, the American style of entrepreneurship and trial, error and choice at the point of marketing and distribution – instead of preintroduction approval – prevails in the large US agriculture sector.

Of course, this different approach to regulation does not signify that variety production in the United States is not significant or that the agriculture sector does not strive for quality improvements as one of its core norms. For example, one unofficial estimate of breeding efforts in 2013 estimates total breeding expenditures in the United States at approximately $2.5 billion and this is comprised of private, public and collaborative breeding efforts. In fact, in the United States, extensive and rigorous variety examinations are occurring in the university research centres, and these centres are financially supported and carefully followed by farmers and various agricultural corporations. Moreover, the government, especially the US Department of Agriculture, plays a huge role in the research and development of varieties, with a sophisticated laboratory and research and testing apparatus integrated within government branches and in collaboration with university and other research centres. In the United States, these investments correlate with a strong commercial rationale for honestly represented but minimally regulated new varieties.

Notes

1 Federal Seed Act, 9 August 1939, 7 U.S.C §§ 1551–1611, online: <www.ams.usda.gov/sites/default/files/media/Federal%20Seed%20Act.pdf> (last accessed 28 January 2016).
2 Ibid.
3 To this author's query, the USDA informally and unofficially offers an approximate cumulative estimate as of 2015 that total utility patents for cultivars has risen to 8,207 and plant variety protection certificates to 10,808. Plant Variety Protection Act, 1970, 7 U.S.C. §§ 2321–2582 [PVPA].
4 Federal Seed Act, supra note 1 at introductory paragraph.
5 Ibid. at s.101(12).
6 The PVPA defines variety as

> a plant grouping within a single botanical taxon of the lowest known rank, that, without regard to whether the conditions for plant variety protection are fully met, can be defined by the expression of the characteristics resulting from a given genotype or combination of genotypes, distinguished from any other plant grouping by the expression of at least one characteristic and considered as a unit with regard to the suitability of the plant grouping for being propagated unchanged. A variety may be represented by seed, transplants, plants, tubers, tissue culture plantlets, and other matter.

Plant Variety Protection Act Regulation, Section 41 on Definitions and Rules of Construction, online: <www.ams.usda.gov/sites/default/files/media/Plant%20Variety%20Protection%20Act.pdf> (last accessed 28 January 2016).

7 Federal Seed Act, supra note 1, at s. 201 and 201(a).
8 Ibid., s. 201(a)6.
9 Ibid., s. 201(a)4.
10 Ibid., s. 201(a) (6)(7).
11 Ibid., s. 201(a)10.
12 See Variety Naming Guidelines to Comply with the Federal Seed Act, online: <www.ams.usda.gov/rules-regulations/fsa/variety-naming> (last accessed 28 January 2016).
13 See <https://dl.sciencesocieties.org/publications/jpr> (last accessed 28 January 2016).

20 Commentary on the Brazilian seed law

Juliana Santilli

Registration waiver for local, traditional or Creole cultivars

Brazilian Law no. 10711 Regulating the National Seed and Seedling System, of 5 August 2003 (commonly known as the Seed Law) aims to 'ensure the identity and quality of materials for multiplication and reproduction of plants produced, sold and used in the national territory.'[1] Despite focusing primarily on the 'formal' system of seeds in the country, this law creates some legal space for varieties developed by local farmers and adapted to the local socioenvironmental conditions – the 'local, traditional and Creole' varieties.

According to the Seed Law, varieties must be registered in the National Cultivar Registry before they can be produced, improved or commercialized. In order to be included in this registry, varieties must be distinct, uniform and stable. In addition, their cultivation and use value must be demonstrated. However, the Seed Law sets forth that 'registration in the National Cultivar Registry of local, traditional or Creole cultivars used by family farmers, Agrarian Reform settlers or indigenous peoples is not mandatory,' since it is very difficult for these local, traditional and Creole varieties to fit the requirements of the National Cultivar Registry (with respect to distinction, homogeneity and stability).[2]

Local, traditional and Creole cultivars are

> varieties developed, adapted or produced by family farmers, agrarian reform settlers, or indigenous peoples with well-established phenotypical characteristics and recognized by local communities as such, and which, according to the Ministry of Agriculture, and considering also their socio-cultural and environmental descriptors (or traits) are not characterized as substantially similar to commercial cultivars.[3]

This definition means that local communities of family farmers, agrarian reform settlers or indigenous peoples must recognize their varieties as 'local, traditional or Creole' and that the Ministry of Agriculture must also consider that they are not substantially similar to commercial cultivars.

The Seed Law does not specify which criteria will be used to distinguish local, traditional and Creole varieties from commercial cultivars. Many family farmers feel that it should be up to the local communities (with the support and participation of official government agencies) to define the necessary criteria for identifying and characterizing varieties that have been developed, produced or adapted to local and specific socioenvironmental conditions as well as the criteria to set them apart from commercial cultivars. After all, the Seed Law requires that sociocultural and environmental descriptors (or traits) be taken into consideration, in addition to the agronomical and botanical descriptors, precisely in order to take into consideration, at the time of defining and characterizing the local varieties, the sociocultural and environmental contexts in which these varieties were developed or adapted by means of natural selection and farmer management.

The Ministry of Agriculture has not issued any regulatory instruments regarding local, traditional or Creole varieties, and the question remains unanswered about whose responsibility it should be to decide which varieties can be considered local, traditional or Creole for the purpose of exempting them from the various legal requirements of commercial varieties. On the other hand, the Ministry of Agrarian Development, which is responsible for the implementation of policies aimed at family farming, issued Directive 51 on 3 October 2007 in order to establish a national register for organizations working with local, traditional or Creole varieties (discussed in more detail later in this chapter). The objective of this register is to provide insurance coverage for farmers who have used traditional, local or Creole varieties (in case there are crop failures) and it does not establish any rules concerning the production, improvement or commercialization of these varieties.

Due to the diversity of agricultural systems in Brazil (which includes agrobusiness, family farming, indigenous and other traditional farming systems), Brazil has two ministries responsible for agricultural and agrarian development policies: the Ministry of Agriculture (Ministério da Agricultura), which is dedicated to policies aimed at supporting agrobusiness, and the Ministry of Agrarian Development (Ministério do Desenvolvimento Agrário), which is responsible for the implementation of policies aimed at strengthening family and small-scale farming. Although both ministries are part of the federal public administration, they frequently promote contradictory public policies since they represent the political and economic interests of very different stakeholders and systems of agricultural production.

Waiver of registration for family farmers, agrarian reform settlers and indigenous peoples multiplying/distributing seed (of local or registered varieties)

The Seed Law also requires all persons (physical and juridical – that is, individuals or companies) that produce, improve, package, store, analyze, trade, import and export varieties to be registered in the National Registry of Seeds and

Seedlings (Renasem). The technical requirements for this registry are extremely burdensome and costly. (There are two registries: one for farmers' varieties and the other one for persons who produce or commercialize the seeds from these varieties; both are legally mandatory.) However, the Seed Law makes an exemption for family farmers, agrarian reform settlers and indigenous peoples, who multiply seeds or seedlings for distribution, exchange or trade among themselves (i.e. with other family farmers, agrarian reform settlers and indigenous peoples). According to the Seed Law, 'family farmers, Agrarian Reform settlers and indigenous peoples who multiply seeds or seedlings for distribution, exchange or trade with each others are not required to register in the National Seed and Seedling Registry.'[4] This mandate means that as long as the distribution, exchange and trade of seed takes place among family farmers, agrarian reform settlers, and indigenous peoples, there is no need for registration. This special treatment applies not only to local, traditional or Creole cultivars but also to registered varieties, as long as they are traded and exchanged among family farmers, according to the Seed Law.

These legal exceptions in favour of family farmers and in regard to local and Creole varieties were inserted into the Seed Law in response to pressure from various social movements and the mobilization of civil society organizations. Farmers' organizations were successful in convincing many congressmen that the rules created to regulate the production, use and trade of commercial varieties should not apply to local/Creole varieties due to their very particular characteristics (dynamism and genetic heterogeneity). The idea of establishing a specific register for local, traditional or Creole varieties was rejected by most family farmers' organizations when the Seed Law was first being discussed at the National Congress in 2002 and 2003. Farmers' organizations felt that such a register could 'freeze' local seeds in time and space, since they are characterized by their evolution and are essentially 'dynamic' varieties. The register would only capture a certain moment or stage in their evolution. They were also afraid that the register could grant exclusive ownership rights (similar to breeders' rights) to those people who registered a local variety, which rightfully should be shared and exchanged by local communities through social networks and according to local rules.

Decree 5,153 of 2004, which established the operational rules for the Seed Law, restricted the operation of this exception as far as family farmers' organizations were concerned. It stated that family farmers' organizations could only distribute (not sell) seeds of local, traditional or Creole cultivars, and that this distribution of seeds could only take place among farmers who are members of these organizations. It meant that, when seeds are distributed through farmers' organizations (associations, cooperatives, unions and so on) and not through individual family farmers, the exemption (of registration) applies only for non-commercial purposes *and* only to local, traditional or Creole varieties. Family farmers, agrarian reform settlers and indigenous peoples can only exchange and trade seeds of registered varieties (among themselves) if they do it individually and not through their organizations. If they act individually, however, they (family farmers, agrarian reform settlers and indigenous peoples) can distribute, exchange and commercialize local or registered varieties among themselves without having to be registered in the National Registry of Seeds and Seedlings.

Many family farmer organizations have argued that this restriction (to the activities that can be developed by farmers' organizations) is illegal, since Decree 5,153 of 2004 is an administrative order and is creating restrictions that do not exist in the Seed Law. They also argue that such restrictions are violating the constitutional right to freedom of association. However, the Brazilian courts have not yet decided on this matter.

The Seed Law does not include a definition of family farmers, agrarian reform settlers or indigenous peoples. However, Law no. 11,326 of 2006, which establishes the National Family Farming Policy, considers a family farmer to be anyone who develops rural activities and meets all of the following criteria:

- does not hold, in any form, land that exceeds four fiscal modules (fiscal modules are established for each Brazilian region and municipality, and they can vary from 5 hectares in the northeast of Brazil to 100 hectares in the Brazilian Amazon);
- uses predominantly the labour of his own family in economic activities developed on his farm;
- has family income predominantly generated from economic activities connected with his farm;
- runs his farm with his own family. However, this definition does not apply to indigenous peoples because their traditional territories have a special legal status.

Some studies carried out in Brazil have demonstrated just how important these 'local' (or 'informal') seed systems are to the social, economic and cultural fabric of the country. According to the Brazilian Association of Seeds and Seedlings (Associação Brasileira de Sementes e Mudas [ABRASEM]), which includes Brazil's largest producers of seeds, Brazilian farmers in the 2006–7 harvest used seeds produced by the 'formal' system in the following proportions: 49 percent in cotton farming; 43 percent in rice; 15 percent in beans; 85 percent in corn; 50 percent in soy; 74 percent in sorghum; and 71 percent in wheat.[5] This data demonstrates that seeds produced in 'local' systems account for 51 percent in cotton farming; 57 percent in rice; 85 percent in beans; 15 percent in corn; 50 percent in soy; 26 percent in sorghum; and 29 percent in wheat. According to ABRASEM, in the 2007–8 harvest, the use of seeds produced by formal systems fell for nearly all crops (except for soy and sorghum: 44 percent for cotton; 40 percent for rice; 13 percent for beans; 83 percent for corn; 54 percent for soy; 88 percent for sorghum; and 66 percent for wheat.[6] In other words, local systems (which are highly dependent on local, traditional or Creole varieties) are responsible for the supply of seeds for the vast majority of Brazilian crops, and the use of seeds produced in the formal system is rapidly growing smaller.

Publicly funded development programs and crop failure insurance schemes

The Seed Law forbids any restrictions on the inclusion of seeds and seedlings of local, traditional or Creole cultivars in publicly funded programs for family farmers.[7] Article 48 of the Seed Law constitutes an important initiative in this regard, since the previous seed law (Law no. 6,570 of 1977) did not acknowledge local seeds, which were treated merely as 'grains,' and it made it difficult to get public support for initiatives aimed at the rescue, improvement and reintroduction of Creole, traditional or local seeds. Legal recognition has made possible government (financial) support for various projects and initiatives undertaken by nongovernmental organizations and farmers.

However, many farmers who used local, traditional or Creole seeds in the 2004–5 and 2005–6 harvests, by using rural credit in the National Program for Strengthening of Family Farming (Programa Nacional de Fortalecimento da Agricultura Familiar), lost their crops due to a severe drought in the central-south region. Their claims for insurance coverage were denied because of their choice to use local, traditional or Creole seeds. Currently in Brazil, farming insurance requires that seeds be included in the agricultural zoning system for climatic risk, which is produced by the Ministry of Agriculture, and only varieties registered in the National Cultivar Registry are eligible to be included in this zoning system. However, the Seed Law determines that 'registration in the National Cultivar Registry of local, traditional or Creole cultivars is not mandatory,' precisely because of the inability of local, traditional or Creole varieties to fit the requirements of the National Cultivar Registry.

In the 2004–5 harvest, the government authorized farm insurance companies to cover the losses (for that harvest only) for rural producers who used local, traditional or Creole varieties that were not included in the zoning system established by the Ministry of Agriculture. In the 2005–6 harvest, the National Monetary Council also authorized payment of farm insurance to farmers who used local, traditional or Creole seeds, extending the benefit to genetically modified soy crops in Rio Grande do Sul. In 2006, the Ministry of Agrarian Development put into effect a national registry of organizations that 'develop recognized work with rescue, management and/or conservation of local, traditional or Creole cultivars.'[8] Currently, this register is regulated by the Ministry of Agrarian Development under Directive 51 of 3 October 2007.

This register is called Cadastro Nacional de Entidades, which means National Register of Organizations (in other words, specific organizations who work with local, traditional or Creole varieties). In order to be registered, the organization must have been in legal existence for at least two years and it must have developed at least two activities that are aimed at rescuing, managing and/or conserving local, traditional or Creole cultivars. The registered organization must report which local, traditional or Creole varieties it has been working with, their basic characteristics and region of adaptation, as well as the technical experts that are responsible for this information. The first stage of the process is for the organization to become registered, and once its registration has been accepted by the Ministry of Agrarian Development it can register the local, traditional or Creole varieties that it has been working with under a different register, the National Register of Local, Traditional and Creole Varieties (Cadastro Nacional de Cultivares Tradicionais, Locais e Crioulas). Thus, the National Register of Organizations must list all organizations working with local, traditional or Creole varieties as well as all local, traditional or Creole varieties with which each organization works. The same variety can be registered by more than one organization because the registration does not give the organization any exclusive right (property or ownership) over the varieties that it has registered, and these varieties can be registered, used and shared by several farmer organizations. The main objective of the register is to provide insurance coverage for those family farmers who use local, traditional or Creole varieties and who will eventually face crop failures.

The register was established by the Ministry of Agrarian Development because farmers' organizations argued that the Seed Law was being violated by government policies. They argued that Article 48 of the Seed Law forbids any restrictions on the inclusion of local, traditional or Creole varieties in publicly funded programs for family farmers and that, if farmers who used local, traditional or Creole varieties were being denied insurance coverage, this provision of the Seed Law was being violated. Therefore, the Ministry of Agrarian Development (which is responsible for insurance policies for family farmers) decided to create a specific register for organizations working with local, traditional or Creole varieties so that they could indicate which varieties they were working with and get a certificate from the Ministry of Agrarian Development that would provide them with insurance coverage, even if the Creole varieties were not included in the zoning system established by the Ministry of Agriculture. That way, the Seed Law, which forbids any restrictions on the inclusion of local, traditional or Creole varieties in publicly funded programs for family farmers, would be more effectively enforced.

According to Directive 51 of 3 October 2007, local, traditional and Creole varieties must meet all of the following requirements in order to be registered:

1. they must be developed, adapted or produced by family farmers, agrarian reform settlers or traditional and indigenous populations and communities;
2. they must have phenotypical characteristics that are well established and recognized by the respective communities;
3. they must have been in use by farmers in one of these communities for more than three years;
4. they cannot be developed by means of genetic engineering or other industrial development processes or laboratory manipulation, be genetically modified or have evolved from hybridization processes that are not controlled by local family farmer communities.

Directive 51 further establishes that local, traditional or Creole cultivars are part of the sociocultural heritage of the local community and are not eligible for patents, ownership or any form of private protection for individuals, businesses or organizations. Furthermore, the directive sets forth that registration does not entitle the organization to any rights of property or ownership of the traditional, local or Creole varieties that the organization has registered in the National Register of Local, Traditional and Creole Varieties.

Many farmers' organizations initially criticized the registration of local, traditional and Creole varieties because they believed it would serve to freeze local seed evolution in time and space, thereby considering these seeds as if they were static and not dynamic. However, most farmers ended up accepting the registration of their local varieties for the sole purpose of accessing insurance coverage. Most family farmers who access public funds desperately need insurance to protect them during difficult times. When natural disasters strike (droughts, floods and so on), farmers need to be sure that they will be provided with insurance coverage even if they use local, traditional or Creole varieties of seeds. In response to the concerns expressed by many farmers' organizations, Directive 51 prevents local, traditional or Creole varieties from being privatized through intellectual property rights, thereby recognizing the collective (and nonexclusive) rights that local communities have over their varieties.

Government officials in Brazil also argue that the system of registration is also a way of producing more information/data on the use of local, traditional and Creole varieties in order to develop public policies aimed at strengthening local and traditional farming systems. The entire system of registration, however, is very recent, having just started in 2009; thus it is too early to say whether it will be effective or not. It has solved the specific problem of access to insurance by farmers who use local, traditional or Creole varieties. However, several issues remain unsolved. For instance, who decides which varieties are local, traditional or Creole and which are not for the purpose of being exempt from registration for produce, use and commercialization? What is the genetic distance necessary to separate a local/Creole variety from a commercial variety?

Most family, traditional and local, small-scale farmers feel that local, traditional and Creole seed systems (which are often called informal systems) should in fact remain out of the scope of the Seed Law, which should apply exclusively to commercial systems. They feel that locally adapted varieties that are used, distributed and traded at the local level, among family farmers, should simply be left out of the Seed Law. Those exceptions that are made for local, traditional and Creole seeds by the Brazilian Seed Law – despite representing an important victory for family and agroecological farming – tend to attenuate the negative effects that this law has on agrobiodiversity but do not alter the general principles and concepts that it is based upon, such as industrial sectorialization and the standardization of agriculture, the denial of the role of farmers as selectors and improvers and so on. These principles and concepts are, in essence, opposed to the environmental logic and sociocultural processes that generate and maintain agrobiodiversity at all of its levels.

Notes

1 Law no. 10711, of 5 August 2003, Regulating the Brazilian National Seed and Seedling System, <www.planalto.gov.br/ccivil_03/Leis/2003/L10.711.htm> (last accessed 10 May 2012).
2 Ibid., Article 11, para. 6.
3 Ibid., Article 2, para. 16.
4 Ibid., Article 8, para. 3.
5 Brazilian Association of Seeds and Seedlings, O mercado de sementes no Brasil, institutional lecture, 21 July 2008, Brasília, 2008, online: <www.abrasem.com.br> (last accessed 10 May 2012).
6 Brazilian Association of Seeds and Seedlings, *Semente: inovação tecnológica* (Anuário 2008), Brasília, June 2008.
7 Law no. 10711, supra note 1, Article 48.
8 Santilli, J. (2011). *Agrobiodiversity and the Law: Regulating Genetic Resources, Food Security and Cultural Diversity*, Earthscan, London, pp. 55–56.

21 Commentary on variety registration regulation in Italy

Alejandro Mejías, Enrico Bertacchini and Riccardo Bocci

Italy provides an excellent case study of how different variety registration requirements can evolve in a country where authority to develop legislation is divided between national and regional levels. In Italy, the regions have played a leading role in developing alternative registration systems linked to nationally established standards in response to conservation-related concerns. An additional factor which makes Italy an interesting case study is that, ultimately, all of the various regional registration schemes and the national law will need to be brought into line with the recent European Union Directives 62/2008, 145/2009 and 60/2010 which derogate EU variety registration standards for "conservation varieties" of seed and seed potatoes, vegetables and fodder plant seed mixtures, respectively. Adding still more complexity to the situation is the fact that in 2013, the Commission (formerly DGSANCO and now DGSANTE) submitted a proposal to European Parliament after five years of negotiation trying to update the overall framework of the European seed legislation. In March 2014, the European Parliament rejected the initiative, and it is still difficult to say if or when it will be reintroduced. Nonetheless, given its potential relevance to the subject matter of this chapter, we include consideration of the rejected proposal.

Italy's first seed regulation and the conditions of registration of varieties in the national register

The first law in Italy to regulate marketing of seed was Royal Decree of 1st July 1926, n. 1361 concerning seed fraud.[1] This decree established some conditions for seed marketing, for examples, that seeds must be labelled indicating a common name, minimum levels of purity (over 95 percent) and germinability (at least 85 percent). It also required the inclusion of information about the origin of the variety.

Later, a law passed in 1938 established a national catalogue of wheat varieties for voluntary certification.[2] This catalogue required three conditions for registration: constant botanical characteristics (*caratteri botanici costanti*), culturally and technically valuable seeds (*accertati pregi colturali e tecnologic*") and ascertained high productivity seeds (*elevata produttività dimostrata*). The National Institute on Improved Seed (Ente Nazionale Sementi Elette) was founded in 1954 for voluntary seed control and certification.[3]

In 1971, Italian Parliament passed national legislation, *Disciplina dell'attività sementiera* (of 25th November 1971 n. 1096). This law established a system for compulsory seed control and plant varieties registration. Only plant varieties that are included in a national list, administered by the Department of Economic Rural Development of the Ministry of Agriculture, can be sold on the open market in Italy. According to the law, to be included on the list, varieties have to be distinct in at least one trait, sufficiently homogenous and stable in the essential traits.

Regional variety registration as part of diversity conservation strategies

Besides having a national variety registration system which regulates the commercialization of seeds, Italy is particularly interesting for having a range of internal regional laws which include various forms of registration linked to conserving, protecting, enhancing and, in some cases, limited commercializing of local and autochthonous varieties (see Table 21.1 on page 352).

Underlying these initiatives is the awareness that there are only a few remaining local or old varieties being grown in Italy today (FAO, 1998). The remaining agricultural species and varieties within each region, and Italy as a whole, are at risk of genetic erosion, requiring measures to encourage conservation and provide incentives for sustainable use of autochthonous genetic resources.

In 1997, Tuscany was the first region to develop a regional law (Regional Law 50/97) on "protection of indigenous genetic resources"[4] (*tutela delle risorse genetiche autoctone*). The Tuscan experience created political and scientific interest in conservation of traditional varieties. Since then, seven regions have issued legislation[5] on this matter and in another four regions draft laws are still under political discussion.[6] Further, after the coming into force of the International Treaty on Plant Genetic Resources for Food and Agriculture (ITPGRFA), Tuscany substituted Regional Law 50/97 with Regional Law 64/2004, "The Protection and Enhancement of Local Breeds and Varieties of Interest to Agriculture, Husbandry and Forestry."

Despite the potential for a proliferation of very different approaches at regional levels, the regional initiatives are quite similar. Most of them include the following basic elements:

- Recognition of local communities as the stewards of the resources (e.g. Lazio and Umbria), or the Region itself (e.g. Tuscany, Emilia Romagna), on behalf of the local communities.
- Establishment of a voluntary, free-of-charge regional register for species, breeds, varieties, populations, cultivars, landraces and clones.
- Establishment of technical-scientific committees to evaluate the nomination proposals of varieties to be inscribed into the regional register.
- Establishment of an on-farm conservation network coordinated by the regional authority. These networks recognize and uphold the figure of the

"custodian farmer" as key actor in the *in situ* on-farm conservation process. Besides farmers, universities, agricultural organizations and research centres, among others, are eligible to join the network.
- Establishment of a regional gene bank for *ex situ* conservation of local and indigenous material.
- Intervention plans for conservation.

The establishment of a regional register is a key part of the Italian regions' strategies for agrobiodiversity conservation. It is worth noting that many of these regional initiatives are not limited to plant varieties, but also extend to animal breeds and forest[7] resources.

Conditions for registration of varieties in regional registries

The conditions for registration of traditional plant varieties tend be fairly similar across the regionally based registries.

Since the objective of these laws is the protection and enhancement of local varieties threatened by genetic erosion, the registers tend to refer to materials that have been in production systems for a long time. Many of the laws refer to indigenous or nonindigenous varieties that have been in production in the area for 50 years. What counts for acceptable documentation/proof of their indigenous nature or the fact that they have been used in an area is not generally defined, and the examination committees are left with considerable discretion in terms of what they can rely upon. Of course, one could argue that the time-length requirement is a kind of "built-in" stability requirement, but it is a potentially very loose requirement. Testimony from farmers – or a written record to the same effect – is deemed acceptable evidence that a variety with similar traits has been used in the area for 50 years or more.

None of the regional laws requires that the plant varieties be distinct, uniform and stable to be registered. Instead, most of the laws only require that the varieties should be identifiable or distinguishable through reference to phenotypic traits. In the case of Tuscany, the scientific commission has established mandatory and facultative characteristics that varieties must comply with in order to be registered.[8] Furthermore, Tuscany requires that applicants establish a connection between the traditional variety in question and the region of origin, to identify the origin and the territory where the variety is being cultivated, the potential and real production of the traditional variety and the resistance to adverse environment of the variety.

All of the regions include risk of genetic erosion as a core prerequisite. However, because of the technical complexity of the matter and lack of scientific consensus on how to define and measure genetic erosion, it is likely that practice is actually varied in terms of how the assessments are made. Definitions of genetic erosion are not usually present in the regional laws; instead, they are left to be handled by the technical-scientific commission involved.[9] At the regional level, Lazio and Emilia Romagna have provided a detailed proposal for defining the basic criteria for considering a variety at risk of genetic erosion. The implementing regulation of the law identifies minimum levels of cultivated land which vary according to the species, and contemplates not only the ecological and agricultural characteristics of the varieties but also, indirectly, natural factors and the production capacity of the farms on the territory. In many instances, the risk of erosion or of disappearance is mainly due to the scarcity of farmers cultivating the crop. The definition of risk, therefore, must also take into account this human factor, which is indirectly linked to the ecological and agricultural properties of the variety.

Rights conferred with respect to registered varieties

The types of rights conferred through the regional registration schemes generally reflect an appreciation of indigenous varieties and breeds as being part of the collective heritage of local communities who are maintaining them.

As noted by Bertacchini (2009), the idea of collective heritage is embedded in the regional laws both in the way some of them refer back to Article 8(j) of the Convention on Biological Diversity,[10] or in the way they state that the region as a collective unit is responsible for the indigenous varieties concerned. The regional laws do not contemplate the institution of any form of individual exclusive rights over registered varieties. The individual or legal person who proposes a variety be registered enjoys no exclusive right over the variety, just as no third party may make a claim for intellectual property rights to it. Rather, inscription in the register and access to the resource accrues first and foremost collective benefits for the community as a whole in terms of conservation and enhancement of the heritage of autochthonous genetic resources. Furthermore, some regional laws (Tuscany and Emilia Romagna) address this point more directly regulating the use of autochthonous genetic resources to create new varieties. Members of the conservation network who intend to apply for a plant breeder right, or a patent on a variety essentially derived from one registered, must request prior consensus to the region or to the responsible agency.

The establishment of on-farm conservation networks, comprised of steward farmers (*coltivatori custody* in Italian), is an innovative approach to dealing with the right to save, replant and exchange seeds for traditional and indigenous varieties which do not fit the requirements of the national registration system.

All of the regional laws allow the members of the network to make the material registered available for conservation or improvement purposes. The laws normally allow farmers within the network to save and to exchange locally, on a nonprofit basis, a small quantity of seed (*modica quantità* in Italian) of registered varieties. Some regional laws allow farmers to sell small quantities[11] of seeds. Bertacchini (2009) notes that these norms, even to a limited extent, reflect the appreciation of the importance of farmers' practices, which in the past have brought about varietal innovation and the continual adaptation of varieties to the territory. The right to save, replant and exchange seeds can be especially important in coping with the risk of extinction of local varieties by putting them to use in agriculture. These activities are also ways of safeguarding and enhancing the cultural heritage and traditional knowledge which are tied in with indigenous varieties.

Status of implementation

This subsection section considers the state of implementation of the regional laws and related registration schemes. Their implementation strongly depends on the availability of financial and technical resources, as well as on the political commitment of the regional authorities. As a result, there are differences in their state of implementation, from region-to-region.

The most developed regulations are in place in Tuscany, Lazio, Emilia Romagna and Marche where registers are already working (there is an annual plan of activities and a budget for each region). For instance, in 2013, Tuscany's register showed 463 arboreal and fruit tree species and 68 herbaceous species, of which 401 and 61 were at risk of genetic erosion, respectively. For the three regions this data is accessible online.[12] Tuscany has also begun to select and register steward farmers as part of the process of supporting the development of networks for conservation and security. The Friuli Venezia Giula legislation is partially operational, with the *ex situ* conservation-related section of the law being implemented; the *in situ*–related sections are not yet operational. Umbria is in the process of conducting a census prior to the establishment of the registry in the context of the Umbrian Rural Development Plan.

Lack of public funds or changes in the institutional contexts may also threaten the implementation of these laws. For example, in 2011, Tuscany formally abolished ARSIA, the public regional body responsible for the implementation of the law. For this reason, there is institutional uncertainty about whether the new responsible body will express the same commitment in continuing the implementation of the regional law provisions.

The EU directives on conservation varieties and their implication for the Italian registration system

In the late 1990s, the European Union became increasingly concerned about the erosion of genetic diversity related to agriculture. Among the many alleged causes for diversity erosion were the strict requirements of EU seed regulation, on the basis that it hindered the conservation through use of landraces and their further evolution/adaption to local and regional agroecological conditions. In December 1997, the EU passed Directive 98/95/CE which introduced a new type of variety called a "conservation variety" that could be marketed within Europe and introduced in the common catalogue. However, directive 98/95/CE could not be implemented as it was. It was still necessary, by way of a subsequent directive, to decide a number of issues, including the conditions of registration of such conservation varieties and the extent to which they could be commercialized.

It was not an easy task to address these outstanding issues. It took 10 years and 14 different draft texts being considered by the Standing Committee on Seeds and Propagating Material for Agriculture before the EU passed Directive 62/2008,[13] which resolved the outstanding issues, attempting to strike a balance between protection of the current seed market and the conservation of agrobiodiversity. One of the most controversial issues was the derogation of the distinctness, uniformity and stability (DUS) requirements for "conservation varieties." The directive establishes that "Member States may adopt their own provisions as regards DUS of conservation," and at the same time, it marks limits to state action (see Table 21.1).

During these 10 years of legal uncertainty at the European level, the Italian legislation was modified to incorporate the new EU provisions concerning derogations to seed regulations registration of conservation varieties. After years of lobbying and pressing the government, an association called Rete Semi Rurali (www.semirurali.net), with the help from a green party parliamentarian, succeeded on convincing the national authorities on passing a decree related to conservation varieties. In 2001, legislative decree n. 212[14] introduced a section dedicated to conservation varieties in the Italian national register. Nevertheless, Italy had to wait another 6 years, until law 46/2007 was passed by the Italian Parliament, to have a more elaborate legal provision concerning the implementation of conservation varieties. That national law, passed by the Parliament to address several miscellaneous international obligations, addressed in one article[15] the creation for a national register of conservation varieties with the intentional of implementing aspects of the International Treaty on Plant Genetic Resources for Food and Agriculture (ITPGRFA) – in particular article 5 (regarding conservation, research, characterization and documentation of phytogenetic resources for food and agriculture), article 6 (regarding sustainable use of phytogenetic resources) and article 9 (regarding farmers' rights). Application Decree of 18th April 2008 followed the law, providing details about the conservation varieties system, drawing both from the texts of the European directive being discussed at that time and from the regional laws' experience. For example, for conservation varieties, the law completely derogated the DUS requirements, without any further reference to minimum conditions to be met for registration of conservation varieties.

A more detailed legal framework has been set out in the form of Legislative Decree D.Lgs. 149/2009, which was enacted to give application *to* the EU directive 62/2008. D.Lgs 149/2009 provides for the acceptance of conservation varieties in the national catalogues of varieties, and the production and marketing of seed and seed potatoes of those varieties. In order to qualify, a variety must be of interest for the conservation of plant genetic resources. Further, such variety cannot be subject to plant variety protection rights; they must also have been deleted from the common catalogue for over two years (if they were ever registered). As a result, a variety is considered eligible for registration as a conservation variety if it is a landrace, an old commercial variety or an old modified commercial variety of interest for the conservation of plant

Table 21.1 Summaries of content of relevant European region, national and Italian regional legislation

European Legislation		
Legislation	*Denomination and requirements for registering a new variety in the European regulation*	*Rights conferred under the legislation (commercialization, nonprofit exchange of material . . .)*
• Council Directive 98/95/EC of 14th December 1998	*Denomination:* Conservation variety. Conditions under which seed may be marketed and therefore included in the common catalogue in relation to conservation *in situ* and the sustainable use of PGR are associated with specific natural and semi-natural habitats and threatened by genetic erosion. Specific conditions shall be established taking into account: • Landraces and varieties which have been traditionally grown in particular regions and threatened by genetic erosion. • The results of unofficial tests and knowledge gained from practical experience during cultivation and the detailed description of varieties shall be taken into account and shall result in exemption from official examination. • Quantitative restrictions shall be established.	Right to market locally the seed. Right to include the variety in the common catalogue.
• Council Directive 2008/62/EC of 20th June 2008	*Denomination:* Conservation variety. *Definition:* Agricultural landraces and varieties which are naturally adapted to the local and regional conditions and threatened by genetic erosion. The varieties must comply with the following requirements:	Right to market locally the seed. Right to include the variety in the common catalogue.

DUS requirements:

Member States may adopt their own provisions as regards DUS requirements fulfilling at least the following requisites:

- The technical questionnaires of Community Plant Variety Office of the EU or the technical questionnaires of UPOV for stability and distinctness.
- For the assessment of uniformity, the directive 2003/90/EC which establishes the minimum conditions for certain agricultural plant species shall apply.

Genetic erosion:

Loss of genetic diversity between and within populations or varieties of the same species over time, or reduction of the genetic basis of a species due to human intervention or environmental change.

No official examination:

No official examination shall be required if the information is sufficient for the decision on the acceptance of the conservation varieties.

Identification of the region of origin:

The Member States shall ensure that the conservation variety must be maintained in that mentioned region.

Certification:

There is a derogation of the certification requirements in respect of minimum of varietal purity and the requirement of official examination if the seed descends from seed produced according to well-defined practices for maintenance of the variety and has sufficient varietal purity.

Marketing conditions:

Seed produced and marketed in its region of origin.

Quantitative restrictions:

Quantity does not exceed 0.5% of the seed of the same species used in that Member State in one season or a quantity necessary to sow 100 hectares, whichever is greater.

(*Continued*)

Table 21.1 (Continued)

Italian National Legislation		
Legislation	*Denomination and requirements for registering a new variety in the national regulation*	*Rights conferred under the legislation (commercialization, nonprofit exchange of material . . .)*
• Law 46/2007 of 6th April "Conversione in legge, con modificazioni, del decreto-legge 15 febbraio 2007, n. 10, recante disposizioni volte a dare attuazione ad obblighi comunitari ed internazionali" • Decree of 18th April 2008 "Disposizioni applicative per la commercializzazione di sementi di varieta' da conservazione"	*Denomination:* Conservation variety. *Definition:* • Autochthonous and nonautochthonous plant varieties that have been integrated into the local agroecosystem for at least 50 years, but which do not appear in any other national register. • Plant varieties that are no longer registered and at risk of genetic erosion. • Plant varieties with some economic, scientific, cultural or landscape interest which are no longer cultivated in Italy, but conserved in national or international botanic gardens, research institutes and gene banks. *Nonofficial testing.* DUS requirements derogated. Genetic modified varieties are excluded. Common/local denomination and synonyms of the variety. Morphological characterization and genetic if it is available. Local area of growth. Historical and cultural evidences and documentation which show the bond between the variety and the area of growth. *Quantitative restrictions:* • The total quantity that each farmer can transfer as seed is the amount necessary to establish a crop of 1,000 square metres for vegetables and 1 hectare for the other agricultural species. • Labelling and identification as a conservation variety.	The right of direct sale is recognized and circumscribed to the local area of production.

Regional Legislation

Region/Regional agency managing the catalogue	*Legislation*	*Denomination and requirements for registering a new variety in the regional regulation*	*Rights conferred under the legislation (commercialization, nonprofit exchange of material . . .)*
Tuscany/Agenzia regionale per lo sviluppo e l'innovazione nel settore agricolo-forestale (ARSIA)	• Regional Law 16th November 2004 n.64 "Tutela e valorizzazione del patrimonio di razze e varieta locali di interesse agrario, zootecnico e forestale." • Regional Decree of 1st March 2007 di attuazione della legge regionale "Tutela e valorizzazione del patrimonio di razze e varieta locali di interesse agrario, zootecnico e forestale."	*Denomination:* Local varieties and races. *Definition:*[17] Species, races varieties, cultivars, clones and populations. • Originated in the Tuscan territory. • Originated outside the region but they have been introduced and integrated in the traditional agriculture of the territory. Species, races, varieties, cultivar, populations and ecotypes: • Derived from phenotypic selection. • Originated in the Tuscan territory but no longer grew and conserved *ex situ* in Tuscany or in other regions or countries. Genetic erosion risk. Quantitative restrictions. Technical link. Historical bond. Morphological and if possible genetic characterization according to international standards.[18] The Scientific Commission must approve the inscription.	The regional regulation allows the nonprofit exchange of modest quantities (established by the ARSIA) of plant genetic materials for recovering, maintaining and reproducing local varieties at risk of genetic erosion to whose are members of the conservation network. In addition, it allows the commercialization of the varieties inscribed in the section of conservation varieties of the register. The additional requirement is the existence of an economic interest in the variety.

(*Continued*)

Table 21.1 (Continued)

Regional Legislation			
Region/Regional agency managing the catalogue	*Legislation*	*Denomination and requirements for registering a new variety in the regional regulation*	*Rights conferred under the legislation (commercialization, nonprofit exchange of material . . .)*
Lazio/Agenzia regionale per lo sviluppo e l'innovazione in agricoltura del Lazio (ARSIAL)	• Regional Law 1st March 2000 n.15 "Tutela delle risorse genetiche autoctone di interesse agrario."	*Denomination:* Autochthonous genetic resources of agricultural interest at genetic erosion risk. *Definition:*[19] Species, races, varieties, ecotypes, clones, spontaneous varieties related to the cultivated specie and populations: • Autochthonous and economically, scientific, environmentally and culturally interesting for the region and at genetic erosion risk. Species, races, varieties and cultivar: • Nonautochthonous but introduced at least 50 years ago. Species, races and varieties: • Autochthonous but no longer grown and conserved in botanic gardens, research institutes and gene banks whether in the country or outside and for which there is a reintroduction interest. Historical, scientific and technical documentation. The variety must be identifiable by a minimum number of traits defined for each variety. The Scientific Commission must approve the inscription. Genetic erosion risk.	The steward farmer member of the conservation network has the right to sell a modest quantity of material (established at the moment of inscription of the local variety) in the local area. The farmer member of the conservation network has the right to replant.

Umbria/not identified yet	• Regional Law 4th September 2001 n.25 "Tutela delle risorse genetiche autoctone di interesse agrario."	*Denomination:* Autochthonous genetic resources of agricultural interest at genetic erosion risk. *Definition:* Species, races varieties, ecotypes, spontaneous varieties related to the cultivated specie, cultivar, clones and populations • Autochthonous and economically, scientific, environmentally and culturally interesting for the region and at genetic erosion risk. • Nonautochthonous but introduced at least 50 years ago and which evolved specific characteristics. Species, races and varieties: • Autochthonous but no longer grew and conserved in botanic gardens, research institutes and gene banks whether in the country or outside and for which there is a reintroduction interest. Genetic erosion risk. The variety must be identifiable by a number of traits defined per every single variety. The Scientific Commission must approve the inscription.	The steward farmer member of the conservation network has the right to sell or exchange a modest quantity of material (established at the moment of inscription of the local variety) in the local area (province).
Friuli Venezia Giulia/ Ente regionale per la promozione e lo sviluppo dell'agricoltura (ERSA)	• Regional Law 22nd April 2002 n.11 "Tutela delle risorse genetiche autoctone di interesse agrario e forestale."	*Denomination:* Autochthonous genetic resources of agricultural interest at genetic erosion risk. *Definition:* Species, races, varieties, cultivars, ecotypes, clones, populations and spontaneous related varieties: • Autochthonous and economically, scientific, environmentally and culturally interesting for the region and at genetic erosion risk. Species, races varieties and cultivars: • Introduced at least 50 years ago and which evolved specific characteristics.	The steward farmer member of the conservation network has the right to sell a modest quantity of material (established at the moment of inscription of the local variety).

(*Continued*)

Table 21.1 (Continued)

Region/Regional agency managing the catalogue	*Legislation*	*Denomination and requirements for registering a new variety in the regional regulation*	*Rights conferred under the legislation (commercialization, nonprofit exchange of material . . .)*
	• Regional Decree 19th July 2004 n. 240 "Regolamento per la tenuta del Registro volontario regionale e per l'iscrizione in esso delle risorse genetiche autoctone della Regione Friuli Venezia Giulia."	Species, races and varieties: • Autochthonous but no longer grew conserved in botanic gardens, research institutes and gene banks whether in the country or outside the country and for which there is a reintroduction interest. Genetic erosion risk defined by decree as permanent reduction in terms of number, uniformity and identification of important morphological traits in the local area. Historical, scientific and technical documentation is required. The variety must be identifiable by a number of traits defined per every single variety. The Scientific Commission must approve the inscription.	The steward farmer member of the conservation network has the right to replant and multiply the material on-farm.
Marche/L'Agenzia per i servizi nel settore agroalimentare delle Marche (ASSAM)	• Regional Law 3rd June 2003 n.12 "Tutela delle risorse genetiche animali e vegetali del territorio marchigiano". • Regional Decree 28th October 2004 n. 116	*Denomination:* Conservation variety. *Definition:* Races varieties, ecotypes, clones, populations, cultivars and spontaneous related varieties: • Autochthonous races, varieties, ecotypes, clones, populations and cultivars. • Nonautochthonous introduced at least 50 years ago.	The commercialization and exchange of plant genetic resources under the conditions of the national and European legislation.

		• Autochthonous but no longer cultivated and conserved in botanic gardens, research institutes and gene banks whether in the country or outside the country and for which there is a reintroduction interest. • Nonautochthonous developed varieties for recovering the genetic diversity of cultivated species of the region which are at risk of disappearing. Genetic erosion risk. Agronomic and morphological distinctions. The Scientific Commission must approve the inscription.	The steward farmer member of the conservation network can reproduce the material under the directives of the ASSAM. The steward farmer member of the conservation network can only grow one variety of each specie.
Campania/Settore Sperimentazione, Informazione Ricerca e Consulenza in Agricoltura (SESIRCA)	• Regional Law 19th January 2007 n.1 "Disposizioni per la formazione del bilancio annuale e pluriennale della regione Campania." • Regional Decree 15th September 2008 n.33 "Salvaguardia delle risorse genetiche agrarie a rischio di estinzione."	*Denomination:* Autochthonous genetic resources of agricultural interest at genetic erosion risk. *Definition:* Species, races, varieties, ecotypes, clones, populations. • Autochthonous. • Autochthonous but no longer cultivated and conserved in botanic gardens, research institutes and gene banks whether in the country or outside the country and for which there is a reintroduction interest. • Nonautochthonous but introduced at least 50 years ago and which evolved specific characteristics. • Derived from phenotypic selection. Historical, scientific and technical documentation is required. Genetic erosion risk. The Scientific Commission must approve the inscription.	The farmer is allowed to regionally exchange plant genetic resources in modest quantities in the local area. Commercialization is allowed under the terms of the European legislation.

(*Continued*)

Table 21.1 (Continued)

Regional Legislation			
Region/Regional agency managing the catalogue	*Legislation*	*Denomination and requirements for registering a new variety in the regional regulation.*	*Rights conferred under the legislation (commercialization, nonprofit exchange of material . . .)*
Emilia-Romagna/ Agricultural General Director Office of the Region	• Regional Law 29th January 2008 n.1 "Tutela del patrimonio di razze e varieta' locali di interesse agrario del territorio Emiliano-Romagno."	*Denomination:* Indigenous genetic resources of agricultural interest. *Definition:* Species, races, varieties, ecotypes, clones, populations. • Autochthonous. • Autochthonous but no longer cultivated and conserved in botanic gardens, research institutes and gene banks whether in the country or outside the country and for which there is a reintroduction interest. • Nonautochthonous introduced and traditionally integrated long time ago. Historical, scientific and technical documentation is required. The Scientific Commission must approve the inscription. Genetic erosion risk.	The members of the conservation network are allowed to regionally exchange plant genetic resources in modest quantities in the local area.
Basilicata	• Regional Law 14th October 2008, n.26 "Tutela delle risorse genetiche autoctone vegetali ed animali di interesse agrario."	*Denomination:* Autochthonous genetic resources at genetic erosion risk. *Definition:* Species, races, varieties, biotypes, ecotypes, clones, populations: • Autochthonous. • Nonautochthonous introduced at least 50 years ago and which evolved specific characteristics. • Autochthonous but no longer cultivated and conserved in botanic gardens, research institutes and gene banks whether in the country or outside the country and for which there is a reintroduction interest. Historical, scientific and technical documentation is required. The variety must be identifiable by a number of traits defined per every single variety. Genetic erosion risk. The Scientific Commission must approve the inscription	The members of the conservation network are requested to make available their registered material for the rest.

Table 21.2 Online access to legislation

European Legislation	
Directive	*Link*
Council Directive 98/95/EC of 14th December 1998.	http://faolex.fao.org/docs/texts/eur19558.doc
Council Directive 2008/62/EC of 20th June 2008.	http://eur-lex.europa.eu/LexUriServ/LexUriServ.do?uri=OJ:L:2008:162:0013:0019:EN:PDF

National Legislation	
Law/Decree	*Link*
Law 46/2007 of 6th April "Conversione in legge, con modificazioni, del decreto-legge 15 febbraio 2007, n. 10, recante disposizioni volte a dare attuazione ad obblighi comunitari ed internazionali."	www.parlamento.it/parlam/leggi/07046l.htm
Decree of 18th April 2008 "Disposizioni applicative per la commercializzazione di sementi di varieta' da conservazione."	www.ense.it/leggiEdisposizioni/decreto18aprile2008.pdf

Regional Legislation		
Region	*Law/Decree*	*Link*
Emilia Romagna	L.R. n. 1/2008, "Tutela del patrimonio di razze e varieta locali di interesse agrario del territorio emiliano-romagnolo."	www.ermesagricoltura.it/wcm/ermesagricoltura/news/2008/01/23_ASS_biodiversita/LR_BIODIVERSITA_2008.pdf
Fruli Venezia	L.R. n.11/2002, "Tutela delle risorse genetiche autoctone di interesse agrario e forestale."	www.consiglio.regione.fvg.it/iterdocs/Serv-LC/ITER_LEGGI/LEGISLATURA_VIII/LEGGI_APPROVATE/205_LR.pdf

(*Continued*)

Table 21.2 (Continued)

Regional Legislation

Region	*Law/Decree*	*Link*
	Regolamento per la tenuta del Registro volontario regionale e per l'iscrizione in esso delle risorse genetiche autoctone della Regione Friuli Venezia Giulia, in applicazione della legge regionale 22 aprile 2002, n. 11.	http://lexview-int.regione.fvg.it/FontiNormative/Regolamenti/D_P_REG_0240–2004.pdf
Lazio	L.R. 15/2000, "Tutela delle risorse genetiche autoctone di interesse agrario."	www.regione.lazio.it/binary/agriweb/agriweb_normativa/L.R._01_03_2000_n_15.1197029105.pdf
Marche	L.R. 12/2003, "Tutela delle risorse genetiche animali e vegetali del territorio marchigiano."	www.sementi.it/normative/regionali/marche/BUR_marche%20n.51%20del%2012.6.03%20-%20risorse%20genetiche.pdf
	Regolamento regionale 28 ottobre 2004, n. 10, Attuazione della L.R. 3 giugno 2003, n. 12: "Tutela delle risorse genetiche animali e vegetali del territorio marchigiano."	www.agri.marche.it/Aree%20tematiche/biodiversit%C3%A0/reg-reg-10_04.pdf
Tuscany	L.R. n 64/2004, "Tutela e valorizzazione del patrimonio di razze e varieta locali di interesse agrario, zootecnico e forestale."	http://germoplasma.arsia.toscana.it/Germo_old/PN_Germo/Download/DPGR%2012R%20attuazione%20L.R.%2064.2004.pdf
	Regolamento di attuazione della legge regionale 16 novembre 2004, n. 64.	http://germoplasma.arsia.toscana.it/Germo_old/PN_Germo/Download/DPGR%2012R%20attuazione%20L.R.%2064.2004.pdf
Umbria	L.R. n. 25/2001, "Tutela delle risorse genetiche di interesse agrario."	www.sementi.it/normative/regionali/Umbria/LR%204–9–2001,%20n.25%20-%20risorse%20genetiche%20autoctone.pdf

Basilicata	L.R. n.26/2008, "Tutela delle risorse genetiche autoctone vegetali ed animali di interesse agrario."	www.regione.basilicata.it/dipagricoltura/default.cfm?fuseaction=linkdoc &doc=4046&link=4047
Campania	L.R. n. 1/2007, "Disposizioni per la formazione del bilancio annuale e pluriennale della regione Campania, legge finanziaria regionale 2007."	www.sito.regione.campania.it/burc/pdf07/burc07or_07/lr01_07/lr01_07.pdf
	Regolamento di attuazione dell'art. 33 della LR 19.01.07, n. 1.	www.consiglio.regione.campania.it/cms/CM_PORTALE_CRC/servlet/Docs?dir=atti&file=ProvLicenziati_8Com_7014.pdf

genetic resources. D.Lgs 149/2009 reintroduced DUS requirements, but eased the standards for uniformity, allowing up to 10 percent off-types. Otherwise the standards are those set by the Community Plant Variety Office (CPVO) or UPOV testing guidelines. The official tests are not mandatory and can be replaced by the results of unofficial tests, knowledge gained from the cultivation of these varieties, and information from authorities or organizations carrying out conservation work.

As for procedural requirements, the ministerial decree implementing the law[16] establishes that the registration process is free of charge.

One of the key characteristics of the conservation varieties system devised by the law is that seed production, selection and marketing must be limited to the region of origin as specified in the application. On the other hand, there is no geographical limitation for the cultivation of the conservation varieties.

Further, there is a derogation of the certification requirements in respect of minimum varietal purity and the requirement of official examination if the seed descends from seed produced according to well-defined practices for maintenance of the variety and has sufficient varietal purity.

Finally, the quantity of seed commercialized per conservation variety shall not exceed 0.5 percent of the seed of the same species used in the country in one season or a quantity necessary to sow 100 hectares, whichever is greater.

In June 2009, the first conservation variety (maize called Nostrano di Storo) was approved by the ministry (decree 11th June 2009) after a request from the Autonomous Province of Trento. By the end of 2011, 15 more conservation varieties were been registered. All these latter varieties have Piemonte as their centre of origin and production. Interestingly, so far, all the conservation varieties registered come from regions which have not enacted regional laws for the protection and enhancement of traditional and indigenous varieties. This result comes mainly from the fact that, even without a regional law, Piemonte and Trentino Alto Adige invested their regional agricultural departments' resources in mapping local biodiversity of interest for agriculture.

Outstanding issues regarding the compatibility of regional laws and the national law, including the derogation for conservation varieties

The national, regional and European norms concerning varieties registration systems have all developed in fits and starts and somewhat out of sync with each other. Therefore it is a challenge to see how all legal provisions will be fully and harmoniously implemented and in particular how the registration systems devised by the different laws will be made complementary.

Of particular interest is to understand how the system devised by regional laws will interact with the system devised by the EU directive on Conservation Varieties and applied in Italy with D.Lgs 149/2009.

On one hand, the aims of the two systems seem to be quite different. On the other hand, they are complementary with regards to the mechanisms devised to sustain the conservation of local and traditional varieties treated by genetic erosion. For instance, many of the regional laws do not establish a right for commercialization of traditional varieties, but instead have set up systems of both *in situ* and *ex situ* conservation at the regional level, and exchanges of seeds as part of those systems. The creation of on-farm conservation networks allows for a free exchange of seeds among registered custodian farmers for traditional varieties that are at risk of genetic erosion. Conversely, the EU Directive on Conservation Varieties, by allowing limited forms of commercialization, is meant to create an economic incentive to address the risk of genetic erosion and loss of conservation varieties in particular.

Further, even if the text of some of the regional laws (i.e. within the country, not the European region) anticipate the commercialization of seeds of traditional and autochthonous varieties, the national law implementing the conservation varieties system will supersede these provisions.

The experiences gained through the implementation of Italian regional laws, identifying and cataloguing traditional varieties and breeds, should provide a useful basis for registering conservation varieties and the legal commercialization of their seeds.

In the current legal framework it is unclear who is actually entitled to commercialize seeds of conservation varieties. For instance, article 19-bis of the Law 1096/1971 (concerning seed control and plant varieties registration) incorporates the derogations for the commercialization of seeds of conservation varieties and states that farmers in the region of origin of the conservation varieties are entitled to sell a small quantity (*modica quantità*) of seeds and propagating material, subject to the application of a ministerial decree. As of June 2013, no such ministerial decree has been drafted and as a result, there is an impasse impeding small farmers' ability to take advantage of the right to sell seed of conservation varieties. In fact, in the current regulatory framework for seed commercialization, only authorized seed producers and seed companies may sell seeds. In many cases, small farmers are not able to comply with the requirements for becoming authorized seed producers. For this reason, unless the forthcoming ministerial decree derogates on the operational requirements for the commercialization of seeds of conservation varieties, the objective of the EU directive in Italy will not be completely achieved.

A few observations regarding the recently rejected EU regulation "on the production and making available on the market of plant reproductive material (plant reproductive material law)"

Being part of the European Union, the Italian scenario for registration requirements and derogations is still likely to change in the near future due to the process of revision of seed marketing directives at the EU level.

In May 2013, the EU parliament submitted to the EU Council a proposal for new EU regulation on plant reproductive material with the aim of consolidating and updating the EU level rules concerning the marketing of plant reproductive material. In March 2014, it was rejected by the EU Parliament, and it is hard to know when or if it will be reintroduced, and in what form.[20] Given that the proposed legislation addressed a number of issues we consider in this chapter, we review its main elements, despite the fact that it was rejected. The proposed legislation appeared to introduce more flexibility in variety registration requirements in four ways. First, the proposed legislation explicitly did not apply to plant reproductive material that was maintained and exchanged in networks of *ex situ* and *in situ* or on farm conservation of genetic resources following national strategies on conservation of genetic resources. Plant reproductive material exchanged in kind between two persons – other than professional operators – was also excluded from the scope of the regulation.

Second, it defined a new category of "nice market reproductive material," that is, material which is made available on the market only in limited quantities by small professional operators. Such material would have been exempted from the requirement of belonging to a registered variety, with the idea that it could be freely commercialized following rules on labelling and packaging defined by the EU Commission.

Third, the proposed legislation provided for the recognition and registration of heterogeneous material (i.e. populations), which did not fulfil the DUS requirements, by empowering the EU Commission to adopt acts for its production and marketability.

Fourth, pursuant to the proposed regulation, the current legal framework for conservation varieties would have been revised to have more relaxed requirements. Old traditional varieties would have continued to be registered on the basis of an officially recognized description without an obligatory DUS examination. The officially recognized description would only need to describe the specific characteristics of the plants and parts of plants which are representative for the variety concerned and make the variety identifiable, including the region of origin.

The quantitative restrictions present in the previous EU Directive would have been abolished. The production of commercialized seed would be limited to regions of origin, but the reproduced material could have been marketed without geographical limitations.

These changes to the existing norms would have represented a step forward in creating a more flexible registration regime, taking into account different interests and needs. It is unfortunate that it was rejected. Regarding the possibilities of marketing seed of heterogeneous materials, the openings contained in the text of the Commission have been included in the Commission Implementing Decision of 18 March 2014, that allows a temporary experiment providing for certain derogations for the marketing of populations of the plant species wheat, barley, oats and maize pursuant to Council Directive 66/402/EEC (http://eur-lex.europa.eu/legal-content/EN/TXT/PDF/?uri=CELEX:

32014D0150&from=EN). It is still unclear the impact of the temporary experiment on the overall legislation, but at least it is a first opening to more diversity in seed marketing.

Notes

1 *Regolamento per l'esecuzione del r.d. 15 ottobre 1925, n. 2033, conv. in legge con l. 18 marzo 1926, n. 562, concernente la repressione delle frodi nella preparazione e nel commercio di sostanze di uso agrario e di prodotti agrari.* URL: www.italgiure.giustizia.it/nir/1926/lexs_11614.html

2 *Legge 28 aprile 1938, n. 546 (gu n. 116 del 23/05/1938) istituzione del _registro nazionale delle varietà elette di frumento_ e disposizioni per la diffusione della coltivazione delle varietà stesse. (pubblicata nella gazzetta ufficiale n. 116 del 23 maggio 1938).* URL: www.italgiure.giustizia.it/nir/1938/lexs_22376.html

3 Formally, ENSE has been abolished in 2010 through the D.L. 78/2010, whose aim has been to rationalize the Italian public agencies and organizations. The mission and activities of the ENSE are now conducted by the INRAN (Istituto Nazionale per la Ricerca Alimentare e la Nutrizione). In 2012 ENSE and INRAN have been merged to the Consiglio per la Ricerca in Agricoltura and now ENSE is named Centro di sperimentazione e certificazione delle sementi (CRA-SCS).

4 For the purposes of this publication we have translated *autoctone* as 'indigenous.' 'Autochtonous' would have been a closer translation, but it is not widely used in current English language. People working in the agriculture and genetic resources will tend to think of farmers' varieties or landraces when they see either of the terms.

5 Lazio (2000), Umbria (2001), Friuli Venezia Giulia (2002), Marche (2003), Emilia-Romagna (2008), Campania (2008) and Basilicata (2008).

6 Regions of Sardegna, Abruzzo, Puglia and Sicilia.

7 This is the case of Friuli Venezia Giula. The Tuscan register also holds a forest section.

8 For further information about the application forms, mandatory traits and facultative ones, see: http://germoplasma.arsia.toscana.it/Germo/modules.php?op=modload&name=MESI_Menu&file=Manager&act=D_2:@102

9 The regional laws name public regional agencies to manage regional catalogues. Moreover, they establish regional scientific-technical commissions in charge of the acceptance or rejection of potential varieties. Those commissions evaluate the eventual risk of erosion of a given variety and determine the area where the materials are allowed to be exchanged. The commissions are usually specialized in animals or plants and are normally composed by one public official from the regional department of agriculture competent in plant or animal genetic resources; one representative of the public regional agency which manages the catalogue; one representative of the farmers; and several scientific and academic experts on the field.

10 Article 8(j) of the CBD states:

> Subject to its national legislation, respect, preserve and maintain knowledge, innovations and practices of indigenous and local communities embodying traditional lifestyles relevant for the conservation and sustainable use of biological diversity and promote their wider application with the approval and involvement of the holders of such knowledge, innovations and practices and encourage the equitable sharing of the benefits arising from the utilization of such knowledge, innovations and practices.

11 Tuscany, Lazio and Friuli Venezia Giulia., in these regions nonprofitable exchange is also recognized as a right.

12 Tuscan register: http://germoplasma.arsia.toscana.it/Germo/modules.php?op=modload&name=MESI_Menu&file=Manager&act=D_1:@201
Lazio register: www.arsial.regione.lazio.it/portalearsial/RegistroVolontarioRegionale/
Marche register: www.assam.marche.it/assam2/_static/_htm/Bio_LR12–03/repertorio/tabella_repertorio.htm
13 The directive is entitled "Providing for certain derogations for acceptance of agricultural landraces and varieties which are naturally adapted to the local and regional conditions and threatened by genetic erosion and for marketing of seed and seed potatoes of those landraces and varieties."
14 *Attuazione delle direttive 98/95/CE e 98/96/CE concernenti la commercializzazione dei prodotti sementieri, il catologo comune delle varieta' delle specie di piante agricole e relativi controlli.*
15 Article 2-bis Law 46/2007 – Provisions for the application of the articles 5, 6 and 9 International Treaty on Plant Genetic Resources for Food and Agriculture, ratified with Law 101, 6th April 2004.
16 Ministry for Agricultural, Food and Forest Policy, Ministerial Decree 17th December 2010, Operational provisions for the implementation of the D.Lgs 149/2009.
17 Tuscany's application form is available at: http://germoplasma.arsia.toscana.it/Germo_old/PN_GERMO/Download/Domanda%20b_n.doc
18 For the specific morphological criteria visit: http://germoplasma.arsia.toscana.it/Germo
19 Lazio's application form is available at: www.arsialweb.it/cms/index.php?option=com_docman&task=doc_download&gid=106&&Itemid=100
20 For a better understanding of the overall process of revision of seed marketing in Europe, see Riccardo, B. (2014). 'Seeds Between Freedom and Rights', *Scienze del territorio* (2): 115–22.

22 Commentary on the Zambian Plant Variety and Seeds Act 1998

Godfrey Mwila

The 1995 Plant Variety and Seeds Act, by way of the 2006 Electoral (Code of Conduct) Regulations (Regulations), creates a mandatory variety registration system for a prescribed list of species.[1] If the variety is not registered, it cannot be marketed within the country. To qualify for registration, varieties must be distinct, uniform and stable (DUS). The act also introduces the concept of quality declared seed (QDS), and the 2006 Regulations establish standards for QDS, which are lower than for other classes of seed.

Background

In recent years, the number of small-scale or resource-poor farmers in Zambia has grown – they currently constitute 70 percent of the countries' farmers. These farmers generally use their own landraces (or farmers' varieties), especially for indigenous traditional food crops such as millet, sorghum, cowpeas, bambara groundnuts and other minor crops. Farmers' varieties for introduced crops such as maize, cassava, sweet potatoes, groundnuts, beans and pumpkins have, to a large extent, also been used by small-scale farmers. However, the extent to which farmers' varieties are used, especially for maize, has been declining due partly to the introduction of new crops and the adoption of improved varieties.

The seed supply system in Zambia includes both informal and formal sectors. The informal sector, which is comprised of local seed systems dominated by the use of traditional varieties and to some extent recycled improved seed, has over the years accounted for well over 70 percent of the national seed supply system, especially for nonhybrid varieties. The formal sector has been mainly dominated by the use of improved and certified varieties, especially hybrids. The major formal sector players are seed companies who have marketed the seed through a network of seed stocklists, agencies and nongovernmental organizations (NGOs) that are scattered throughout the country.

Despite the promotion of maize production among small-scale farmers through fertilizer and seed subsidies (and despite farmers' appreciation of the benefits of improved seed), traditional crop production systems, using farmers' local varieties and methods of seed provision, has continued to be, and remains, the backbone of subsistence agriculture over the years.

Policy shift

Zambian agriculture in general underwent significant transformation after the country attained political independence in 1964. Major changes in the Zambian agricultural policy, which were preceded by a major shift in the economic policy from state control to market liberalization, started taking place in the mid-1990s. This was when an agricultural policy statement for the year 2000 and beyond was put in place, setting out basic guiding principles for the operation of the agricultural sector as it evolves from a highly controlled and regulated industry to one that was expected to be fully liberalized and market driven. The most visible effect of the liberalization policy was the withdrawing of subsidies on fertilizers and improved seeds since 1991. Key objectives of the new agricultural policy framework included the need to ensure household and national food security through the annual production of dependable and adequate supplies of foodstuffs and to ensure that the existing resource base was maintained and improved upon. Strategies put into place to achieve these goals included the diversification of crop production, improved use of available natural resources, helping farmers deal with natural disasters and an emphasis on sustainable farming systems.

The development of the seed system was also influenced by these policy changes, which moved from a system that only promoted the formal sector to one that sought to promote and integrate the informal sector into the seed supply system. The objectives of promoting the informal sector included the need to facilitate increased availability and accessibility by strengthening the local seed systems, which were largely based on local traditional crop varieties.

Development of seed laws

The initial seed laws regulating the seed industry in Zambia were embodied in the Federal Seeds Act, which was put into place in 1965.[2] This act was later replaced by the Agriculture (Seeds) Act in 1967.[3] The 1967 act provided regulations regarding the release of varieties as well as regarding the testing, inspection and import of seeds. It also outlined the necessary conditions for the sale of seeds within the country. Under this act, only varieties that have been adequately tested under Zambian conditions and that are entered into the official variety list are eligible for marketing in the country. By and large, the Agriculture (Seeds) Act was designed to promote the formal sector and improved seed, providing the regulatory framework for variety release, seed certification, production and distribution. From the start, the conditions for variety release that were included in the act did not cater to farmers' varieties or local traditional varieties.

As a means of improving the administration of the Agriculture (Seeds) Act and broadening stakeholder involvement, a National Variety Release Committee was formed in 1984 to scrutinize the release of crop varieties for commercial use in Zambia. Around the mid-1990s, it gradually became apparent that the current seed law was inadequate to provide a balance between the commercial and subsistence farming systems, therefore prompting the need to review and amend the existing legislation. Amendments to the 1967 Agriculture (Seeds) Act resulted in the 1998 Plant Variety and Seeds Act.[4] This act provided for increased private sector participation in seed quality control by way of licensing and the introduction of minimum standards for QDS. These standards were designed to promote seed production at community levels and facilitate the development of seed provision systems in outlying areas. It was expected to contribute to the promotion of the informal seed sector.

According to the Food and Agriculture Organization (FAO), the purpose of QDS is to offer an alternative quality control system that can be used for those areas, crops and farming systems in which highly developed seed quality control schemes may be difficult to implement.[5] These minimum standards offer an opportunity for accommodating varieties of crops, such as farmers' varieties, that may not easily fit within a conventional seed quality control scheme. It is said to be an open scheme that meets the needs of farmers in a flexible way without compromising the basic standards of seed quality. Adoption of the QDS standards and approaches in Zambia has allowed small-scale farmers to become more involved in the production of certified seed through their farmers' associations. For example, groundnut seed for a registered groundnut variety called Chalimbana has been produced by small-scale farmers and distributed through a rural and smallholder-based farmers' association in Zambia. A few similar cases involving seed crops that are not adequately attractive to large commercial seed companies such as common beans have also taken place.

The FAO's QDS guidelines have ensured that QDS systems could be extended to materials that were not necessarily meet the standard of DUS. To date, these guidelines have not been implemented in Zambia with respect to farmers' varieties that did not already satisfy the DUS standards for registration. While it might be challenging to do so in practice, it would certainly be a useful approach, given that the use of farmer-saved seed, either of the traditional local varieties or of recycled seed of improved varieties, by most small-scale farmers has generally been acknowledged to be part of, and a significant component of, the local seed systems and a significant component of the informal seed sector.

Other provisions of the reviewed act include the decentralization of seed testing, which has led to the need to establish satellite seed-testing laboratories and has been useful in supporting the development of the seed provision systems in rural areas. The participation of the private sector in seed quality control has included licensing to carry out seed inspections and testing in accordance with conditions set out in the act.

Most seed companies in Zambia have concentrated on hybrid maize, which has been the most economically viable seed. This decision has left most of the traditional crops (e.g. groundnuts, beans, sorghum, millet and cowpeas) unattended to and, therefore, has led to a shortage of seed/planting material for these crops. The role of the informal sector in filling this gap, especially in outlying areas, and integrating with the formal sector has been seen as critical not only for a viable and sustainable seed industry in Zambia but also for improving the productivity of the majority of small-scale farmers. In this regard, a number of government, NGO and donor-supported programs were put in place to assist and develop seed provision systems in rural areas. Activities under these programs have included seed training to facilitate seed production at community levels and seed distribution in disaster situations.

The development of the rural seed provision system has created enthusiasm among smallhold seed growers. However, this system does not include seed multiplication for farmers' varieties but, rather, uses improved varieties for which it does not even have a source of breeder/basic seeds. There is a need to implement some of the measures included in the agricultural policy that relate to addressing the constraints in seed production and distribution among smallholder farmers and facilitating their increased involvement. The latter can be accomplished, for instance, through the promotion of participatory plant breeding with the aim of evaluating the products of plant breeding and local varieties. Measures that have been considered as alternatives for sustaining these seed production initiatives include the need to support the multiplication of prebasic and basic seed of varieties developed through participatory plant breeding and making them available for seed production.

Notes

1 Plant Variety and Seeds Act, 1998, Chapter 236 of the Laws of Zambia.
2 Federal Seeds Act, 1965.
3 Agriculture (Seeds) Act, No. 14 of 1967.
4 Plant Variety and Seeds Act, supra note 1.
5 Food and Agriculture Organization (FAO), *Plant Production and Protection* (Rome: FAO, 2006) at 185.

23 Commentary on the regulation on production management of farm households' plant varieties in Vietnam

Nguyen Van Dinh and Nguyen Ngoc Kinh

The need for a legal document about managing the informal seed system

Vietnam is an agricultural country with 70 percent of the population living by agriculture. The crop production sector plays an important role in agriculture remarkable contributions coming from the plant breeding sector. In years of subsidization with big government investment capital, an official seed system (involving companies, public research institutes, and universities) produced a large quantity of seed for farmers. However, this official system has its limitations. For example, it only provides farmers with 10–20 percent of the rice seed that they need annually. This shows the important role of the farmers' informal variety system, which still supplies the vast majority of seed for production, even for rice, which is a very important commercial crop in Vietnam.

The same is true for most other plant varieties, with the exception of maize, and some vegetables and flowers, vegetable species and fruit trees: seeds provided by farmers (selling, exchanging or saving their own seed) still accounts for the largest proportion produced and used annually. This illustrates the important role of farmers in maintenance, preservation and selection of seed used for their process of agricultural production themselves.

There are many policies and guidelines in Vietnam that affect the informal seed system inadvertently, in its pursuit of policy goals not directly related to those seed systems. While the role of government agencies and state-owned companies researching, selecting, breeding or producing seeds is addressed throughout most of these regulations, until quite recently none of them directly addressed or provided support for farmers' roles in these processes. The Ministry of Agriculture's Decision, 'Regulation on Production Management of Farm Households' Plant Varieties, 2008'[1] was designed to address this situation, to encourage the effective development of the informal seed system.

In the early stages of consideration of such a legal instrument, it was considered that it should be developed with the following elements:

- It should be consistent with, and complement, existing laws in Vietnam.
- It should benefit farmers partly by contributing to raising their own awareness, and that of national researchers and agriculture extension workers, about farmers' dynamic role in their informal seed systems, as a precursor to further developing their own capacities and the advantages to be gained by informal seed systems.
- It should cover all aspects influencing the development of the informal seed system, including such areas as economics, politics, society, culture, environment and traditional knowledge.
- It should regulate the informal seed system to improve the local residents' quality of life.
- It should contribute to developing the local residents' ability to conserve and stock local genetic resources.

In November 2006, the Department of Cultivation suggested issuing a decision at the ministerial level. By the end of 2007, the draft Decision on 'Regulations on Managing Informal Seed Production' was completed thanks to the cooperation of core agencies like the Vietnam Seed Association, Department of Cultivation, Department of Legislation of the Ministry of Agriculture and Rural Development, and many other authorities and unions.

The process of developing the law was relatively short but intense. In March 2007, the first draft was given to all members of the multistakeholder task force that was established to guide a five-year national research and capacity building project supported by the internationally organized Genetic Resources Policy Initiative (GRPI). The draft was also presented for feedback at the meeting of the National Seed Consulting Council and at the conference held by the Seed Association. Based on feedback received, a second draft entitled 'Regulations on Encouraging the Development of the Informal Seed System' was developed. This second draft included articles covering the following seven areas: (1) object and scope of regulation, (2) definition of the informal seed system, (3) purpose of encouraging the development of the system, (4) role of the system in saving, conserving and exploiting native genetic resources, (5) role of the system in researching, selecting and breeding new varieties, (6) improving the quality of varieties bred from informal seeds and (7) organizing and implementing the responsibilities of six agencies.

In November 2007, a conference with representatives of the seed program, seed associations and the production department was organized to discuss the title and scope of the draft and whether the word 'policy' would be used. Participants also looked at the purpose of producing seeds in the informal system and which phase should get policy support (production for own consumption, for exchange within the community or in cooperation with commercial enterprises). Among other things, they discussed the level of farmer participation in research as well as the role of research agencies. In addition, they examined the issue of registration for seed protection: if specialty seeds belong to the community, how should support for the informal seed system be provided, what should be the source of finances and where should participants in the seed system be registered?

After the conference, a third draft was developed, which included nine articles that were more detailed in terms of management and division of responsibilities. This draft was circulated for comments and, after more discussion and work, a fifth draft was issued as the official Decision (No. 35/2008/QĐ-BNN) of the Ministry of Agriculture and Rural Development, entitled 'Regulation on Production Management of Farm Households' Plant Varieties'[2] (hereinafter, the 'Decision').

Summary of the Decision

The Decision consists of a total of 10 articles. (Because it is short, it is reproduced in its entirely at the end of this commentary.) The Decision's objectives are to motivate farmer households to be involved in protecting, maintaining and making reasonable use of local genetic resources, making good choices and breeding newly-found varieties and to create favourable conditions for households production of high quality and low production price seed to meet the production demand.

The Decision is applied to farm households, cooperative teams and organizations that utilize, circulate and transact plant varieties on the market.

Very significantly, the Decision formally recognizes the important contributions that farm households can make to collecting, maintaining and conserving local plant genetic resources, to developing new plant varieties, and to seed purification and seed production. The Decision complements this recognition by formally opening up the possibility of various forms of government financial support to farmer households (and collectives) for such activities. This represents a very important step forward; in the absence of such a legal recognition of their role, and explicit recognition of their ability to receive support (and to make proposals for support in their own name), funds from the listed sources would not have been available to support the related activities.

The Decision also clarifies that farm households may apply for plant variety protection rights under the national plant variety protection law, and may register new varieties pursuant to the national plant ordinance, provided the varieties satisfy the conditions for protection that are set out in those acts. The Decision does not create alternative criteria for either plant variety protection of for variety registration, so to be recognized for plant variety protection rights, the household variety would need to satisfy the conditions of distinctness, uniformity and stability as established in the plant variety protection law.[3] However, it does state that the associated costs of testing and registration should be lower or covered by the government.

The Decision clarifies that farmer households do not have to comply with seed production standards as set out in the national seed ordinance when the seed is for their own use, or for exchange among farmers within the local district administrative unit. However, if they are commercializing seed, they must also comply with the seed ordinance production conditions.

Very importantly, the Decision specifies various forms of support that must be provided from different government departments, research agencies and agricultural extension, including strengthening capacity of farming households in genetic resources identification, collection, maintenance and use; seed purification and developing new plant varieties; testing performance of farm household varieties; developing pilot demonstration models of farmer households' seed production; and developing and submitting the expenditures for informal seed activities to the provincial people's committee for approval.

MINISTRY OF AGRICULTURAL AND RURAL DEVELOPMENT	**SOCIALIST REPUBLIC OF VIETNAM** **Independence – Liberty – Happiness**

REGULATION

On production management of farm households' plant varieties

(Issued together with Decision No. 35/2008/QĐ-BNN dated 15/02/2008 by Minister of Ministry of Agriculture and Rural Development)

Article 1. Application scope and objects

1 Application scope
   This document defines management contents for activities including maintenance, conservation and exploitation of local genetic resources and plant varieties; new plant variety breeding; seed production and exchange in communities or on market circulation.

2 Application objects
   This regulation applies to farm households, cooperative teams, cooperatives, and clubs (hereafter referred to farm households) involved in activities specified in Clause 1 of this Article.

Article 2. Farm households' plant varieties

Farm households' plant varieties are varieties developed and produced by farmers for use, exchange or circulation on market.

Article 3. Objectives of farm households' plant variety management

1 Encourage farm households participating in effective maintenance, conservation and exploitation of local plant genetic resources and new plant variety development.
2 Create favorable conditions for farm households' production of high quality and low production price seed to meet the production demand.

Article 4. Collection, conservation and exploitation of local plant genetic resources and varieties

1 Farm households allowed to participate in activities including collection, maintenance, conservation, exploitation and utilization of local plant genetic resources and varieties, as specified in regulation on plant varieties.
2 Implementation of activities specified in Clause 1 of this Article:
   a Farm households develop proposals and investment projects submitted for People's Committee of provinces and cities under the central government for approval and implementation;
   b Farm households participate in projects of other organizations;
   c Farm households shall register for recognition of mother stocks and stock nurseries as prescribed in Decision No. 67/2004/QĐ-BNN dated 24/11/2004 by Minister of Ministry of Agriculture and Rural Development on selection, recognition, management and utilization of mother stocks and stock nurseries of industrial crops and perennially fruit trees.
   d Expenditure for collection, maintenance, conservation, exploitation and utilization of local plant genetic resources and varieties is supported by state budget approved by authorized agency, including:
      - Central state budget supports projects approved by ministerial and governmental levels.
      - Provincial budget supports projects approved by People's Committee of provinces and cities under the central government.

Article 5. New plant variety development and seed purification

1 Farm households are allowed to apply scientific projects on new plant variety development and seed purification to Science and Technology Department of provinces and cities under the central government or to Science, Technology and Environment Department of Ministry of Agriculture and Rural Development (MARD) as specified in Decision No. 36/2006/QĐ-BNN dated 15/5/2006 issued together with Regulation on management of scientific and technological projects by Minister of MARD.

2 Plant varieties developed by farm households can directly or authorize other local or central organizations to conduct varietal testing/evaluation, trial production and recognition as specified in Decision No. 95/2007/QĐ-BNN dated 27/11/2007 by Minister of MARD on recognition of new agricultural plant varieties.

Article 6. New plant variety protection

Farm households who are breeders of new plant varieties shall register for new plant variety protection as specified in Intellectual Property Law and Government Decree No. 104/2006/NĐ-CP dated 22/9/2006 on concrete regulation and guidelines to enforce some articles of Intellectual Property Rights on Plant variety rights.

Article 7. Budget sources supporting farm household breeding activities

1 Farm households producing seed in concentrated seed production areas under provincial seed programs in the period 2006 to 2010 shall be supported by provincial budgets for infrastructure and equipment to serve seed production and processing as specified in Decision No. 17/2006/QĐ-TTg dated 20/1/2006 by Prime Minister on continuation of Decision No. 225/1999/QĐ-TTg dated 10/12/1999 on programs of crop varieties, animal breeds and forest varieties to 2010 and Ministerial Circulation No. 15/2007/TTLT-BTC-BNN&PTNT dated 8/3/2007 by Ministry of Finance and MARD on guidelines of governmental budget use and management for programs of crop varieties, animal breeds and forest varieties.
2 Budget for testing, trial production, new plant variety recognition, mother stock and stock nursery evaluation and selection, and registration of new plant varieties developed by farm households shall be supported by governmental budget. Concrete financial support shall be approved by the Chairman of People's Committee of provinces and cities under the central government based on Provincial Agriculture and Rural Development Department's proposals.

Article 8. Production, exchange and circulation of farm households' seed

1 Farm households producing seed for households' use or exchange in district administrative unit shall not be obligatory to provide all production conditions as specified in Clause 1 of Article 36 of Ordinance on Plant varieties. If planting materials of industrial crops

and perennial fruit trees using vegetative propagation should use mother stocks or stock nursery.

2 Farm households producing seed for commercialization purposes:
   a Seed in the list of main crops shall need to provide all conditions as specified in Clause 1 of Article 36 of the Ordinance on Plant varieties.
   b Industrial crops and perennial fruit trees shall need to provide all conditions as specified in Clause 1 of Article 38 of the Ordinance on Plant varieties.
   c Shall execute regulation on quality announcement, testing and certification and labeling.

Article 9. Implementation

1 Crop Production Department
   - Lead organization for activity management and instruction of farm households' plant varieties nation-wide.
   - Organizing Science and Technology Board for evaluating testing results, proposing recognition and release to production of new plant varieties developed by farm households.
   - Supervising mother stock and stock nursery evaluation and selection, recognition and management of fruit trees and perennially industrial crops nation-wide.
   - Supervising National Testing Centre of plant varieties, plant-derived products and fertilizers and Office for protection of new plant varieties to guide and help locals and farm households in implementing testing and trial production activities, recognition and registration for plant variety protection.

2 Science, Technology and Environment Department
   - Guiding locals to help farm households applying, developing and implementing projects on plant genetic resource collection, conservation and utilization, purifying seed, and developing new plant varieties submitted to authorized bodies for approval.
   - Cooperating with Crop Department in control and evaluating testing results of farm household new developed plant varieties.

3 Research institutes under MARD
   - Consulting farm households about plant genetic resource and variety conservation, maintenance and exploitation, seed purification, and new plant variety development research.

4 National Centre of Agricultural and Fishery Extension
   – Supervising the agricultural extension network to support farm households with technical training in farm household breeding activities and in setting up pilot demonstration model of farm household purified and newly developed plant varieties.

5 Agriculture and Rural Development Departments of provinces and cities under the central government
   - Directly managing and guiding farm household breeding activities in the locality.
   - Developing and submitting Provincial People's Committee for budget approval of farm household breeding activities as specified in Article 7 of this Decision.
   - Supervising local agricultural extension for training and developing pilot demonstration model of farm households' seed production.

Article 10. Implementation provisions

1 Organizations and individuals involved in activities of local plant genetic resource and variety conservation, maintenance, and exploitation, of new plant variety development, seed production and circulation in community or markets shall be administratively sanctioned or criminally executed according to violating degree.
2 If in the course of implementation, any problems or new matters arise, organizations and individuals involved in farm household breeding activities should report to MARD by written document for prompt solution.

Notes

1 Regulation on Production Management of Farm Households' Plant Varieties, Decision No. 35/2008/QĐ-BNN dated 15/2/2008.
2 Ibid.
3 Decree No. 104/2006/ND-CP* On Detailed Regulations to Implement Some Articles in the Intellectual Property Law, Chapter on Plant Variety Rights.

24 Commentary on the registry of native crops in Peru

Law 28477 and the registry of native potatoes

Manuel Ruiz Muller

Registries of native crops and potatoes in Peru

The move towards creating a registry for native crops was strongly stimulated by the idea that these crops needed some form of legal protection, especially from misuse, misappropriation and biopiracy in general. Policy discussions had already been under way for a considerable amount of time, when Law 28477 Declaring Crops, Native Breeds and Usufruct Wildlife Species are Part of the Nation's Natural Heritage was enacted in 2005.[1] Soon thereafter, in 2008, a regulation created the National Registry for Native Potatoes.

Law 28477 establishes a closed list – in an annex – of crops, breeds and wild species that are recognized as part of the natural patrimony of the nation. This list includes a wide range of species and crop varieties, including native potatoes as well as domesticated animal breeds and wild animals. The law places on the Ministry of Agriculture (in coordination with local and regional governments and public and private entities) the responsibility of registering, disseminating, conserving and promoting genetic materials of these crops and breeds as well as the production, commercialization and internal and international consumption of domesticated breeds of animals and wild animals (indicated in the annex), according to sustainability criteria.

In this context, there are a few comments that can be made in regard to the practical implications of Law 28477. First, the law gives special recognition and legal status to those crops, breeds and animals that are contained in the list – they are all part of the natural patrimony of the nation. What does this special recognition mean? Essentially it means that the state has a special interest in these crops and breeds, which is reflected in a series of measures conducive to their conservation, wider use, registration and so on. This may include a series of measures, projects and activities to realize these goals. The law does not specify what this registration will specifically entail in legal terms. For example, whether registration may mean the reconition of constitutive, exclusionary rights or simply involve a declaratory tool with no specific or enforceable right attached to it. Second, the law also describes a range of measures that will be adopted to ensure the appropriate conservation and promotion of the listed crops, breeds and animals. Finally, the law does not recognize nor grant specific rights over the listed crops and breeds. There are no specific beneficiaries or rights holders. Rather, it is the state, in the name of the nation (in accordance with the National Constitution of 1993), which takes on the responsibility of ensuring that its economic, cultural and political interests (specifically in biodiversity) are safeguarded.

Specific rights over seeds are granted through the protection of new plant varieties under Supreme Decree 035–2011-PCM, which is a type of system modelled after the International Convention for the Protection of New Varieties of Plants (UPOV, in its 1991 version) and derived from Decision 345 of the Andean Community on the Protection of New Varieties of Plants (1993).[2] In addition, Law 27262 and its regulation on Seeds also determines how certified seeds can be commercialized throughout the country. Both these legal regimes are mostly concerned with modern, high-yielding varieties and not traditional, local and native crops. However, the seed regime does recognize "non-certified" seeds as a special category of genetic resource that will require special regulations and is mosty related to native and ancestral cultivars that may be legally commercialized throughout the country.

In contrast with the more classical seed protection regimes, the National Registry for Native Potatoes was created through Ministerial Resolution 0533–2008-AG in July 2008. This registry was created in response to the express recognition of the potato's critical importance and contribution as a key component of the Peruvian people's diet as well as a good portion of the world's diet and food security in general. More specifically, in the preamble of this resolution, the state acknowledges that it is necessary to create mechanisms that facilitate access to information regarding native Peruvian potatoes, by means of a registry that contains reliable genetic, morphological and anatomical indicators of this tuber. Such information may become the technical basis for the potato's international recognition and protection.

The registry is overseen by the Ministry of Agriculture. Its implementation, maintenance and updating is the responsibility of the National Institution for Agricultural Innovation (INIA). INIA, in turn, is responsible for developing and enacting the necessary complementary provisions and guidelines that may be required for the appropriate operation of the registry. Ministerial Resolution 0533–2008-AG also provides that INIA, under the supervision of the Ministry of Agriculture, honours various other agreements with other research institutions (such as the Agrarian University or the International Potato Center) to support the implementation and continued updating of the registry.

The registry does not grant rights, however. Registration per se is carried out by INIA. Its main functions are twofold. On the one hand, it provides useful technical data and information regarding native Peruvian potatoes to any interested party (researchers, farmers, universities and so on). On the other hand, and equally importantly, it may serve as a 'defensive mechanism' or order to ward off attack from patents or other intellectual property rights that may invoke novelty or inventiveness or seek to claim origin in regard to unique and endemic potato crops of the Peruvian Andes. Data, information, registration dates, knowledge and other details contained in the registry may assist in invalidating claims put forward in patent or in plant breeders' rights applications. The idea is that the registry stimulates small farmers and native seed producers to identify themselves as conservers and producers of these crops. Though no monetary compensation or IP-type exclusive rights are envisioned, social recognition becomes important through the registry.

Policy, social and legal framework

The initial interest for agrobiodiversity can be traced to the early 1990s with the development of policies and laws that implemented the principles of access to genetic resources and benefit sharing, conservation of agrobiodiversity and protection of traditional knowledge under the Convention on Biological Diversity (CBD).[3] At this time, Peru enacted a new set of laws including an environmental code, a biodiversity conservation law, a biosafety law, a protected areas law and, immediately afterwards, a national biodiversity strategy, a law for the promotion of medicinal plants, a law for the protection of traditional knowledge and a series of regulations.

All of these pieces of legislation made a direct and sometimes indirect reference to the need to conserve agrobiodiversity and enhance the capacity of farmers and institutions to make better and more efficient use of their components (seeds, soils, agroecosystems, animal breeds and so on). In terms of specific projects, the Global Environment Facility's In Situ Conservation of Native Crops and Wild Relatives Project, which ran from 1997 to 2006, was a pioneering and important awareness-raising effort, through which, probably for the first time, ideas regarding how to effectively protect native crops and their wild relatives first emerged. The Genetic Resources Policy Initiative (GRPI, led by IPGRI at the time), which was set up from 2003 to 2007, also served to draw attention to agrobiodiversity and native crops and especially to policy and legal elements regarding conservation and sustainable use. Other initiatives such as the Potato Park, the Andean Project for Peasant Technologies, the Science and Technology Coordinator of the Andes, Association Arariwa, El Centro IDEAS, and a wide range of localized projects have all contributed in different ways, from providing different perspectives and approaches to revaluing agrobiodiversity.

Although the importance of potatoes for world, national and local needs is widely recognized, 'new' crops such as mashua (*Tropaeolum tuberosum*), maca (*Lepidium meyeni*), arracacha (*Arracacia xanthorriza*), yacon (*Smallanthus sochifolius*), Oca (*Oxalis tuberosa*) and others have emerged with considerable potential to satisfy food and nutritional needs (even health needs) of a wider population in Peru. They are often called 'underutilized crops,' a term that gives little indication of their importance at the local/community level or for food security purposes.

At the same time as these developments, more and more interest was being placed in the 1990s on a highly emotional and politically charged phenomenon: biopiracy, or the misappropriation of Peruvian biodiversity through intellectual property theft and other means. While the CBD recognized national sovereignty over biodiversity, more and more cases were being identified in which Peruvian products in all fields (natural products, pharmaceuticals, bioremediation, agroindustry and so on) were subject to intellectual property rights and thus appropriated through indirect means. Regardless of the validity of this concept in strictly technical terms, it became very significant politically in terms of Peru's vast wealth of biodiversity.

The potential role of these native crops has been in a way rediscovered by social and natural scientists, and economists and lawyers have realized that incentives and specific regulations are required to support the conservation, development, protection and wider utilization of these seeds and crops. It is this recognition and multidisciplinary interest that has led policy makers and decision makers to develop specific laws and measures. The role and drive provided by a true 'gastronomic boom' that started in Peru in the late 199s to native crops has also been instrumental to the revaluing of local and native agribiodiversity, seeds and small farmers' activities.

Some legal issues and final remarks

The notion of registering biodiversity or native crops has always been very appealing for countries with high biodiversity. Probably since the initial discussions in Peru regarding genetic resources and traditional knowledge in the early 1990s, registering biodiversity has become almost synonymous with protecting biodiversity – the idea being that the act of registration automatically grants certain rights. This is not necessarily so in technical terms. A lot depends on the objective of the register and whether it is of a constitutive or declarative nature. In the first case, the registry creates and grants rights (which would need to be defined in content, scope, beneficiaries and so on). In the latter case, the registry does not create a right but simply recognizes the existence of a seed, maybe its location or its developers or some of the related knowledge, in addition to providing other useful technical or economic information related to the seed.

These discussions regarding the registry and its objectives are important because they may define who owns, controls or has certain rights (in opposition to or the exclusion of others). Both Law 27844 and Ministerial Resolution 0533–2008-AG do not create legal rights. In the case of Law 27844, it recognizes or stresses a situation that is also recognized in the Peruvian Constitution of 1993 (and other national laws), which is that natural resources, including genetic resources, are the patrimony of the nation, and the state exercises rights over these resources in the nation's name and representation.

Property rights over the list of crops and breeds in the annex to Law 27844 and the national registry cannot be granted for a simple reason: many of the crops, breeds and varieties are shared between communities (and even countries) and are also widely distributed. Thus, assigning a specific right to a specific person would be of no practical use whatsoever.

Notes

1 Law 28477 Declaring Crops, Native Breeds and Usufruct Wildlife Species Part of the Nation's Natural Heritage, 22 March 2005, online: <www.wipo.int/wipolex/en/details.jsp?id=6684> (last accessed 15 June 2012).

2 International Convention for the Protection of New Varieties of Plants, 2 December 1961, available at <www.upov.int/en/publications/conventions/index.html> (last accessed 15 June 2012).

3 Convention on Biological Diversity, 31 I.L.M. 818 (1992).

25 Commentary on the draft proposal for the establishment of a native seeds registry in Costa Rica

Jorge Cabrera Medaglia

Introduction

Costa Rican seed law requires that varieties are appropriately registered and that they meet specific technical requirements before they can be commercialized. In practice, this implies that the seeds sold by farmers – known nationally as 'native' (*nativas*) or 'local' (*locales*) seeds – could not be legally commercialized in the either formal or informal markets if they are not registered with the National Seeds Office.

With the objective of promoting the conservation and use of local varieties and making it legal for farmers to trade and sell seeds,[1] proposals have been made to modify the legal framework regarding variety registration in Costa Rica. These proposals are part of a broader effort to support the conservation and sustainable use of plant genetic resources and for the implementation of farmers' rights. They are also consistent with, and motivated by, Costa Rica's obligations pursuant to the Food and Agriculture Organization's (FAO's) International Treaty on Plant Genetic Resources for Food and Agriculture[2] (ITPGRFA), especially with respect to articles 5, 6 and 9 regarding conservation, sustainable use and farmers' rights.

Costa Rican legal framework for the commercialization of seeds

Costa Rican legislation regarding seeds is varied and has undergone several changes over time. Among them, the following may be mentioned:

- Law No. 6289 of December 4th, 1978, and its amendments (Seeds Law): The National Seeds Office is created as an entity associated to the Ministry of Agriculture and Livestock; among whose functions are included the registration of several categories of seeds. The Office will not authorize the commercialization of a seed variety not compliant with the respective regulations and not properly registered.[3] The production, processing and commercialization of seeds without observing that which is declared in the Law and its regulations, will be considered an infraction of the Law.[4] Conditions of registration are: stability, uniformity, distinctiveness and agronomic value or proved use (*valor comprobado*).[5] The traditional DUS requirements and one additional condition of agronomic value or proved use are those to be fulfilled by the varieties.

- The Regulation of the Seeds Law was approved by Executive Decree No. 12.907-A of July 7th, 1981, published in *La Gaceta*, No. 180 of September 21st, 1981. In the Regulation are reaffirmed some provisions contained by the Law related to the obligation to register the seeds before its commercialization.[6] The commercialization of seed varieties not registered in the Registry of Commercial Varieties is considered an infraction of the Law.[7] The Registry of Commercial Varieties was reformed – responding to different reasons and for several purposes – but the obligations to register the seed varieties at said registry before commercializing them was retained. The procedures and requirements for the registration of commercial varieties do not represent major technical or legal impediments to registering varieties that come from participatory plant breeding programs. In the past different varieties of bean have been registered (*Gibre y Curré*, both red grains), obtained through this mode of genetic enhancement.[8]
- Regulation for the Import, Export and Commercialization of Seeds; issued by the National Seeds Office (not published as an Executive Decree). This Regulation was published in *La Gaceta*, No. 73 of April 18th, 2005. It reaffirms the obligation to register as set out the Seeds Law (following the same conditions for registration) and regulates the requirements for the national commercialization of the seed (as well as for the import and the export of seeds).
- The Executive Decree No. 32.487-MAG (*La Gaceta*, No. 146 of July 29th, 2005): This Decree emphasizes the authority of the National Office to establish regulations to ensure that every imported seed complies with the quality standards of the seed reproduced in the country, regardless of whether the seed is commercialized or not.
- The General Regulation for the Certification and Quality Control of Seeds, approved by the Board of Directors of the National Seed Office in Session No. 599, on December 9th, 2009. This provision regulates the process of certification through the different categories of seeds and their quality control.
- Additionally, there are some Central American technical regulations about this matter, such as the RTCA 65.05.34:06, Central American Regulation for the Production and Commercialization of Certified Seeds of Soybean and Basic Grains. This technical regulation contains provisions for the registration of seeds intended to harmonize requisites and conditions for the registration of commercial varieties among the different Central American countries[9] as part of the efforts to establish a Central American Custom Union. However, the regulation does not provide the conditions for registration, but leaves them to be determined according to the laws of each country (as described earlier for Costa Rica).
- Finally, the Law for the Development, Promotion and Encouragement of the Organic Agricultural and Livestock Activity, No. 8591, published in

La Gaceta, of August 14th, 2007; recognizes the concept of native, local or traditional seeds (section j), in the following manner,

> Seeds corresponding to cultivated and developed varieties by local farmers and communities. Regardless of their origin, they are adapted to agricultural practices and local ecosystems. They are regulated by what is provided in article 82, and following articles, of the Biodiversity Law; Law No. 7788 of April 30th of 1998.[10]

Law No. 8591 also provides that:

> The State through the competent authorities, will promote, encourage and protect the right of the people and of the agricultural organizations to access, use, trade, multiply and save native seeds with the intention of preserving the native genetic heritage in benefit of current and future generations of organic growers. The Ministry of Agriculture and Livestock shall ensure the compliance of this provision, in accordance with that has been established in the Convention on Biological Diversity, approved by the Law. No. 7416 of June 8th of 1994 and the Biodiversity Law, Law No. 7788 of April 30th of 1998.[11]

Law No 8591 provides a justification for amendments to the current variety and seed registration regimen in Costa Rica to promote, encourage and protect farmer´s rights with respect to trade native seeds.

Draft proposals

Background of the draft proposals

At the end of the last decade, the FAO sponsored a number of projects in Costa Rica to improve seed law related to quality control, registering process, and so on. The National Seed Office also, by its own volition, wanted to amend the national legal framework to provide appropriate recognition of the farmers' practice to exchange and commercialize seeds in informal markets. This approach was also consistent with the national legal framework in place regarding the conservation and sustainable use of traditional or native varieties, as well as the country's obligations under the ITPGRFA and the Convention on Biological Diversity.

Under the policies and laws described in the preceding section, no traditional or local varieties have been ever registered in Costa Rica, because they could not fulfil the legal requirements. On the other hand, there has also never been an administrative or criminal sanction against anyone for commercializing the seed of native or local varieties. Nonetheless, it was considered desirable to reform to the applicable laws to expressly recognize the legal right to commercialize these kinds of seeds.

Initially, the strategy was to include the amendment as part of a broader reform of the regulation (No. 12.907-A of 1981) passed under the Seeds Law.

However, due to some comments received from other national ministries (especially those in charge of the issuance of technical regulations), a decision was taken to include the modifications into a new regulation (called the Technical Regulation for the Registry of Commercial Varieties) that would replace the 2005 regulation.

Main content

In relation to the issue of native varieties, the Draft Technical Regulation for the Registry of Commercial Varieties contains a provision that reads:

> The registration in the Registry of Commercial Varieties could have the following exceptions, subjected to previous evaluation by the National Seeds Office:[12]
>
> - **Local, traditional or native varieties:** for these varieties, the National Seeds Office will **provide a database, where interested parties might include** information about these materials, as a mechanism of support in the conservation of plant genetic resources and allowing to officially cataloguing them as known genetic resources. For this purpose the following information shall be provided:
>
>   I. Person, community or organization.
>   II. Name of the variety.
>   III. Origin of the variety (if it's known).
>   IV. General morphological description allowing identification of the variety.
>   V. General features of the variety (agronomic or culinary characteristics; or known properties).
>   VI. Geographical distribution (locations where the variety is distributed).

The draft regulation also proposes to modify the concept of seeds commercialization in the following manner:

> Seed Commercialization: the offering for sale, the possession for sale, the sale and any other commercial operation (cession, delivery or transfer) with purpose of commercial exploitation; of seeds to third parties, for remuneration or not. Excluded from this definition are: the delivery of seed samples with research purposes, the cession or exchange among farmers of local, traditional or native seeds.[13]

The main purpose of the reform is to create a legal space for the sale of seeds of native varieties (mostly in informal markets) through two complementary mechanisms: (1) creating an express exception to the obligation of registration before commercialization and replacing that obligation with a voluntary registry pursuant to alternative criteria; and (2) excluding from the legal definition of seed commercialization the exchange of native, traditional or local seeds among farmers. This last mechanism on its own would not, in principle, allow the exchange of farmer's seeds in open markets or to persons who cannot be considered as farmers. However, the two parts read together seems to support this possibility.

Pending challenges

The proposed technical regulation constitutes an important step forward to support the farmer's rights under the ITPGRFA by providing a legal basis for the commercialization of native or local seeds. However, the proposed reform if approved, still presents some potential shortcoming and challenges:

- The exception to register native seeds – as it is written in the proposal – leaves to the discretion of the National Seeds Office to allow their commercialization (article 12 of the draft technical regulation indicates that the exceptions are subjected to previous evaluation by the National Seeds Office), which – at least theoretically – could lead to the prohibition and restriction of the commercialization of this kind of seeds if the national authorities so decide (based on technical and other reasons).
- The database created by the proposal – with the purpose of improving the existing information to support the conservation of plant genetic resources – is for voluntary use of the farmers, which does not guarantee that this information will be effectively obtained and used for conservation and other goals. Although the proposed regulation does not mention it expressly, this registry should be public due to the general Costa Rican laws concerning access to information (including environmental information).[14] In the past, efforts to create a similar system for the registration of traditional knowledge under the Biodiversity Law[15] encountered the opposition of some indigenous and local communities on the basis that the registered information would need to be public.
- Finally, despite the interpretation offered earlier regarding a whole reading of the proposed technical regulation, the fact that the definition of commercialization excludes exchanges among farmers could lead to the interpretation that sales made by farmers to nonfarmers are outside the scope of the exception of the definition of commercialization.

Other complementary reforms

In addition to the proposed technical regulations, there is also a draft proposal for a new Seeds Law being promoted by the National Seeds Office which is under discussion at the Agricultural Affairs Commission of the Legislative Assembly. The proposed new Seeds Law has the explicit objective of promoting the conservation and use of plant genetic resources, the promotion of food security and the achievement of environmentally sustainable production. However, the proposal does not include the development of new criteria for the commercialization of native or traditional varieties.

It is proposed that the National Seeds Office should be appointed as the national competent authority on plant genetic resources for food and agriculture (PGRFA), in accordance with Costa Rica's obligations under the ITPGRFA. Until recently, no national competent authority for the implementation of ITPGRFA had been appointed. The main mandate of the national competent authority will be ensure, coordinate, promote and guide the conservation and sustainable use of these resources. It is also proposed that the National Commission on Plant Genetic Resources (CONAREFI for its acronym in Spanish) should be created (now its legal basis is only an executive decree), which will integrate all the relevant sectors and will act as an advisor entity to support the National Seeds Office, including in the implementation of the ITPGRFA.

Conclusion

These reforms (both the draft technical regulations and the Draft Seeds Law) aim to consolidate a long-standing practice related to the sale of native or local seeds in informal markets.

Most importantly, they propose the creation of a voluntary registration system where farmers could register their native seeds, with the aim of increase the information on plant genetic resources and improving their conservation and sustainable use, and to legalize at least some forms of exchanges of those seeds between farmers and possibly their commercialization. This effort is complementary to the provisions on plant genetic resources included in the proposed Draft Seeds Law.

It is not at all clear if these proposed policies will be accepted and implemented. However, in the meantime, they represent concrete efforts to promote farmers' rights in the country.

Notes

1 There is no official data concerning the size of the informal markets of seeds.
2 Ratified by Costa Rica by Law No. 8539 of August 2006.
3 Article 15.
4 Article 26.
5 Described in article 56 of the 2005 Regulations.
6 Article 74 and following articles.
7 Article 85.
8 PITTS *Technical Report of the Gibre and Curré varieties for inclusion in the Register of Commercial Varieties of the National Seed Office*, November 2006. However, the fact that two varieties have been registered does not necessarily mean that there are others varieties developed through participatory plant breeding out that were not registered because they do not satisfy the criteria for registration.
9 In the process of the Central American Customs Union in the field of seeds, there are two main objectives: the harmonization of the registration of commercial varieties and

of the regulations of quality control (seed certification, verification of quality standards, etc.). Therefore, several technical regulations have been adopted in the framework of the Central American Customs Union essentially with the purpose of facilitate the mutual recognition of national registries or to harmonize requisites and conditions for the registry of seeds.

10 Law No. 8591, article 5j.
11 Law No. 8591, article 20.
12 Article 12.
13 Article 2.
14 As established in the Constitution and the General Public Administration Law.
15 Article 82.

26 Commentary on the registration of traditional varieties in Benin

Raymond Sognon Vodouhe and Michael Halewood

In 1989, by way of decree, the government of Benin endorsed the idea of creating a national catalogue of plant species and varieties.[1]

In 2008, the Economic Community of West African States (ECOWAS) adopted regional regulations for the harmonization of national seed regulation. As part of that effort, it created that the West African Catalogue of Plant Species and Varieties.[2]

In 2011, the government of Benin published the first version of the *Catalogue Beninois des Espèces et Variétés Végétales* (CaBEV; MAEP, 2011). The CaBEV includes three lists of varieties. The first list is for newly bred varieties that are distinct, uniform and stable (DUS) and have demonstrable value for cultivation and use (VCU) over other existing, registered varieties. The seeds of these varieties can be produced and commercialized in any of the member states that subscribe to the regional harmonization law.

The second list of varieties is for those that are DUS, but not necessarily VCU. Seeds of these varieties can be produced in Benin (and other regional member states) but must be commercialized outside the region.

The third list of varieties contains traditional varieties. Traditional varieties and lists of traditional varieties are not mentioned in the regional harmonization law. However, the regional harmonization law does not prohibit the creation of such additional national variety lists linked to solely domestic purposes. Traditional varieties can be included in the CaBEV list if they are identifiable; they do not have to be DUS or VCU. They can be taken off the list if and when they lose their identifiable characteristics.[3] In practice, so far, all of the work to identify and register varieties, including traditional varieties, in Benin has been undertaken by INRAB (Institute National des Recherches Agricoles du Benin). Efforts of the National Registry Commission have had to be sporadic, given the lack of predictable funds and other demands on the Commission members' time. As of January 2015, the Commission has successfully registered the traditional varieties set out in Table 26.1.

It is prohibited to produce and commercialize the seed of species and varieties that have not been registered and included in CaBEV. In theory, this prohibition extends to the seed of traditional varieties as well as the seed of varieties that have been bred by formal sector organizations.

Table 26.1 Traditional varieties included in Benin national catalogue

Crop	*Variety names*
Pennisetum glaucum (pearl millet)	Nkouadi, Zongo
Sorghum bicolor	Koussoubakou, Tokogbessenou, Chorossoya, Sinou, Souarou, Wawiro, Mahi, Spaya, Natissoya, Chabicouma
Oryza sativa (rice)	Gambiaka
Vigna ungiculata (cowpea)	Kpodjiguegue
Manihot esculenta (cassava)	Odongbo. Adoborou, Kpaki Swan, Ahotonon
Dioscorea rotundata and *D. alata*	Laboko (*D. rotundata*), Morokorou (*D. rotundata*), Ahimon (*D. rotundata*), Guirissi Baka (*D. rotundata*), Kokoro Gbanou (*D. rotundata*), Tabane (*D. rotundata*), Singou (*D. rotundata*), YakanougoSoussou (*D. rotundata*), Tantoumani (*D. rotundata*), Agogo (*D. rotundata*), Douba Yessirou (*D. rotundata*), Tara (*D. rotundata*), Tikenianti (*D. rotundata*), Koussoussouka (*D. rotundata*), Guiwa (sankounou Aloungan, Sakata) *D. Alata*

Export of traditional varieties is only possible with the prior approval of the Minister of Agriculture after receiving advice from the National Committee of Seeds and Plants. Exploitation of such traditional genetic resources must be for the benefit of local populations.[4]

Notes

1 Decret No 89–378 du octobre 1989.
2 Regulation C/REG.4/05/2008 on Harmonization of the Rules Governing Quality Control, Certification and Marketing of Plant Seeds and Seedlings in ECOWAS Region, Sixtieth Ordinary Session of Minsters, Abuja, 17–18 May 2008 available at www.coraf.org/wasp2013/wp-content/uploads/2013/07/Regulation-seed-ECOWAS-signed-ENG.pdf (last accessed May 29, 2015).
3 Management of varieties, Titre II, Gestion et protection des variétés, Chapitre 1, Gestion des Variétés, paras 17, 18.
4 Protection of varieties, Titre II, Gestion et protection des variétés, Chapitre 2, Protection des Variétés, paras 22, 23, 24.

References

MAEP (Ministère de l'Agriculture, de l'Enlevage et de la Pêche) (2011). *Catalogue Béninois des Espèces et Variétés Végétales* (CaBEV). Ministère de l'Agriculture, de l'Elevage et de la Pêche, Tome.

Index

Note: Page numbers with *f* indicate figures, those with *t* indicate tables, and those with *b* indicate boxes.